Inotropic Stimulation and Myocardial Energetics

Hj. Just, Ch. Holubarsch, H. Scholz (eds.)

Inotropic Stimulation and Myocardial Energetics

Springer-Verlag Berlin Heidelberg GmbH

CIP-Titelaufnahme der Deutschen Bibliothek

Inotropic stimulation and myocardial energetics/Hj. Just... (eds.).

ISBN 978-3-662-07910-2   ISBN 978-3-662-07908-9 (eBook)
DOI 10.1007/978-3-662-07908-9
NE: Just, Hanjörg [Hrsg.]

Basic Res. Cardiol. ISSN 0300-8428
Indexed in Current Contents.

Copyright © 1989 by Springer-Verlag Berlin Heidelberg
Originally published by Dr. Dietrich Steinkopff Verlag GmbH & Co. KG, Darmstadt in 1989

Medical editor: Sabine Müller – English editor: James C. Willis – Production: Heinz J. Schäfer

Preface

Inotropic stimulation of the myocardium, as well as vasodilation and diuresis as essential principles in the treatment of congestive heart failure have recently met with considerable criticism and reevaluation.

It is generally agreed that unloading of the heart, either through vasodilation and/or diuresis, improves the working conditions of the dilated, failing heart. It reduces myocardial oxygen consumption through reduction of chamber radius and, thereby, wall tension as the major determinants of myocardial oxygen consumption.

Inotropic stimulation, quite in contrast, does not conserve oxygen. It rather consumes energy and that may be disadvantageous in situations of compromised oxygen supply and energy metabolism of the working myocardium. However, under conditions of sufficient oxygen supply and metabolic support inotropic stimulation may bring about increased pumping and subsequent improvement of myocardial failure.

In recent years it could convincingly be demonstrated that vasodilation leads to symptomatic improvement of congestive heart failure, improvement of exercise tolerance, and it prolongs life – especially in the case of ACE-inhibitors and the combination of hydralazine with long-acting nitrates. Quite in contrast, equally beneficial effects could not be demonstrated for inotropic agents in congestive heart failure. Only for the cardiac glycosides has it been shown that beneficial effects can be achieved, especially if atrial fibrillation with absolute arrhythmia is present. The influence of the cardiac glycosides on the latter represents an effect which is independent of the inotropic action. As far as inotropic stimulation as a therapeutic factor by itself in acute and/or chronic congestive heart failure is concerned, statistical proof is lacking that therapeutic benefit can be achieved.

On the other hand, clinical experience shows that inotropic stimulation may be of considerable value, especially in cardiogenic shock, where beta- and alpha-adrenergic catecholamines find widespread application. It is not clear, however, to what extent vasoactivity of these substances accounts for the therapeutic effect. Pathophysiological studies suggest inotropic stimulation to be highly desirable *after* ventricular unloading, accomplished either through vasodilation or diuresis.

Evaluation of inotropic drugs has met with difficulty, because virtually all inotropic drugs exhibit vasoactivity. Beta-adrenergic agents are characterized by vasodilation. This group of drugs includes substances such as isoproterenol, dobutamine, dopexamine or fenoterol. Partial agonists, such as xamoterol or prenalterol will also have to be included. The balance between beta-1 (inotropic) and beta-2 (vasodilating) properties may vary dependent on dosage (dobutamine) and the individual situation. In chronic congestive heart failure with beta-1 receptor downregulation, beta-2 effects, and hence vasodilation, may dominate; Noradrenalin induces arterial vasoconstriction.

Stimulation of the alpha-receptors will probably induce some inotropism, but the predominant effect is vasoconstriction. This evidently will counteract any myocardial stimulant effect and must be considered undesirable in heart failure, unless we are dealing with situations with inadequate peripheral vasoconstriction (early shock, septic shock). Dopaminergic receptors seem to exhibit inotropic properties to a certain degree. Their predominant action, however, is a peripheral vasoconstriction through secondary noradrenaline release and renal vasodilation brought about by specific dopaminergic receptors in the renal parenchyma. Dopamine has been widely used in intensive care medicine. More recently, oral preparations have been tested in congestive heart failure, for example levodopa and ibopamine. The latter drug is a prodrug and is transformed into epinine, the active substance. Here inotropic stimulation and renal vasodilation are achieved simultaneously.

Phosphodiesterase inhibitors are characterized by vasodilator activity as well as inotropic stimulation. It is this group of drugs which has recently received the most attention (amrinone, milrinone, enoximone). The vasodilating properties are so pronounced that hypotension may occur. The very pronounced vasodilation is apt to mask oxygen-consuming properties of inotropic stimulation. The effectiveness of these drugs seems to be seriously impaired in chronic heart failure, as soon as receptor downregulation and, probably, impairment of receptor/effect–coupling occurs. Here not only beta-1-adrenergic activity, but also inotropism mediated through phosphodiesterase inhibition seems to be inhibited.

Adenylate cyclase activators such as forskoline, have recently been studied. Experience has been disappointing; they do not seem to offer a new approach. Histamine agonists are also dependent on a functioning receptor mechanism at the membrane level. Experience with these drugs has been insufficient so far. Calcium agonists are currently not considered drugs for the treatment of heart failure, because they combine inotropic stimulation and vasoconstriction. This can, at most, be considered desirable in cases of early septic shock, but applicability on a wider range cannot be claimed.

The evaluation of inotropic drugs with regard to their influence on myocardial oxygen consumption and myocardial contractile economy has so far been widely neglected. It may well be that the clue to a more logical evaluation of inotropic drugs rests with the definition of the "energetic profile". In the past it has been very difficult to differentiate vasoactivity and inotropism in regard to myocardial energetics. Recently this differentiation has become possible.

For these reasons we have choosen to attempt a reevaluation of inotropic stimulation. For some of the currently used drugs an oxygen-wasting effect can be shown. This applies to sympathomimetic amines as well as to the phosphodiesterase inhibitors. On the other hand, cardiac glycosides and some new inotropic drugs which have additional "calcium-sensitizing" effects, seem to be able to bring about inotropic stimulation with proportionate cost in terms of oxygen consumption.

The second Gargellen-Conference held June 16, 1988, in Gargellen/Montafon, Austria, was dedicated to the evaluation of inotropic drugs under the aspect of myocardial energetics. The previous Gargellen-Conference of June 1986 addressed the contractile process and the energetic costs of myocardial contraction and thereby laid the basis for this symposium.

We are very grateful to our sponsors, Dr. Karl Thomae GmbH, Biberach/Riss (represented by Dr. Benedikter), Bayer AG, Leverkusen (represented by Dr. Kubitz), and Fisons AG (represented by Dr. Roloffs). We also gratefully acknowledge the support of Gödecke AG, Freiburg; Beiersdorf AG, Hamburg; Winthrop AG, Wedel/Holstein; and of our publishers Dr. D. Steinkopff-Verlag, Darmstadt, for expert help and efficiency.

We hope that this book will help to improve our understanding of inotropic stimulation substances and thereby improve our therapeutic armamentarium for the frequent and serious syndrome of congestive heart failure.

Hj. Just
Ch. Holubarsch
H. Scholz

Contents

III. Clinical experience with positive inotropic substances

IV. Energetic aspects in the intact heart

Contents   IX

**Use of a conductance catheter to detect increased left ventricular inotropic state
by end-systolic pressure-volume analysis**
Leatherman, G. F., T. L. Shook, S. M. Leatherman, W. S. Colucci 247

**Separation between vasodilation and positive inotropism by assessment of myocardial
energetics in patients with dilated cardiomyopathy**
Holubarsch, Ch., G. Hasenfuss, M. Allgeier, H. W. Heiss, Hj. Just 257

Subject index 267

1. Cellular and subcellular mechanisms, energy turnover

Introduction:
Mechanisms of positive inotropic effects

H. Scholz

Abteilung Allgemeine Pharmakologie, Universitäts-Krankenhaus Eppendorf, Hamburg, FRG

Summary

This review deals with the principal mechanisms which are known to play a role in positive inotropism: 1) The myoplasmic Ca^{2+} concentration may be increased by increases in cyclic AMP. Beside receptor-mediated stimulation (isoprenaline) or direct stimulation (forskolin) of the adenylate cyclase, the cyclic AMP may be increased by phosphodiesterase inhibition; 2) Cyclic AMP-independent activation of Ca^{2+} channels can be brought about by alpha-adrenergic agents (phenylephrine) or so-called calcium agonists; 3) Only a small increase in myoplasmic Na^+ concentration can greatly enhance the force of contraction by an increase in the intracellular Ca^{2+} concentration. This is possible by inhibition of the Na^+/K^+-ATPase (glycosides) or by prolongation of the open state of Na^+ channels (DPI 201-106); 4) A direct inhibition of the Na^+/Ca^{2+} exchange has been discussed for amiloride; 5) A prolongation of the action potential induced by K^+ channel-inhibiting agents such as 4-amino-pyridine may increase the myoplasmic Ca^{2+} concentration by a prolongation of the slow Ca^{2+} inward current; 6) An increased Ca^{2+} sensitivity of the contractile proteins has been demonstrated for a number of compounds in vitro; the contribution of such an effect to the overall positive inotropism is unknown because a calcium sensitizer without any effects on calcium or sodium movements is not yet available.

The increase in the force of myocardial contraction is ultimately due to one of two mechanisms (Fig. 1): an increase in intracellular free Ca^{2+} concentration, $\langle Ca^{2+}\rangle_i$, to interact with the contractile proteins (① in Fig. 1) or to an increased sensitivity of the myofilaments for Ca^{2+} (② in Fig. 1), or both (9). It will be discussed in detail in this section how these variables might be affected by various inotropic drugs.

1. Increased myoplasmic Ca^{2+} concentration

Mechanisms leading to an increase in myoplasmic Ca^{2+} concentration are marked ①–⑤ in Fig. 1. These include:

1) Increase in intracellular cAMP level
   - Stimulation of adenylate cyclase
   - Inhibition of phosphodiesterase
2) Cyclic AMP-independent activation of Ca^{2+} channels
   - Alpha-adrenergic agents
   - Ca-agonists
3) Increase in myoplasmic Na^+ concentration
   - Inhibition of Na^+/K^+-ATPase
   - Prolongation of the open state of Na^+ channels
4) Direct inhibition of Na^+/Ca^{2+} exchange
5) Inhibition of K^+ channels.

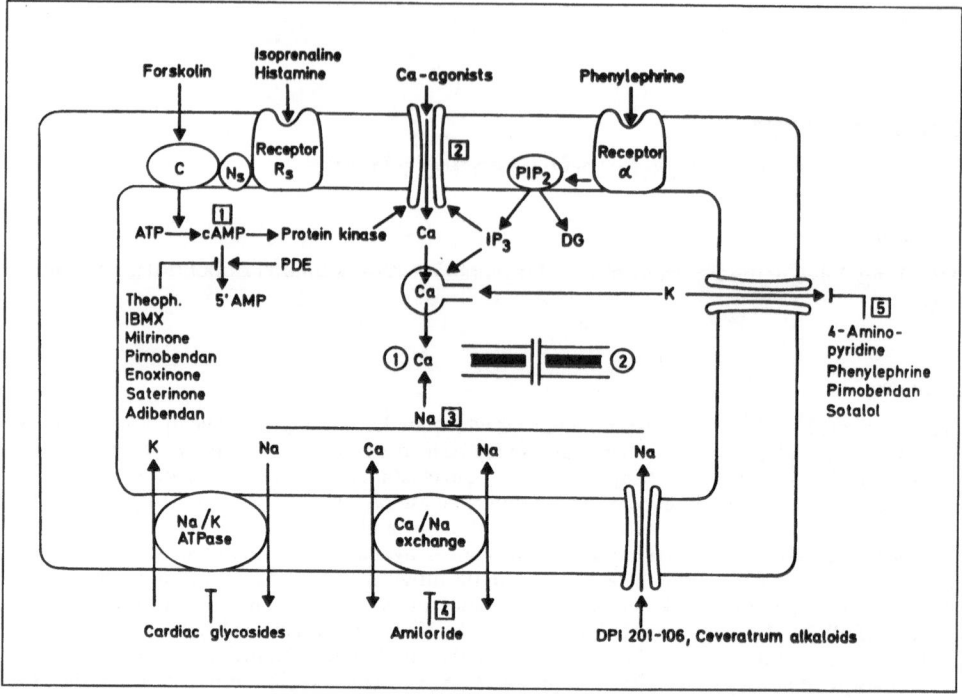

Fig. 1. Schematic summary of possible mechanisms of positive inotropic effects of various positive inotropes. All drugs increase the cellular Ca^{2+} concentration (①) or increase the Ca sensitivity of the contractile proteins (②). Mechanisms leading to an increase in $\langle Ca^{2+}\rangle_i$ are marked (1–5)

An increase in intracellular cAMP level (1 in Fig. 1) may be brought about by a receptor-mediated (e.g., isoprenaline, histamine) or a direct (forskolin) stimulation of adenylate cyclase (increase in cAMP formation) or by an inhibition of phosphodiesterase (decrease in cAMP breakdown). The main effect of cAMP is a phosphorylation of functional proteins and an increase in the slow Ca^{2+} inward current. This leads to an increased Ca^{2+} release from the sarcoplasmic reticulum, either because it served as a greater trigger for Ca^{2+}-dependent Ca^{2+} release or because it increases the filling of these stores with Ca^{2+} that can be released during subsequent beats.

In principle, it does not matter whether the increase in cAMP is produced by a stimulation of adenylate cyclase or by an inhibition of phosphodiesterase. It is, however, important to note that PDE inhibitors act independently of beta-adrenoceptors and, hence, also in the presence of beta-adrenoceptor blocking drugs and that they produce not only positive inotropic but also vasodilator effects. New positive inotropic agents preferentially inhibit PDE III; this is shown for pimobendan in Fig. 2.

Undesired effects of cAMP increasing agents include positive chronotropic and arrhythmogenic effects [8]. Moreover, beta-adrenoceptor agonists decrease the number of beta-adrenoceptors and thus tolerance develops rapidly [2]. Evidence is increasing that tolerance develops also with other cAMP-increasing agents including PDE-inhibitors. This might be due to an effect on the coupling between beta-adrenergic or other receptors and the catalytic subunit of the adenylate cyclase [3, 5, 7].

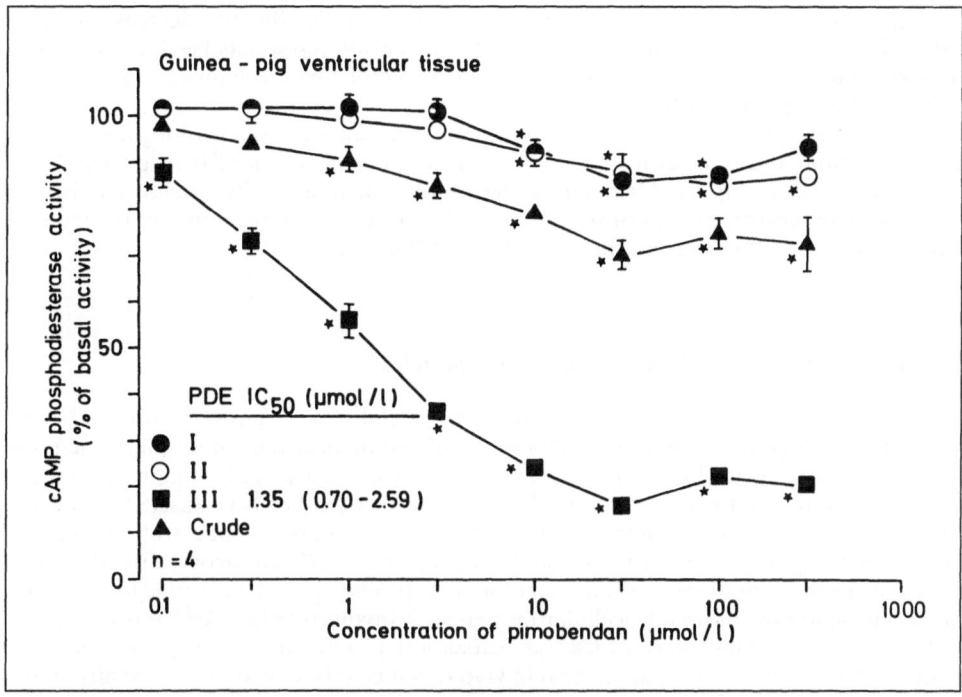

Fig. 2. Effects of pimobendan on the activity of cAMP-phosphodiesterase (PDE) I–III and on crude cAMP-PDE in guinea pig ventricular myocardial tissue. Substrate concentration 1 µmol/l cAMP. The experiments were performed with methods described [1]

An increase in slow Ca^{2+} inward current without elevation of the cAMP concentration (② in Fig. 1) can be brought about by alpha-adrenergic agents such as phenylephrine or by the so-called Ca-agonists Bay K 8644 and CGP 28392. Alpha-receptor agonists exert an increase in phosphatidylinositol turnover. It is presently widely discussed whether or not this effect is causally related to the alpha-adrenergic positive inotropic effect (see [10]).

The transsarcolemmal movement of Ca^{2+} and hence the inotropic state of cardiac muscle is also controlled by the intracellular Na^+ concentration (③ in Fig. 1). The link in the interrelationship between Na^+ and Ca^{2+} is the Ca^{2+}/Na^+ exchange system in which transsarcolemmal Ca^{2+} movements are coupled to opposite movements of Na^+. Thus, Na^+ can exit from the cell in exchange for Ca^{2+}, or Ca^{2+} can exit in exchange for Na^+, depending upon the prevailing Na^+ and Ca^{2+} electrochemical gradients. It is important to note that only a very small increase in $\langle Na^+ \rangle_i$ is required to achieve a large increase in $\langle Ca^{2+} \rangle_i$ and, hence, in force of contraction. Quantitatively, a 1 mM increase in intracellular Na^+ activity is accompanied by about a 100% increase in force of contraction.

As yet two mechanisms have been detected which are likely to produce an increase in myocardial force of contraction via an increase in $\langle Na^+ \rangle_i$:
1) Inhibition of Na^+/K^+-ATPase (e.g., cardiac glycosides)
2) Prolongation of the open state of Na^+ channels (DPI 201-106, ceveratrum alkaloids).

A direct inhibition of Na^+/Ca^{2+} exchange ($\boxed{4}$ in Fig. 1), leading to a decrease in Ca^{2+} efflux, and thus an increase in $\langle Ca^{2+}\rangle_i$ and force of contraction has been discussed for amiloride and amiloride analogs [6]. The importance of this mechanism, however, requires further experimentation.

K^+ channel-inhibiting agents such as 4-aminopyridine lead to a prolongation of the action potential and thus, at least theoretically, to a prolongation of the slow Ca^{2+} inward current ($\boxed{5}$ in Fig. 1). A similar mechanism has also been discussed to contribute to the positive inotropic effect of sotalol, phenylephrine or pimobendan, but the biological significance of this mechanism is as yet not well defined.

2. Increased Ca^{2+} sensitivity of the contractile proteins

An increased Ca^{2+} sensitivity of the contractile proteins (Ca^{2+} sensitization) can be revealed by studying the contractile behavior of isolated myocardial contractile structures or by measuring intracellular Ca^{2+} signals with the Ca^{2+}-sensitive bioluminescent protein aequorin in intact cardiac muscle. A Ca^{2+} sensitization has been discussed, for instance, for sulmazole, pimobendan, methylxanthines at millimolar concentrations, alpha-adrenoceptor agonists, and DPI 201-106 (for references see [9]). In theory, this effect can contribute to the overall inotropic action of a drug, provided the drug is able to cross the intact cell membrane or an intracellular messenger is involved in the Ca^{2+} sensitizing effect. However, the importance of Ca^{2+} sensitization as a positive inotropic mechanism in intact cardiac muscle or even more so in vivo remains to be elucidated. A positive inotropic "Ca^{2+} sensitizer" without any other effect, e.g., without any effect on myocardial Ca^{2+} or Na^+ movements, is still lacking.

References

1. Bethke T, Brunkhorst D, v. der Leyen H, Meyer W, Nigbur R, Scholz H (1988) Mechanism of action and cardiotonic activity of a new phosphodiesterase inhibitor, the benzimidazole derivative adibendan (BM 14.478), in guinea-pig hearts. Naunyn Schmiedebergs Arch Pharmacol 337:576–582
2. Bristow MR, Ginsburg R, Minobe W, Cubicciotti RS, Scott Sageman W, Lurie K, Billingham ME, Harrison DC, Stinson EB (1982) Decreased catecholamine sensitivity and β-adrenergic-receptor density in failing human hearts. N Engl J Med 307:205
3. Feldman AM, Copelas L, Gwathmey JK, Philips P, Warren SE, Schoen FJ, Grossman W, Morgan JP (1987) Deficient production of cyclic AMP: pharmacologic evidence of an important cause of contractile dysfunction in patients with end-stage heart failure. Circulation 75:331
4. Honerjäger P (1982) Cardioactive substances that prolong the open state of sodium channels. Rev Physiol Biochem Pharmacol 92:1–74
5. Kessler PD, Van Dop C, Feldmann AM (1988) G proteins: transmembrane signal processors in the heart. Heart Failure 3:239
6. Luciani S, Floreani M (1985) Na^+-Ca^{2+} exchange as a target for inotropic drugs. Trends Pharmacol Sci 6:316
7. Schmitz W, Scholz H, Erdmann E (1987) Effects of alpha- and beta-adrenergic agonists, phosphodiesterase inhibitors and adenosine on isolated human heart muscle preparations. Trends Pharmacol Sci 8:447–450
8. Scholz H, Meyer W (1986) Phosphodiesterase-inhibiting properties of newer inotropic agents. Circulation 73 (Suppl III):III-99–III-108

9. Scholz H (1986) Positive inotropic agents: Different mechanisms of action. In: Erdmann E, Greeff K, Skou JC (eds) Cardiac Glycosides 1785–1985; Steinkopff, Darmstadt, 181–188
10. Scholz J, Schaefer B, Schmitz W, Scholz H, Steinfath M, Lohse M, Schwabe U, Puurunen J (1988) Alpha-1 adrenoceptor-mediated positive inotropic effect and inositol trisphosphate increase in mammalian heart. J Pharmacol Exp Ther 245:327–335

Author's address:

Prof. Dr. Hasso Scholz, Abteilung Allgemeine Pharmakologie, Universitäts-Krankenhaus Eppendorf, Martinistrasse 52, D-2000 Hamburg 20 (FRG)

Adrenoceptors in myocardial regulation: concomitant contribution from both alpha- and beta-adrenoceptor stimulation to the inotropic response*

J. B. Osnes, H. Aass, T. Skomedal

Department of Pharmacology, University of Oslo, Norway

Summary

The studies presented deal with the α_1-adrenoceptor mediated inotropic effects of noradrenaline obtained by exclusive ("pure") α_1-adrenoceptor stimulation or by concomitant stimulation of α_1- and β-adrenoceptors in myocardium. The pure β-adrenergic effects of noradrenaline were also quantified. Interactions between the two receptor systems were studied. The pure α_1- and β-adrenergic effects of noradrenaline, respectively, were achieved separately in the presence of high concentrations of appropriate receptor blockers. The experiments were performed on isolated ventricular myocardium from rat, rabbit, and man.

The pure α_1-adrenergic inotropic effects were about 35–50% of control (basal) and half the pure β-adrenergic effects both in rat and rabbit myocardium. Ventricular myocardium from man exhibited an α_1-adrenergic inotropic effect of the same magnitude (50% of control [basal]) as did rabbit papillary muscle.

Determination of the α_1-adrenergic inotropic component during concomitant β-adrenoceptor stimulation was associated with difficulties. Several experimental approaches on rat and rabbit myocardium are presented and discussed. Some types of experimental approaches obviously underestimate the α_1-adrenergic component. The methods regarded as reliable revealed an α_1-adrenergic inotropic effect of about 20–30% during combined adrenoceptor stimulation by noradrenaline. Concomitant β-stimulation reduced the α_1-adrenergic effect by about 50%, while α_1-stimulation attenuated the β-effect to a lesser degree (about 20–25%). A model is presented on a mutual attenuation of the functional expression of the two receptor systems. Thus the α_1-adrenergic inotropic component increases when the β-adrenergic stimulation is antagonized and the β-effect largely compensates for the loss of the α_1-effect during myocardial α_1-adrenoceptor blockade.

1. Background and scope

Since 1948, there has been no doubt that β-adrenoceptors are involved in the inotropic effect of adrenergic stimulators in myocardium [3]. It was several years however, before it was acknowledged that also stimulation of myocardial α_1-adrenoceptors was able to elicit an inotropic effect ([18], reviews: [4, 6, 11, 13]). But still some questions exist on the role of this part of the adrenergic regulation of heart function. For some years an α_1-adrenergic inotropic effect of noradrenaline was doubted, at least in some species [9, 14–17]. An interaction or cooperation between α_1- and β-adrenergic effects during concomitant stimulation of the two receptor systems by noradrenaline was a further challenge of investigation. The present paper deals with both of these aspects.

* Supported by the Norwegian Research Council for Science and the Humanities and the Norwegian Council on Cardiovascular Diseases.

2. Separate α_1- and β-adrenergic inotropic effects of noradrenaline

As the most successful studies on α_1- and β-adrenergic effects had been performed by pure and separate receptor stimulations, we adapted the same approach to the studies on the effects of noradrenaline.

Pure β-adrenergic effects were achieved by noradrenaline given in the presence of and after preincubation with a relatively high concentration of the α_1-blocker prazosin (10^{-7} mol/l) to eliminate α_1-receptor stimulation. Pure α_1-receptor stimulation was achieved by noradrenaline in the presence of relatively high concentrations of a non-selective β-receptor blocker without partial agonistic activity (5×10^{-6} mol/l propranolol or 10^{-6} mol/l timolol). We studied the effects on isolated papillary muscles from rat and rabbit heart and on ventricular strips from human heart.

The maximal adrenergic effects are illustrated in Fig. 1. Clear α_1-adrenergic inotropic effects (about 35–50% of control [basal]) appeared in all three myocardial preparations. Remarkable is that the effect in man is of the same magnitude (about 50% of control [basal]) as that in rabbit, which has been claimed to be the animal of choice for studying α_1-adrenergic heart effects [7]. Both in rat and rabbit myocardium the pure α_1-adrenergic inotropic effect is about half the separately determined pure β-adrenergic effect.

Another aspect which came out of these and similar studies was the qualitative difference between the two inotropic effects of noradrenaline. α_1-Adrenoceptor stimulation increased isometric contraction almost proportionally to or slightly more than relaxation while β-adrenoceptor stimulation elicited a disproportionally higher increase of relaxation (e.g. [1]). This was also reflected in the time-course of the contraction relaxation cycle: α_1-adrenoceptor stimulation caused a slight increase of the cycle duration or left it unchanged while β-adrenoceptor stimulation markedly shortened both the contraction and the relaxation phase.

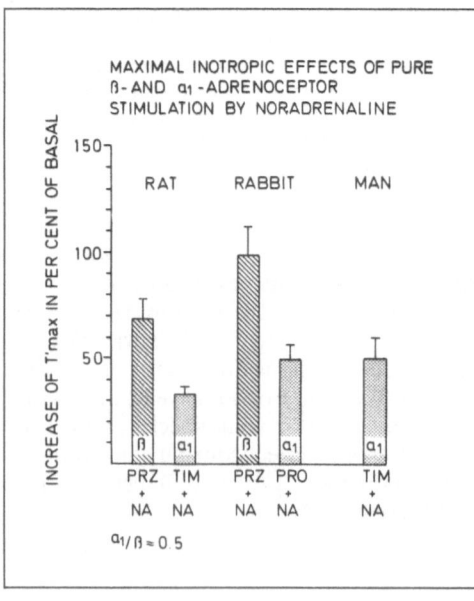

Fig. 1. Maximal inotropic effects of pure β- and α_1-adrenoceptor stimulation by noradrenaline (NA) in rat and rabbit heart papillary muscles and in heart ventricular strips of man (α_1-effect only). The isolated preparations were mounted isometrically in an organ bath containing a physiological salt solution. The muscle strips were driven electrically at 1 Hz. Methodological details are described elsewhere [1, 2, 5]. The inotropic effects are expressed as percent increase of maximal rate of force development ($T'_{max} = (dT/dt)_{max}$). Pure β-adrenergic stimulation was achieved in the presence of 10^{-7} mol/l prazosin (PRZ). Pure α_1-adrenergic stimulation was obtained in the presence of 10^{-6} mol/l timolol (TIM) (rat, man) or 5×10^{-6} mol/l propranolol (PRO) (rabbit). The blockers were added 15 min (rat) or 45 min (rabbit, man) before noradrenaline. SEM is indicated

3. Is there an α_1-adrenergic inotropic effect of noradrenaline during concomitant β-adrenoceptor stimulation?

Although the existence of a pure α_1-adrenergic inotropic effect of noradrenaline indicates that it operates also during β-adrenoceptor stimulation, this must be shown directly. Some reports on lacking effect of α-receptor blockers on the response to noradrenaline had thrown some doubt on a role for the α-adrenergic inotropic effect (for further comments and references see [5]).

Attempts to determine an α_1-adrenergic inotropic effect in the presence of β-adrenoceptor stimulation can be done in several ways. But, as is evident from the following, it is not straight forward to obtain reliable results.

3.1 Attempts to determine an α_1-adrenergic component by "subtraction"

A conventional way to quantify an α_1-adrenergic inotropic component would be to determine the maximal effect of noradrenaline and subtract the maximal effect obtained in the presence of a high concentration ($1-1.2 \times 10^{-7}$ mol/l) of the potent α_1-adrenoceptor blocker prazosin that is able to eliminate α_1-adrenoceptor stimulation. This approach is not intended to shift the dose-response curve to higher concentrations of agonist, but is instead expected to reduce the maximum. The results which we obtained in rat [5] and rabbit papillary muscles, did not seem to reveal significant α-adrenergic components (Fig. 2). Only a slight decrease of maxima was caused by prazosin. Thus it might be tempting to exclude a significant role of α-adrenoceptors in the inotropic response to noradrenaline alone in the absence of β-receptor blockade. But a prerequisite for such a conclusion would be that the β-adrenergic effect is the same in the presence and absence of the α_1-receptor blockade. This is not necessarily the case. In fact, as shown below, the

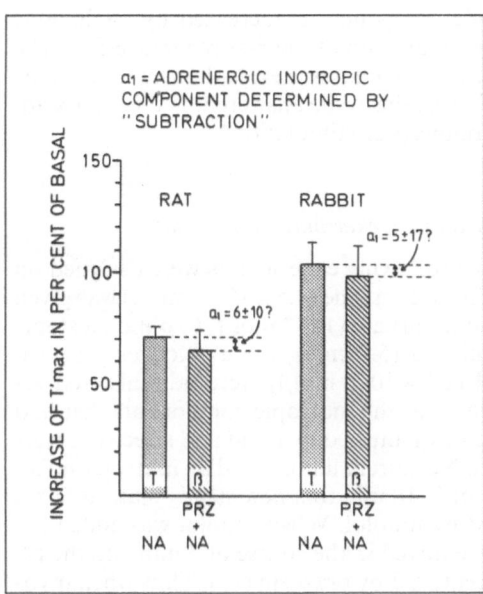

Fig. 2. Maximal inotropic effects of noradrenaline (NA) in the absence (T) and in the presence (β) of 1.2×10^{-7} mol/l (rat) or 10^{-7} mol/l (rabbit) prazosin (PRZ). An α_1-adrenergic component was calculated as the difference between the two maximal effects (further discussion in the text). The experiments were performed on rat and rabbit heart papillary muscles. Experimental conditions are given in the legend to Fig. 1

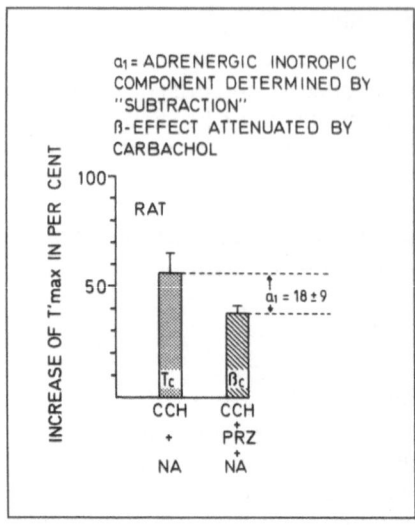

Fig. 3. Maximal inotropic effects of noradrenaline (NA) in the presence of 10^{-5} mol/l carbachol (CCH), T_c and in the presence of both carbachol and 1.2×10^{-7} mol/l prazosin (PRZ), β_c. An α_1-adrenergic component was calculated as the difference between the two maximal effects. The experiments were performed on rat heart papillary muscles (see legend to Fig. 1)

β-effect turned out to be reduced by concomitant α_1-adrenergic stimulation. The further experimental approaches intended to circumvent this difficulty.

Similar experiments were performed on rat papillary muscles which were exposed to carbachol (10^{-5} mol/l) to attenuate the β-adrenergic effects beyond the β-adrenoceptor [8]. In this case a high concentration of prazosin (1.2×10^{-7} mol/l) significantly reduced the maximal inotropic effect [5]. This "substraction method" revealed an α_1-adrenergic component of 18% (Fig. 3). Thus, at least during an antagonized/attenuated β-adrenergic effect, stimulation of α_1-adrenoceptors elicits an inotropic effect.

The results indicate that the α_1-adrenergic component is increased by cholinergic stimulation either directly or indirectly through reduction of the β-adrenergic effect. The latter explanation would imply that β-stimulation attenuates the α_1-adrenergic inotropic component. Thus the challenging question still remains to be answered: Is there an α-adrenergic inotropic effect also during full β-adrenoceptor stimulation?

3.2 Adrenergic inotropic components determined by blocker induced reversal

In an attempt to answer the question above the following experiments were designed on rat heart papillary muscle: A maximal dose of noradrenaline (4.5×10^{-6} mol/l) was given in the presence of cocaine (3×10^{-5} mol/l) and ascorbate (10^{-4} mol/l) to obtain a stable inotropic response. After the effect had stabilized (5–7 min), the α_1-blocker prazosin (1.2×10^{-6} mol/l) and the β-blocker timolol (1.2×10^{-5} mol/l) were administered sequentially. The rationale for this approach was that the inotropic components that had developed in the presence of each other, could be quantified by rapid and selective reversal by receptor blockers at high concentrations. Noradrenaline exerted an inotropic effect of 68%. Prazosin reversed 18% in the course of 8–10 min to a new steady state at 50%. This remaining effect was completely reversed by timolol. When timolol was added before prazosin, an inotropic effect of 50% was removed in the course of 5 min and the residual inotropic effect (18%) was completely removed by prazosin (12). Thus when myo-

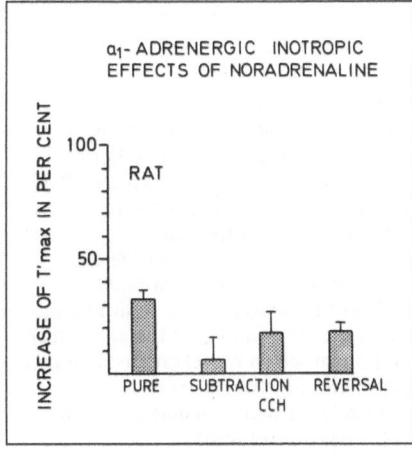

Fig. 4. α_1-Adrenergic inotropic effects of noradrenaline in rat heart papillary muscles in the absence or presence of β-adrenoceptor stimulation and determined by different methods. "Reversal" represents the component reversed by 1.2×10^{-6} mol/l prazosin added at steady-state maximal response 5–7 min after addition of noradrenaline (further explanation in the text and Figs. 2 and 3)

cardial α_1- and β-adrenoceptors were maximally stimulated, the α_1-adrenergic inotropic component amounted to 18% (Fig. 4), while the β-adrenergic component was 50%. The relative contribution from α_1-adrenoceptor stimulation to the total inotropic effect of noradrenaline thus was about 25%.

This "reversal method" disclosed a definite α_1-adrenergic component which had developed concomitantly with full β-stimulation. This is in sharp contrast to the "subtraction method" (see above) which obviously underestimated this component. A reasonable explanation is that the pure β-adrenergic component which was subtracted (65%) (Fig. 2), was larger than the β-adrenergic component during combined adrenoceptor stimulation (50%). The α_1-adrenergic component during combined stimulation was about half the size of the inotropic effect of pure α_1-adrenoceptor stimulation by noradrenaline (Fig. 4).

Thus during combined adrenoceptor stimulation in rat myocardium there was a mutual attenuation of the functional expression of the two receptor systems.

3.3 α_1-Adrenergic component during low degree of receptor blockade

The large inotropic effect of pure α_1-adrenoceptor stimulation in rabbit papillary muscle made it worthwhile to see if a relatively low concentration of prazosin could displace the dose-response curve sufficiently to disclose an α_1-adrenergic component and simultaneously permit it to be fully expressed by increasing the concentrations of noradrenaline. From theoretical considerations and preliminary experiments 3×10^{-9} mol/l prazosin was chosen for α_1-receptor blockade. The dose response curve thus obtained for the inotropic effect of noradrenaline on rabbit papillary muscle exhibited two components partially separated. The larger component (about 80% of total) had not been shifted by the blocker and represented, obviously, the β-adrenergic effect. The smaller component (about 20% of total) had been shifted to higher agonist concentrations by prazosin and was thus the α_1-adrenergic component. This response component had then developed in the presence of full β-receptor stimulation. The maximal inotropic effect was 144% and thus the α_1-adrenergic inotropic effect during combined receptor stimulation was about

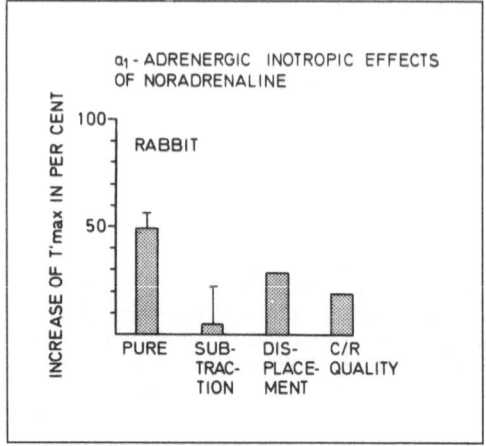

Fig. 5. α_1-Adrenergic inotropic effects of noradrenaline in rabbit heart papillary muscles in the absence and presence of β-adrenoceptor stimulation and determined by different methods. "Displacement" indicates the α_1-adrenergic component of the dose-response curve that had been displaced by 3×10^{-9} mol/l prazosin. "C/R (contraction/relaxation) quality" indicates the α_1-adrenergic component which was estimated from an analysis of the effects on the contraction-relaxation cycle (further explanation in the text and in the legend to Fig. 6)

30% (Fig. 5). These results are in sharp contrast to the apparently lacking α_1-adrenergic component found by the "subtraction method" (Fig. 2) which, as it does for rabbit myocardium, gives an underestimation. This means that results obtained in the absence and presence of the α_1-blocker cannot easily be combined as in Fig. 2.

The α_1-adrenergic component during combined receptor stimulation seems in rabbit myocardium to also be about half the size of the pure α_1-adrenergic effect obtained separately (Fig. 5). The two adrenergic inotropic components are obviously not independent of each other and they seem to exert a mutual attenuation in rabbit myocardium as well.

3.4 α_1-Adrenergic component disclosed by analyzing the quality of the contraction relaxation cycle

α_1- and β-adrenoceptor stimulations change the contraction relaxation cycle differently: α_1-stimulation increases the contraction and the relaxation almost proportionally, or the contraction slightly more than the relaxation ([1], reviews: [11, 13]). β-Stimulation increases the relaxation more than the contraction and causes an earlier onset of relaxation in each cycle. These differences can be illustrated by plotting the maximal rate of tension development (T'_{max}) against the maximal onset-rate of relaxation ($T_{min} = d^2T/dt^2$). This has been done in Fig. 6 for rabbit papillary muscles exposed to increasing concentrations of noradrenaline. The pure α_1- and β-adrenergic stimulation, respectively, exhibit markedly different relationships between effects on contraction and relaxation [11]. During combined receptor stimulation by noradrenaline in the absence of blockers the relationship between contraction and relaxation followed an intermediate course. The latter curve is assumed to be the result of varying contributions from the α_1- and β-adrenergic components. Accordingly, the relative contribution from these two components could be calculated from the slopes of the curve. In the lower concentration range of noradrenaline the α_1-adrenergic effect contributed (by about 40%), in the range of EC_{50} (by about 30%) and at maximal inotropic effect the α_1-effect contributed (by about 18%) to the total response (Fig. 6). As the maximal inotropic effect of noradrenaline without blockers was

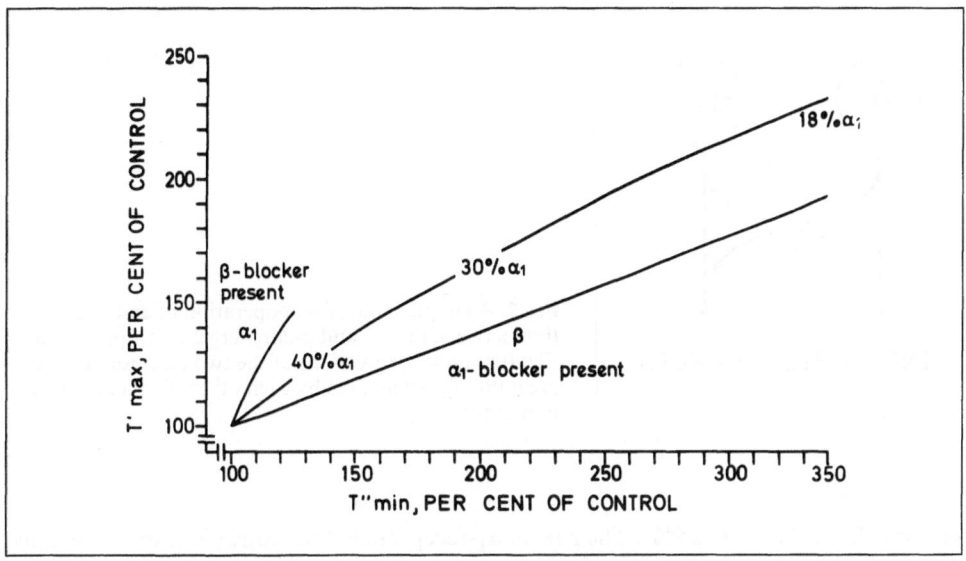

Fig. 6. Analysis of the effects of noradrenaline on the contraction-relaxation cycle of rabbit heart papillary muscle. Corresponding values for contraction (T'_{max}, ordinate) and maximal onset-rate of relaxation (T_{min}, abscissa) after α_1-adrenergic (to the left), β-adrenergic (to the right) and combined stimulation (in the middle) by increasing concentrations of noradrenaline. The curves are based on dose-response relationships presented elsewhere [1]. α_1: α_1-adrenergic stimulation by increasing concentrations of noradrenaline in the presence of 5×10^{-6} mol/l propranolol (added 45 min before noradrenaline). β: β-adrenergic stimulation by increasing concentrations of noradrenaline in the presence of 10^{-7} mol/l prazosin (added 45 min before noradrenaline). *Intermediate curve:* Combined α_1- and β-adrenoceptor stimulation by increasing concentration of noradrenaline in the absence of blockers. The relative contributions from α_1-adrenoceptor stimulation are indicated (further explanation in the text)

104% (Fig. 2), the relative contribution of 18% corresponded to an α_1-adrenergic inotropic effect of about 19% in the presence of maximal β-stimulation. This is clearly lower than the pure α_1-effect and comparable to that obtained in the presence of the low concentration of prazosin (Fig. 5).

4. Conclusion

From an overall evaluation of our studies on rat and rabbit heart papillary muscles an α_1-adrenergic inotropic effect is also evoked during combined adrenoceptor stimulation by noradrenaline. Thus there is a concomitant contribution from both α_1- and β-stimulation. During maximal stimulation the relative contribution from α_1-receptors is about 20–25%. An α_1-adrenergic component may, however, avoid detection by some usually applied methods. The "reversal method" seems to be an appropriate way to disclose both adrenergic inotropic components. There is a mutual attenuation of the functional expression of the two receptor systems (Fig. 7). Concomitant β-stimulation reduces the α_1-adrenergic inotropic effect by about 50%, while α_1-stimulation attenuated the β-effect

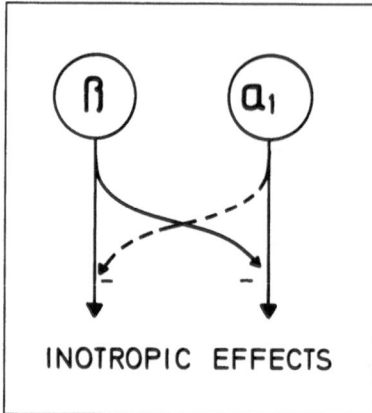

Fig. 7. A simple model for cooperation of and interaction between the α_1- and β-adrenergic inotropic effects. The functional expressions of the two receptor systems are mutually attenuated by each other (further explanation in the text)

INOTROPIC EFFECTS

relatively less (about 20–25%). The role of α_1-receptors in the inotropic response is thus increased when β-adrenergic stimulation is antagonized, e.g., by receptor-blockers, by cholinergic agents [5, 8] and probably by adenosine (10). The β-effect largely compensates for the loss of the α-effect during myocardial α_1-adrenoceptor blockade.

This cooperation of and interaction between the α_1- and β-adrenoceptors provide the heart muscle with a high degree of flexibility in its adrenergic response during various conditions. A future challenge would be to further explore the functional significance and the molecular mechanisms of this interplay.

References

1. Aass H, Skomedal T, Osnes J-B (1983) Demonstration of an alpha adrenoceptor-mediated inotropic effect of norepinephrine in rabbit papillary muscle. J Pharmacol Exp Therap 226:572–578
2. Aass H, Skomedal T, Osnes J-B, Fjeld NB, Klingen G, Langslet A, Svennevig J, Semb G (1986) Noradrenaline evokes an α-adrenoceptor-mediated inotropic effect in human ventricular myocardium. Acta pharmacol et toxicol 58:88–90
3. Ahlquist RP (1948) A study of the adrenotropic receptors. Am J Physiol 153:586–600
4. Benfey BG (1982) Function of myocardial α-adrenoceptors. Life Sci 31:101–112
5. Christiansen HB, Horgmo GI, Skomedal T, Osnes J-B (1987) Enhancement of the α-adrenergic inotropic component of noradrenaline by simultaneous stimulation of muscarinic acetylcholine receptors in rat myocardium. Eur J Pharmacol 142:93–102
6. Endoh M (1982) Adrenoceptors and the myocardial inotropic response: Do alpha and beta receptor sites functionally coexist? In: Kalsner S (ed) Trends in autonomic pharmacology, vol 2. Urban & Schwarzenberg, Baltimore Munich, pp 303–322
7. Endoh M, Blinks JR (1988) Actions of sympathomimetic amines on the Ca^{2+} transients and contractions of rabbit myocardium: Reciprocal changes in myofibrillar responsiveness to Ca^{2+} mediated through α- and β-adrenoceptors. Circ Res 62:247–265
8. Endoh M, Motomura S (1979) Differentiations by cholinergic stimulation of positive inotropic actions mediated via α- and β-adrenoceptors in the rabbit heart. Life Sci 25:759–768
9. Endoh M, Schümann HJ, Krappitz N, Hillen B (1976) α-Adrenoceptors mediating positive inotropic effects on the ventricular myocardium: some aspects of structure activity relationships of sympathetic amines. Japn J Pharmacol 26:179–190

10. Endoh M, Yamashita S (1980) Adenosine antagonizes the positive inotropic action mediated via β-, but not α-adrenoceptors in the rabbit papillary muscles. Eur J Pharmacol 65:445–448
11. Osnes J-B, Aass H, Skomedal T (1985) On adrenergic-regulation of heart function: role of myocardial α-adrenoceptors. In: Refsum H, Mjøs OD (eds) α-Adrenoceptor blockers in cardiovascular disease. Churchill Livingstone, Edinburgh, pp 69–102
12. Osnes J-B, Christiansen HB, Horgmo GI, Schiander IG, Skomedal T (1987) Alpha-adrenergic stimulation contributes to the inotropic effect of noradrenaline. Biomed Biochim Acta 46(8/9):S417–S420
13. Scholz H, Brückner R, Mügge A, Reupcke C (1986) Myocardial alpha-adrenoceptors and positive inotropy. J Mol Cell Cardiol 18:79–87
14. Schümann HJ (1980) Are there α-adrenoceptors in the mammalian heart? Trends in Pharmacological Sciences 1:195–197
15. Schümann HJ, Endoh M, Wagner J (1974) Positive inotropic effects of phenylephrine in the isolated rabbit papillary muscle mediated both by alpha- and beta-adrenoceptors. Naunyn-Schmiedeberg's Arch Pharmacol 284.133–148
16. Schümann HJ, Wagner J, Knorr A, Reidemeister JC, Sadony V, Schramm G (1978) Demonstration in human atrial preparations of α-adrenoceptors mediating positive inotropic effects. Naunyn-Schmiedeberg's Arch Pharmacol 302:333–336
17. Wagner J, Schümann HJ, Knorr A, Rohm N, Reidemeister JC (1980) Stimulation by adrenaline and dopamine but not by noradrenaline of myocardial α-adrenoceptors mediating positive inotropic effects in human atrial preparations. Naunyn-Schmiedeberg's Arch Pharmacol 312:99–102
18. Wenzel DG, Su JL (1966) Interactions between sympathomimetic amines and blocking agents on the rat ventricle strip. Arch Int Pharmacodyn 160:379–389

Author's address:

Dr. Jan-Bjørn Osnes, Department of Pharmacology, P.O. Box 1057 Blindern, N-0316 Oslo 3, Norway

Studies on the cellular mechanisms of action of positive and negative inotropic agents

A. M. Watanabe, F. Green, Z. Ahmad

Department of Medicine, Indiana University School of Medicine, Indiana, USA

Summary

We have reviewed the mechanism by which drugs that elevate cyclic AMP level modify myocardial contractility. We have presented preliminary evidence about the mechanism by which muscarinic agonists antagonize the effects of these drugs. Finally, we suggest that the protein phosphorylation experiments, particularly if done in dispersed myocytes, could be an efficient and cost-effective method of screening drugs which may act by elevating intracellular levels of cyclic AMP.

An important mechanism by which many drugs increase myocardial contractility is by activating the cyclic AMP – dependent protein kinase system [1]. There are multiple ways by which drugs can activate this system. First, agents might act on receptors located in the plasma membrane. When agonist drugs interact with these receptors, adenylate cyclase is activated thus leading to increases in intracellular cyclic AMP levels. Forskolin, the diterpene compound, acts directly on the catalytic subunit of adenylate cyclase to increase its activity and thus leas to elevations in intracellular levels of cyclic AMP. A third mechanism by which drugs can increase cyclic AMP is by inhibiting phosphodiesterase which converts cyclic AMP into its inactive metabolite 5'-AMP. All of these different classes of drugs lead to elevations in cyclic AMP levels and as a result, cyclic AMP dependent protein kinase is activated. Cyclic AMP-dependent protein kinase, in turn, phosphorylates proteins within the myocardial cell, and this leads to alterations in myocardial contractility [1].

When cyclic AMP dependent protein kinase is activated, a variety of different proteins within myocardial cells are phosphorylated [1]. Several of these proteins are thought to be involved in regulating myocardial contractility, including the slow calcium channel, troponin I, and phospholamban [1]. Various other proteins which are involved in regulating the metabolic state of the myocardial cell are also phosphorylated.

Among the phosphodiesterase inhibitors, there are a variety of different compounds that can inhibit the enzyme and lead to phosphorylation of various proteins by cyclic AMP-dependent protein kinase [2]. This includes theophylline, isobutyl-methylxanthine, amrinone, milrinone, and a number of the newer, positive inotropic agents, many of which are discussed in this volume. It appears that the agents must inhibit type 3-phosphodiesterase in order to be effective positive inotropic agents by the mechanism of activating cyclic AMP-dependent protein kinase. Interestingly, all of these agents, which work via the cyclic AMP dependent protein kinase mechanism, are antagonized by muscarinic agonists. This will be discussed later.

By perfusing isolated perfused ventricles with radiolabeled, inorganic phosphate, it is possible to radiolabel the ATP pool. When hearts that have been treated in such a manner are administered inotropic agents that act via the cyclic AMP-dependent protein kinase system, proteins which are phosphorylated incorporate radiolabeled ^{32}P. Thus, it is possible to detect and quantify such protein phosphorylation by autoradiographic techniques [3–5].

In our laboratory we have been studying the phosphorylation of several proteins, particularly phospholamban, troponin I, and to a lesser degree, sarcolemmal proteins which are phosphorylated when intracellular cyclic AMP levels are elevated [3–5]. Our studies have shown that phosphorylation of phospholamban correlates very well with the positive inotropic effects of agents that elevate cyclic AMP levels in the myocardial tissues. Phosphorylation of phospholamban correlates very well with the time-course of positive inotropic effects of various drugs that elevate cyclic AMP. Also, the concentration response of phosphorylation of phospholamban correlates well with the physiological repsonse of the heart to different concentrations of the drugs [3]. Finally, various biochemical parameters such as calcium uptake and calcium ATPase activity of sarcoplasmic reticulum are modified by these drugs in a time-course and concentration response that correlates well with phosphorylation of phospholamban. These types of studies strongly suggest that the phosphorylation of phospholamban contributes to the inotropic effects of drugs that elevate cyclic AMP levels.

All of the different compounds previously mentioned which elevate intracellular myocardial cyclic AMP levels, regardless of the mechanism by which that elevation occurs, lead to phosphorylation of phospholamban [3]. In addition, muscarinic agonists antagonize phosphorylation of phospholamban mediated by all of these different agents [5]. Much of our effort in recent years has been directed toward understanding the mechanism by which muscarinic agonists antagonize the phosphorylation of proteins in myocardial cells. Three possible mechanisms could account for this effect of muscarinic agonists: 1) inhibition of adenylate cyclase; 2) inhibition of cyclic AMP dependent protein kinase; and 3) activation of phosphatases. Inhibition of cyclic AMP dependent protein kinase has been evaluated in several other laboratories, and there is no evidence to indicate that activation of muscarinic receptors by choline esters has any effect on cyclic AMP-dependent protein kinase activity in myocardial cells. Muscarinic receptors are coupled to adenylate cyclase via a G protein which has been designated G_i. When muscarinic receptors are activated by agonists, they inhibit adenylate cyclase activity via G_i and this leads to a reduction in the amount of cyclic AMP generated by the enzyme [6]. Thus, when muscarinic receptors are stimulated, the steady state levels of cyclic AMP in myocardial cells are reduced. This certainly is a mechanism by which muscarinic agonists antagonize the physiological effects and the phosphorylation of proteins mediated by agents that elevate cyclic AMP by activating adenylate cyclase.

Our studies indicate, however, that this mechanism cannot totally explain the inhibitory action of muscarinic agonists on positive inotropic drugs that act via the cyclic AMP dependent protein kinase system. First, muscarinic agonists powerfully antagonize the positive inotropic effects and phosphorylation of proteins induced by agents that elevate cyclic AMP levels without activating adenylate cyclase; for example, phosphodiesterase inhibitors. Secondly, muscarinic agonists powerfully inhibit phosphorylation of proteins induced by some positive inotropic agents, such as phosphodiesterase inhibitors and forskolin, without substantially lowering intracellular cyclic AMP levels [5]. Similar observations have been made with adenosine [7]. Because of these types of observations, we have sought additional mechanisms for the effect of muscarinic agonists, and we have recently begun to examine the effects of muscarinic agonists on phosphoprotein phosphatase activities.

We have begun to examine the effects of muscarinic agonists on type-1 phosphatase activity and on inhibitor 1, which is a protein that regulates the activity of type-1 phosphatase. We have evidence that agents that elevate cyclic AMP levels, including isoproterenol and forskolin, increase inhibitor 1 activity, presumably by leading to phosphorylation of inhibitor 1, mediated by cyclic AMP dependent protein kinase. This increase in activity of inhibitor 1 would be expected to decrease the activity of phosphatase type 1, which is regulated by inhibitor 1. The muscarinic agonist acetylcholine antagonizes the effects of both isoproterenol and forskolin on inhibitor 1 activity.

We also have recent, preliminary evidence of the effect of agents that elevate cyclic AMP on type-1 phosphatase activity. Isoproterenol and forskolin both decrease type-1 phosphatase activity. Whether or not this decrease in activity is entirely due to the effect of these agents on inhibitor-1 activity is unknown. This is a possibility but, in addition, there could be other mechanisms by which these agents that elevate cyclic AMP levels regulate type-1 phosphatase activity. The muscarinic agonist acetylcholine antagonizes the effect of isoproterenol and forskolin on type-1 phosphatase activity. In addition, acetylcholine alone stimulates type-1 phosphatase activity.

Thus, in summary, the agents which elevate cyclic AMP levels in myocardial cells increase inhibitor-1 activity and decrease type-1 phosphatase activity. This would be expected to lead to increased phosphorylation of proteins. The muscarinic receptor agonist acetylcholine antagonizes the effect of isoproterenol and forskolin on inhibitor-1 activity and type-1 phosphatase activity. In addition, acetylcholine alone stimulates type-1 phosphatase activity. This would be expected to decrease the state of phosphorylation of various proteins. Thus, muscarinic agonists regulate protein phosphorylation by at least two mechanisms: 1) by inhibiting the activity of adenylate cyclase, which would lead to reduced steady state levels of intracellular cyclic AMP and thus diminish activation of cyclic AMP dependent protein kinase; and 2) by regulating protein phosphorylation by modulating the activity of type 1 phosphatase.

Finally, a few words about assessing protein phosphorylation as a method for evaluating the mechanism of action of new inotropic agents. Many different inotropic agents are thought to act by elevating cyclic AMP levels and thus activating cyclic AMP dependent protein kinase. A standard way of assessing such agents is by measuring steady-state tissue cyclic AMP levels. However, this approach is associated with significant difficulties. First, the radioimmunoassay for cyclic AMP is relatively difficult to perform and it can be difficult to generate consistently reproducible data, particularly if the drugs in question elevate cyclic AMP levels only modestly. A second, and perhaps more important problem, is the possibility that cyclic AMP may be compartmentalized within cells. For example, cyclic AMP may be membrane associated and cytoplasmic, and different drugs might affect only one or the other of these compartments. If total tissue cyclic AMP levels are measured, it is not possible to detect such compartmentalization. An approach to getting around such problems has been the measurement of cyclic AMP-dependent protein kinase activity. However, this assay is also difficult to perform and, depending on the species (in which there may be type-1 or type-2 cyclic AMP-dependent protein kinase), it may be more or less difficult to assay the activity of this enzyme.

Another approach to assessing the mechanism of action of newer inotropic agents is to evaluate the phosphorylation of proteins such as phospholamban. This can be done in isolated perfused ventricles, or more simply and efficiently in dispersed myocytes which can be prepared from hearts of small animals such as rats or guinea pigs. The myocytes can be prelabeled with radioactive inorganic phosphate. They are then exposed to the different drugs of interest. The cells are then solubilized in SDS, separated by PAGE and the protein phosphorylation evaluated by autoradiography. By using this method, we have assessed a number of inotropic agents, including amrinone, milrinone, UDCG 115, and UDCG 212, and we have compared such phosphorylation with that induced by isoproterenol and IBMX. All of these agents cause phosphorylation of phospholamban, particularly in guinea pig myocytes, which contain a substantial amount of phosphodiesterase type 3 which is inhibited by these agents. Interestingly, cells made from rat hearts, which do not contain significant amounts of phosphodiesterase type 3, do not demonstrate the phosphorylation of phospholamban in reponse to the agents, except at high concentrations of the drug. Other investigators have reported that guinea pig ventricles are also much more responsive to the positive inotropic effects of these agents than are rat ventricles.

It would be important in the future to attempt to correlate the phosphorylation of different proteins, such as phospholamban, with the inotropic effects of new drugs. If the correlation appeared to be strong – for example, in terms of time response and concentration response – this would provide supportive evidence that the drugs are acting by elevating intracellular cyclic AMP levels.

References

1. Watanabe AM, Lindemann JP (1984) Mechanisms of adrenergic and cholinergic regulation of myocardial contractility. In: Sperelakis N (ed) Physiology and pathophysiology of the heart. Nijhoff M., pp 377–404
2. Scholz H, Meyer W (1986) Phosphodiesterase-inhibiting properties of newer inotropic agents. Circulation 73(3 Pt 2):III99–108
3. Lindemann JP, Jones LR, Hathaway DR, Henry BG, Watanabe AM (1983) Beta-adrenergic stimulation of phospholamban phosphorylation and Ca^{2+}-ATPase activity in guinea pig ventricles. J Biol Chem 358:464–471
4. Lindemann JP, Watanabe AM (1985) Phosphorylation of phospholamban in intact myocardium: role of Ca^{2+}-calmodulin-dependent mechanisms. J Biol Chem 260:4516–4525
5. Lindemann JP, Watanabe AM (1985) Muscarinic cholinergic inhibition of B-adrenergic stimulation of phospholamban phosphorylation and Ca^{2+} transport in guinea pig ventricles. J Biol Chem 260:13122–13129
6. Watanabe AM, McConnaughey MM, Strawbridge RA, Fleming JW, Jones LR, Besch HR Jr (1978) Muscarinic cholinergic receptor modulation of beta adrenergic receptor affinity for catecholamines. J Biol Chem 253:4833–4836
7. Bohm M, Bruckner R, Newman J, Nose M, Schmitz W, Scholz H (1988) Adenosine inhibits the positive inotropic effect of 3-isobutyl-1-methylxanthine in papillary muscles without effect on cyclic AMP or cyclic GMP. Br J Pharmacol 93:729–738

Author's address:

August M. Watanabe, M.D. Department of Medicine, Indiana University School of Medicine, Indianapolis, Indiana, 46240, USA

On the mechanism of positive inotropic effects of alpha-adrenoceptor agonists

W. Schmitz, C. Kohl, J. Neumann, H. Scholz, J. Scholz

Abteilung Allgemeine Pharmakologie, Universitäts-Krankenhaus Eppendorf, Universität Hamburg, FRG

Summary

The positive inotropic effect of the alpha$_1$-adrenoceptor agonist phenylephrine is accompanied by an increase in the presumed second messengers inositol 1,4,5-trisphosphate (1,4,5-IP$_3$) and inositol 1,3,4,5-tetrakisphosphate (1,3,4,5-IP$_4$). Both 1,4,5-IP$_3$ and 1,3,4,5-IP$_4$ sensitize myocardial contractile proteins in chemically skinned fibers. In addition to the Ca^{++} releasing effect of 1,4,5-IP$_3$ from the sarcoplasmic reticulum the Ca^{++}-sensitizing effect of the inositol phosphates may play a role in alpha$_1$-adrenergic positive inotropism. In isolated heart muscle preparations from patients with endstage heart failure (due to dilated cardiomyopathy) beta-adrenergic as well as alpha$_1$-adrenergic effects are reduced compared to preparations from healthy hearts. The reduced beta-adrenergic effects can in part be explained by an increased content of signal transducing G$_1$-proteins. It is tempting to investigate whether other G proteins are also altered in severe congestive heart failure.

Introduction

The positive inotropic effects of catecholamines are predominantly mediated by myocardial beta-adrenoceptors. In addition, a significant part of the positive inotropic effect is brought about by stimulation of alpha$_1$-adrenoceptors (for reviews see [35, 10, 34]. Alpha$_1$-adrenoceptor-mediated positive inotropic effects can also be detected in isolated human heart muscle preparations [36, 9, 1, 21].

Unlike the beta-adrenergic response the alpha$_1$-adrenoceptor-induced positive inotropic effect is not accompanied by an increase in myocardial cAMP content (for review see [35, 10]). Instead, it may involve an cAMP-independent increase in slow calcium inward current (I$_{si}$) and/or a sensitization of the contractile proteins for calium [8, 3, 13]. More recent evidence suggests that the presumed second messenger inositol 1,4,5-trisphosphate (1,4,5-IP$_3$) is involved in alpha$_1$-adrenoceptor-mediated effects. Stimulation of alpha$_1$-adrenoceptors via coupling by a G-protein activates phospholipase C (PLC) which hydrolyzes phosphatidylinositol bisphosphate (PIP$_2$), generating the two presumed second messengers diacylglycerol (DG) and 1,4,5-IP$_3$ [2]. The G-protein involved in phospholipase C activation in the heart is of unknown nature [33, 5]. DG activates protein kinase C (PKC). The functional role of PKC in heart muscle contraction is as yet not fully understood [26]. 1,4,5-IP$_3$ has been shown to release Ca^{++} from intracellular Ca^{++} stores in pancreatic cells, hepatocytes, and a variety of other cell systems including smooth muscles [37, 2, 11, 20, 12]. In cardiac and skeletal muscle fibers some authors failed to observe an 1,4,5-IP$_3$-induced Ca^{++} release from the sarcoplasmic reticulum [24, 31]. However, there is evidence that 1,4,5-IP$_3$ is able to release Ca^{++} from skeletal or cardiac sarcoplasmic reticulum [17, 14, 27] and that 1,4,5-IP$_3$ can thereby increase the force of contraction [39, 40, 27].

In the present study we investigated the influence of the alpha$_1$-adrenoceptor agonist phenylephrine on force of contraction and on the formation of 1,4,5-IP$_3$ and its phosphorylation product inositol 1,3,4,5-tetrakis (1,3,4,5-IP$_4$) phosphate which is also assumed to possess second messenger function [18] in the heart. In chemically skinned porcine cardiac muscle fibers we studied whether or not 1,4,5-IP$_3$ and/or 1,3,4,5-IP$_4$ can account for the alpha$_1$-adrenoceptor-mediated increase in Ca^{++}-sensitivity of the contractile proteins. In addition the effects of phenylephrine on isolated human heart muscle preparations were investigated. In another series of experiments we studied whether alterations of the amount of G-(or N-)proteins (GTP or nucleotide binding regulatory proteins) may be involved in the reduced responsiveness to positive inotropic stimulants in heart muscle preparations from severe failing human hearts.

Methods

Force of contraction, inositol phosphates, and phospholipids were measured in electrically driven (1 Hz) left auricles isolated from rats as described previously [32, 35a]. In some experiments inositol 1,4,5-trisphosphate, inositol 1,3,4-trisphosphate and inositol 1,3,4,5-tetrakisphosphate were measured after separation on Partisil SAX 10 columns by high performance liquid chromatography (HPLC) with slight modifications as described by Irvine et al. [19]. These experiments were performed in the presence of 10 mmol/l LiCl. Chemically skinned porcine cardiac muscle fibers were prepared and Ca^{++}-sensitizing effects were measured according to Herzig et al. [16]. Force of contraction in human right ventricular trabeculae (paced at 0.5 Hz) and content of G$_i$-protein in human heart membrane preparations isolated from the explanted hearts from patients undergoing urgent heart transplantation were measured as described [22, 25]. All experiments with phenylephrine were performed in the presence of the beta-adrenoceptor antagonists propranolol (1 μmol/l) to exclude interference from beta-adrenoceptor stimulation. The values presented are means ± SEM. Significant differences were estimated by Student's t-test for paired or unpaired observations. A p-value of less than 0.05 was considered significant.

Results

Effects of phenylephrine on myocardial force of contraction, 1,4,5-IP$_3$ and 1,3,4,5-IP$_4$ content

Figure 1 (left panels) shows that the alpha$_1$-adrenoceptor agonist phenylephrine (10 μmol/l) increases force of contraction to about 160% of predrug value (Fig. 1 A) and simultaneously increases IP$_3$ to about 180% of control (Fig. 1 B) whereas phosphatidylinositol bisphosphate (PIP$_2$), the precursor of IP$_3$, is decreased to about 40% of control (Fig. 1 C). In contrast, the positive inotropic effect of the beta-adrenoceptor agonist isoprenaline was not accompanied by changes in IP$_3$ or PIP$_2$ (Fig. 1, right panels) indicating that the effects of phenylephrine cannot be mimicked by any positive inotropic intervention and thus are not unspecific effects.

To determine if another presumed second messenger, namely 1,3,4,5-IP$_4$, the phosphorylation product of 1,4,5-IP$_3$, is generated in the heart and whether it increases in response to alpha$_1$-adrenoceptor stimulation, we performed HPLC analyses of tissue ex-

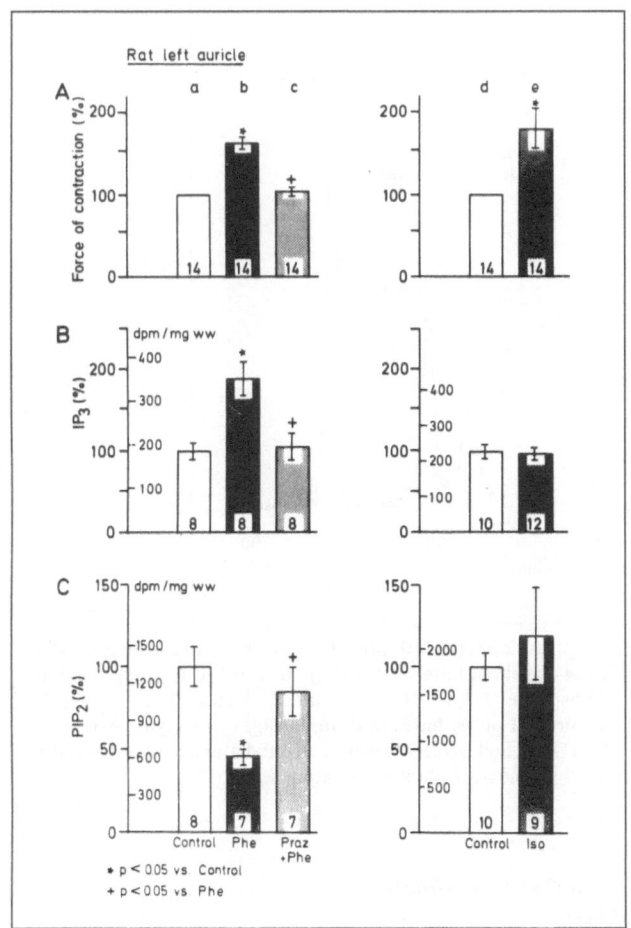

Fig. 1A–C. Effects of phenylephrine (Phe; 10 μmol/l) in the presence of propranolol (1 μmol/l) on force of concentration (**A**), inositol triphosphate content IP$_3$; **B** and phosphatidyl inositolbisphosphate (PIP$_2$) in electrically driven (1 Hz) rat left auricles (left panels). The effects of isoprenaline (Iso; 1 μmol/l) are shown for comparison (right panels). The effects of Phe were antagonized by prazosin (Praz; 0,11 μmol/l). Ordinates: force of contraction in percent of predrug value (**A**), IP$_3$ content (**B**) and PIP$_2$ content (**C**) in percent of control and in dpm per mg wet weight. The time of incubation with Phe or Iso was 15 min. The numbers in the columns denote the number of experiments

tracts from isolated rat electrically driven left auricles without and with stimulation by phenylephrine (10 μmol/l). The original chromatograms in Fig. 2 show that beside 1,4,5-IP$_3$ its phosphorylation product 1,3,4,5-IP$_4$, as well as the dephosphorylation product of 1,3,4,5-IP$_4$, the biologically inactive 1,3,4-IP$_3$ can be detected in the heart. All three inositol phosphates increase in response to phenylephrine.

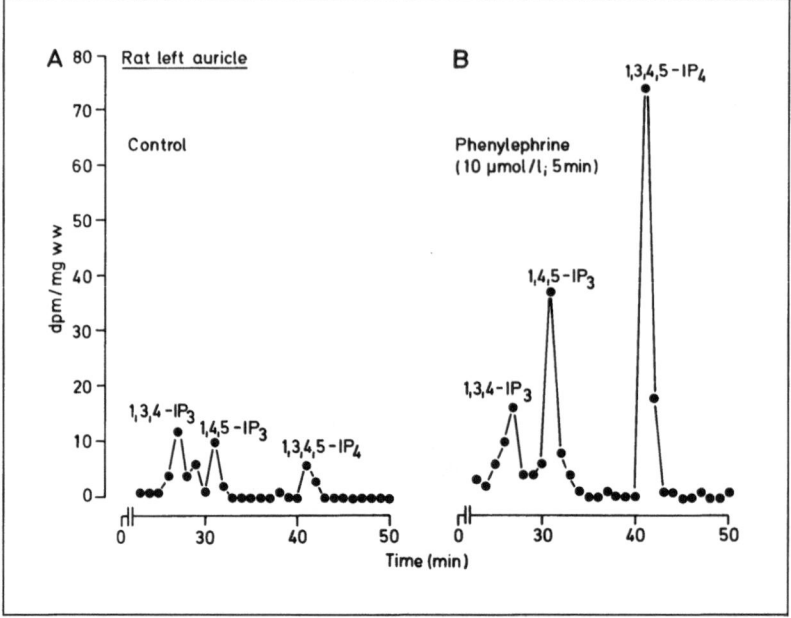

Fig. 2 A, B. HPLC analysis of the effect of phenylephrine (10 μmol/l); in the presence of propranolol (1 mol/l) on the content of inositol 1,3,4-trisphosphate (1,3,4-IP$_3$), inositol 1,4,5-trisphosphate (1,4,5-IP$_3$) and inositol 1,3,4,5-tetrakisphosphate (1,3,4,5-IP$_4$) in isolated, electrically driven (1 Hz) rat left auricles. Ordinate: content of the inositol phosphates in dpm per mg wet weight. Abscissae: elution time in min; 1 ml fractions of the ^{3}H-labelled products were collected and counted for radio-activity. Note that phenylephrine increased all inositol phosphates analysed

Effects of 1,4,5-IP$_3$ and 1,3,4,5-IP$_4$ on Ca^{++}-sensitivity
in skinned porcine cardiac muscle fibers

We investigated whether or not 1,4,5-IP$_3$ and 1,3,4,5-IP$_4$ increase the Ca^{++}-sensitivity of chemically skinned porcine cardiac muscle fibers. The effects of 1,4,5-IP$_3$ and 1,3,4,5-IP$_4$ (100 μmol/l each) on a skinned cardiac fiber pre-contracted with 2.3 μmol/l Ca^{++} are illustrated in Fig. 3 A. Ca^{++} was added at zero time for 15 min. 1,4,5-IP$_3$ applied for 15 min increased the Ca^{++}-induced force development. A further increase in force was obtained if 1,3,4,5-IP$_4$ was applied in the continued presence of 1,4,5-IP$_3$. The effects of 1,4,5-IP$_3$ and 1,3,4,5-IP$_4$ were readily reversible within 15 min upon washout with drug-free bathing solution. Figure 3 B summarizes the results obtained on 28 fibers isolated from three pig hearts. 1,4,5-IP$_3$ or 1,3,4,5-IP$_4$ alone (a + b) caused about a 10% sensiti-zation of the contractile proteins, i.e., they increased force to 110% of the Ca^{++}-induced force. When 1,3,4,5-IP$_4$ (100 μmol/l) was applied in addition to 1,4,5-IP$_3$ (100 μmol/l) a further increase in force to 128% of control was observed (c). These results were observed applying 1,3,4,5-IP$_4$ 15 min after 1,4,5-IP$_3$ in the continued presence of 1,4,5-IP$_3$. Similar results were obtained when 1,4,5-IP$_3$ and 1,3,4,5-IP$_4$ (100 μmol/l each) were applied si-multaneously (data not shown). Inositol 1-monophosphate (IP$_1$; 100 μmol/l) investigated for comparison did not affect Ca^{++}-induced force development (100.2 ± 1.3% of control; n = 6). Since 1,4,5-IP$_3$ and 1,3,4,5-IP$_4$ used were tri-lithium and tetra-potassium salts, re-

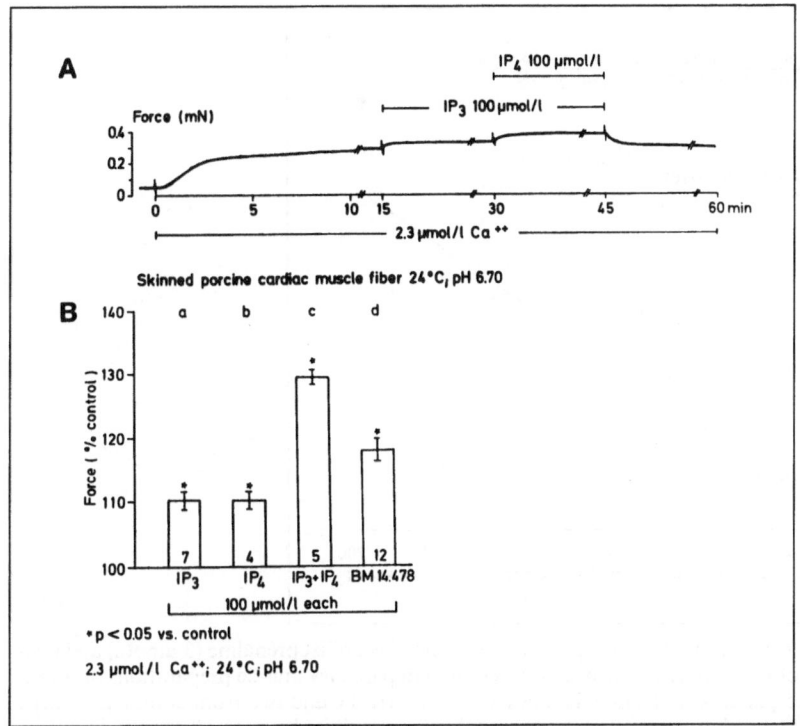

Fig. 3 A. Original recording of the effects of inositol 1,4,5-trisphosphate (IP_3) and inositol 1,3,4,5-tetrakisphosphate (IP_4) on Ca^{++}-induced isometric force development of a chemically skinned porcine cardiac muscle fiber. A submaximal contraction was induced by 2.3 µmol/l Ca^{++} applied for 15 min and was further increased by IP_3 (100 µmol/l) for an additional 15 min. IP_4 (100 µmol/l) applied for 15 min in the continued presence of IP_3 further increased contraction. The effects of IP_3 and IP_4 were readily reversible within 15 min upon removing IP_3 and IP_4 by changing the solution against IP_3- and IP_4-free bathing solution

Fig. 3 B. Effects of IP_3 (a) and IP_4 (b) alone or in combination (c; 100 µmol/l each) on Ca^{++}-induced isometric force development. For comparison the effect of the Ca^{++}-sensitizing agent BM 14.478 (d; 100 µmol/l) is shown. The data are expressed as percent force induced by 2.3 µmol/l Ca^{++} (% control). The Ca^{++}-induced force amounted to 0.27 ± 0.03 mN (n=13). The numbers in the columns denote the numbers of experiments. Asterisks denote significant differences vs control

spectively, we investigated whether 300 µmol/l LiCl or 400 µmol/l KCl had any influence on the Ca^{++}-induced force; neither salt affected force development (data not shown). For comparison the well-known Ca^{++}-sensitizing positive inotropic agent BM 14.478 (7,7-dimethyl-2-((4-pyridyl)-6,7-dihydro-3H,-5H pyrrolo (2,3-f)benzimidazol-6-one; 100 µmol/l) was investigated. As shown previously [15a] BM 14.478 increased force to 117% of control (Fig. 3 Bd). Furthermore, cAMP (100 µmol/l) had a relaxing effect on the skinned fibers, i.e., it reduced Ca^{++}-induced force by about 50% in accord with published data [30]. Thus, with the model used, both Ca^{++}-sensitizing and Ca^{++}-desensitizing effects can be detected.

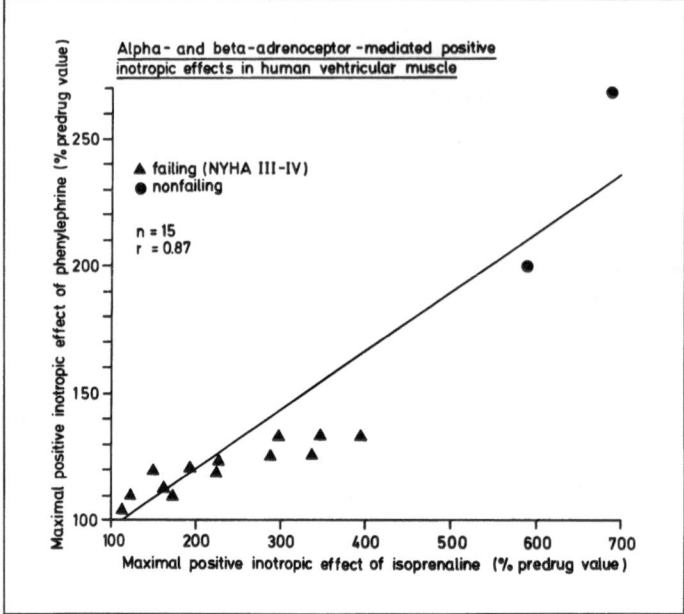

Fig. 4. Correlation between the maximal positive inotropic effect of isoprenaline (3 µmol/l) and phenylephrine (30 µmol/l) in electrically driven (0.5 Hz) human papillary muscle preparations of trabulae isolated from 15 patients: 13 from failing hearts NYHA III–IV and two from nonfailing hearts. Preparations with reduced responsiveness to isoprenaline also exhibited only small positive inotropic effects in response to phenylephrine

Alpha₁-adrenoceptor-mediated positive inotropic effects in human heart muscle preparations

It is well known that cAMP-mediated positive inotropic effects are strongly reduced in preparations from human hearts with endstage heart failure (see [34]). Figure 4 shows that the maximal positive inotropic response to the alpha₁-adrenoceptor agonist phenylephrine correlates with the beta-adrenoceptor-mediated positive inotropic effects in preparations from 13 failing hearts (NYHA III–IV) and from two nonfailing hearts. There is a pronounced phenylephrine-induced (about 190% and 270% of predrug value) as well as a pronounced isoprenaline-induced (600–700% of predrug value) increase in force of contraction in the nonfailing hearts. The effects of both substances are strongly reduced in the failing heart.

G-proteins in the human heart

An altered amount of signal transducing G-proteins could explain the reduced responsiveness of the failing human heart to positive inotropic drugs. Figures 5A and B show an increased labelling of a 40 kDA protein, most probably the α subunit of G_i, in failing hearts from patients with idiopathic dilated cardiomyopathy as compared to nonfailing hearts. In inflammatory heart disease (myocarditis) no change in G_i-content was ob-

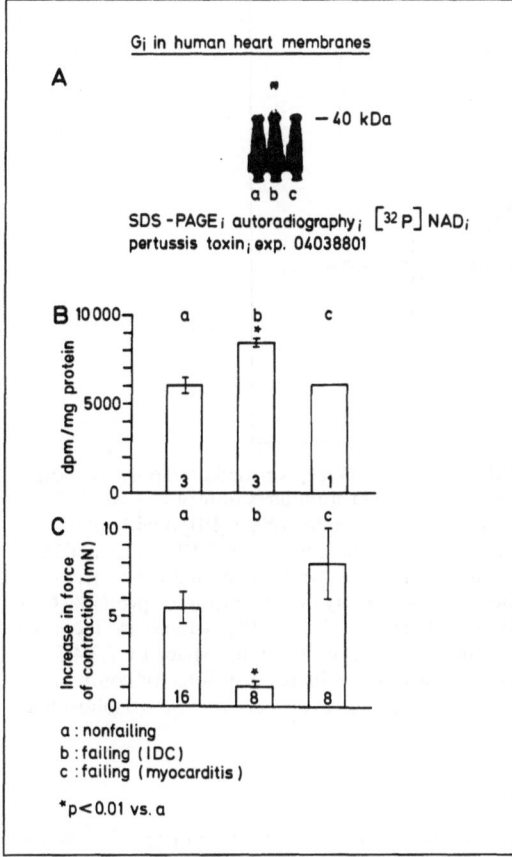

Gᵢ in human heart membranes

A

− 40 kDa

a b c

SDS -PAGE; autoradiography; [^{32}P] NAD;
pertussis toxin; exp. 04038801

B

C

a : nonfailing
b : failing (IDC)
c : failing (myocarditis)

*p<0.01 vs. a

Fig. 5 A. Autoradiography of SDS poly-acrylamide gel electrophoresis of radioactive products resulting from pertussis toxin catalyzed ADP-ribosylation of cardiac membranes from nonfailing (a) or failing human hearts obtained from patients with idiopathic dilated cardiomyopathy (b; IDC) and from a patient with myocarditis (c). Identical amounts of protein were loaded on each track

Fig. 5 B. Histograms of levels of radioactivity incorporated at 49 kDa. Peaks were cut out and counted for radioactivity and dpm were referred to protein (25 µg) applied to each lane. Experiments were performed in duplicate. Numbers in the columns denote the number of different patients

Fig. 5 C. Increase in force of contraction 5 min after application of isoprenaline (1 µmol/l) in millinewton (mN). Ventricular trabeculae were obtained from the same hearts as in Fig. 3 B. Numbers in the columns denote the number of different trabeculae. Predrug values of force of contraction were 1.84 ± 0.34 (a), 1.62 ± 0.28 (b), and 1.51 ± 0.27 mN (c). Asterisks indicate significant difference vs control (a)

served. The increase in Gᵢ coincided with a reduced contractile response to β-adrenoceptor stimulation with isoprenaline. In myocarditis the positive inotropic effect of isoprenaline was not reduced.

Discussion

The positive inotropic effect of the alpha$_1$-adrenoceptor agonist phenylephrine was accompanied by an increase in the presumed second messengers 1,4,5-IP$_3$ and 1,3,4,5-IP$_4$. It was shown previously that an increase in IP$_3$ precedes the alpha$_1$-adrenoceptor-mediated increase in force of contraction [28, 32, 35a] fulfilling a prerequisite for a possible second messenger role. However, the methods used in these studies could not distinguish between the biologically active isomer 1,4,5-IP$_3$ and the inactive or (much less active) isomer 1,3,4-IP$_3$. In addition, in these studies the IP$_3$ fraction may have been contaminated by the phosphorylation product of 1,4,5-IP$_3$ the 1,3,4,5-IP$_4$. Attempts to directly detect 1,3,4,5-IP$_4$ in the heart have failed so far [41, 29]. However, Renard and Poggioli [29] described an 1,4,5-IP$_3$-3-kinase activity in the heart which produced 1,3,4,5-IP$_4$ from

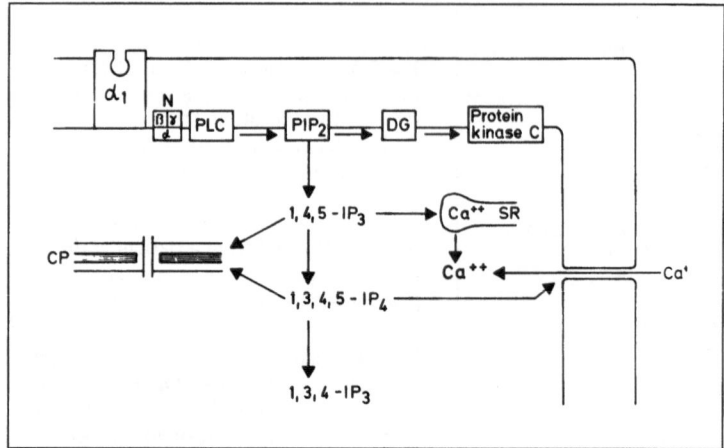

Fig. 6. Hypothetical scheme of the mechanism of the alpha$_1$-adrenoceptor-mediated positive inotropic effect. Occupation of the receptor (α_1) via a N (or G)-protein stimulates phospholipase C (PLC) leading to an enhanced hydrolysis of phosphatidyl inositol 4,5-bisphosphate (PIP$_2$) with an increased production of diacylglycerol (DG) and inositol 1,4,5-trisphosphate (1,4,5-IP$_3$). DG activates protein kinase (1,4,5-IP$_3$ may increase cytosolic Ca^{++} concentration by releasing Ca^{++} from the sarcoplasmic reticulum [SR]) and may increase the Ca^{++}-sensitivity of the contractile proteins (CP). 1,4,5-IP$_3$ is phosphorylated to inositol 1,3,4,5-tetrakisphosphate (1,3,4,5-IP$_4$) which also increases the Ca^{++}-sensitivity of CP. An increase of Ca^{++}-entry from the extracellular space by 1,3,4,5-IP$_4$ has been found in sea urchin eggs. All these effects can lead to an increase in force of contraction. 1,3,4,5-IP$_4$ is dephosphorylated to the inactive (or much less active) inositol 1,3,4-trisphosphate (1,3,4-IP$_3$)

exogenously applied 1,4,5-IP$_3$. In the present study we provide direct evidence for the existence of 1,3,4,5-IP$_3$ in the heart.

It is still a matter of debate whether or not 1,4,5-IP$_3$ releases Ca^{++} from the cardiac sarcoplasmic reticulum [17, 24, 14, 27] and hence whether or not it may function as a second messenger for the alpha$_1$-adrenoceptor-mediated increase in force of contraction. Concerning the physiological role of 1,3,4,5-IP$_4$ it has been shown that in combination with 1,4,5-IP$_3$ it somehow facilitates Ca^{++} entry from the extracellular into the intracellular space in sea urchin eggs [18] and activates Ca^{++}-dependent K$^+$-channels in the lacrimal gland [23]. In the heart the effects of 1,3,4,5-IP$_4$ have not been investigated. The present finding that both 1,4,5-IP$_3$ and 1,3,4,5-IP$_4$ sensitized the contractile proteins of chemically skinned porcine cardiac muscle fibers may provide a physiological role in the regulation of myocardial force of contraction of these inositol phosphates. These data are in line with the Ca^{++} sensitizing effects of alpha$_1$-adrenoceptor agonists found in the rabbit heart with the aequorin technique [3, 13]. A Ca^{++}-sensitizing effect of 1,4,5-IP$_3$ has also been found in the skinned skeletal muscle fibers [38]. One might argue that the concentration of the inositol phosphates to produce Ca^{++}-sensitization are rather high (100 µmol/l). However, the endogenously produced concentrations of 1,4,5-IP$_3$ or 1,3,4,5-IP$_4$ are not known and the concentrations necessary to elicit Ca^{++}-release from the SR in skeletal or cardiac muscle are comparably high [27, 39].

In summary it is reasonable to conclude that the alpha$_1$-adrenoceptor-mediated increase in 1,4,5-IP$_3$ as well as in 1,3,4,5-IP$_4$ provides a prerequisite for a possible second messenger role of these inositol phosphates in $alpha_1$-adrenoceptor-mediated positive inotropism. The positive inotropic effect of phenylephrine may partly be brought about by the Ca^{++}-sensitizing effects of 1,4,5-IP_3 and 1,3,4,5-IP_4 (Fig. 6).

Human heart muscle preparations

The present data demonstrate that the well-known reduced beta-adrenoceptor-mediated positive inotropic effect coincided with a reduced $alpha_1$-adrenergic positive inotropic effect in isolated heart muscle preparations from patients with severe congestive heart failure. The reduced responsiveness to beta-adrenergic agonists may in part be explained by beta-adrenoceptor down-regulation in severe heart failure [6, 7]. In addition the about 40% increase in G_i content can also explain the reduced responsiveness of beta-adrenoceptor agonists as well as of phosphodiesterase inhibitors [15]. The increased amount of the inhibitory signal transducing G_i protein could keep the adenylate cyclase activity reduced, resulting in cAMP production insufficient to produce maximal positive inotropic effects even in the presence of phosphodiesterase inhibitors. The fact that an increased G_i content as well as a reduced beta-adrenergic response has not been detected in myocarditis indicates that the increase in G_i does not occur in heart failure from any etiology. However, it may constitute an etiolocigal factor in dilated cardiomyopathy.

The increase in adenylate cyclase-coupled G_i cannot explain the reduced cAMP-independent effect of the $alpha_1$-adrenoceptor agonist phenylephrine. Since down-regulation of $alpha_1$-adrenoceptors in severe congestive heart failure has not been found [4] it is tempting to investigate whether other G-proteins are also altered in heart failure.

Acknowledgements. This work was supported by the Deutsche Forschungsgemeinschaft.

References

1. Aass H, Skomedal T, Osnes JB (1983) Demonstration of an alpha-adrenoceptor-mediated inotropic effect of norepinephrine in rabbit papillary muscle. J Pharmacol Exp Ther 226:572–578
2. Berridge MJ, Irvine RF (1984) Inositol trisphosphate, a novel second messenger in cellular signal transduction. Nature 312:315–321
3. Blinks JR, Endoh M (1986) Modification of myofibrillar responsiveness to Ca^{++} as an inotropic mechanism. Circulation 73 (suppl III):III-85–III-98
4. Böhm M, Diet F, Feiler G, Kemkes B, Erdmann E (1988) α-adrenoceptors and α-adrenoceptor-mediated positive inotropic effects in failing human myocardium. J Cardiovasc Pharmacol 12:357–364
5. Böhm M, Schmitz W, Scholz H (1987) Evidence against a role of a pertussis toxin-sensitive guanine nucleotide-binding protein in the $alpha_1$-adrenoceptor-mediated positive inotropic effect in the heart. Naunyn Schmiedebergs Arch Pharmacol 335:476–479
6. Bristow MR, Ginsburg R, Umans V, Fowler M, Minobe W, Rasmussen R, Zera P, Menlove R, Shah P, Jamieson S, Stinson EB (1986) β_1- and β_2-adrenergic-receptor subpopulations in nonfailing and failing human ventricular myocardium: coupling of both receptors subtypes to muscle contraction and selective β_1-receptor down-regulation in heart failure. Circulation. 59:297–309
7. Brodde OE, Schüler S, Kretsch R, Brinkmann M, Borst HG, Hetzer R, Reidemeister C, Warnecke H, Zerkowski HR (1986) J Cardiovasc Pharmacol 8:1235–1242
8. Brückner R, Scholz H (1984) Effects of α-adrenoceptor stimulation with phenylephrine in the presence of propranolol on force of contraction, slow inward current and cyclic AMP content in the bovine heart. Br J Pharmacol 82:223–232

9. Brückner R, Meyer W, Mügge A, Schmitz W, Scholz H (1984) α-adrenoceptor-mediated positive inotropic effect of phenylephrine in isolated human ventricular myocardium. Eur J Pharmacol 99:345–347
10. Brückner R, Mügge A, Scholz H (1985) Existence and functional role of alpha$_1$-adrenoceptors in the mammalian heart. J Mol Cell Cardiol 17:639–645
11. Burgess GM, Godfrey PP, McKinney JS, Berridge MJ, Irvine RF, Putney JW (1984) The second messenger linking receptor activation to internal Ca release in liver. Nature 309:63–66
12. Dawson AP, Irvine RF (1984) Inositol (1,4,5)-trisphosphate promoted Ca^{2+} release from microsomal fractions of rat liver. Biochem Biophys Res Commun 120:858–864
13. Endoh M, Blinks JR (1988) Actions of sympathomimetic amines on the Ca^{2+} transients and contractions of rabbit myocardium: Reciprocal changes in myofibrillar responsiveness to Ca^{2+} mediated through alpha- and beta-adrenoceptors. Circ Res 62:247–265
14. Fabiato A (1986) Inositol (1,4,5)-trisphosphate-induced release of Ca^{++} from the sarcoplasmic reticulum of skinned cardiac cells. Biophys J 49:190a
15. Feldmann MD, Copelas L, Gwathmey JK, Phillips P, Warren SE, Schoen FJ, Grossman W, Morgan JP (1987) Deficient production of cyclic AMP: pharmacologic evidence of an important cause of contractile dysfunction in patients with end-stage heart failure. Circulation 75:331
15a. Freund P, Müller-Beckmann B, Strein K, Kling L, Rüegg JC (1987) Ca^{2+}-sensitizing effect of BM 14.478 on skinned cardiac muscle fibres of guinea-pig papillary muscle. Eur J Pharmacol 136:243–246
16. Herzig JW, Feile K, Rüegg CJ (1981) Activating effects of AR-L 115 BS on the Ca^{2+} sensitive force, stiffness and unloaded shortening velocity (V$_{max}$) in isolated contractile structures from mammalian heart muscle. Drug Res 31:188–191
17. Hirata M, Suematsu E, Hashimoto T, Hamachi T, Koga T (1984) Release of Ca^{++} from a non-mitochondrial store site in peritoneal macrophages treated with saponin by inositol 1,4,5-trisphosphate. Biochem J 223:229–236
18. Irvine RF, Moor RM (1986) Microinjection of inositol 1,3,4,5-tetrakisphosphate activates seaurchin eggs by a mechanism dependent on external Ca^{2+}. Biochem J 240:917
19. Irvine RF, Anggard EE, Letcher AJ, Downes CP (1985) Metabolism of inositol 1,4,5-trisphosphate and inositol 1,3,4-trisphosphate in rat parotid glands. Biochem J 229:505–511
20. Joseph SK, Thomas AP, Williams RJ, Irvine RF, Williamson JR (1984) Myo-inositol 1,4,5-trisphosphate. A second messenger for the hormonal mobilization of intracellular Ca^{2+} in liver. J Biol Chem 259:3077–3081
21. Kohl C, Schmitz W, Scholz H, Scholz J, Tóth M, Döring V, Kalmár P (1988) Evidence for alpha$_1$-adrenoceptor-mediated increase of inositol trisphosphate in the human heart. J Cardiovasc Pharmacol: in press
22. Meyer W, Neumann J, Nose M, Schmitz W, Scholz H, Scholz J, Starbatty J, Steinkraus V, Döring V, Kalmár P, Rödiger W, Klöppel G, Hanrath P (1988) Inotropic response in CHF: myocarditis vs dilated cardiomyopathy. Am Heart J 115:1346–1348
23. Morris AP, Gallacher DV, Irvine RF, Petersen OH (1987) Synergism of inositol trisphosphate and tetrakisphosphate in activating Ca^{2+} dependent K$^+$ channels. Nature 330:653–655
24. Movsesian MA, Thomas AP, Selak M, Williamson JR (1985) Inositol trisphosphate does not release Ca^{2+} from permeabilized cardiac myocytes and sarcoplasmic reticulum. FEBS Lett 185:328–332
25. Neumann J, Schmitz W, Scholz H, von Meyerinck L, Döring V, Kalmár P (1988) Increase of myocardial G$_i$-proteins in human heart failure. Lancet: in press
26. Nishizuka Y (1986) Studies and perspectives of protein kinase C. Science (Wash. DC) 233:305–312
27. Nosek TM, Williams MF, Zeigler ST, Godt RE (1986) Inositol trisphosphate enhances calcium release in skinned cardiac and skeletal muscle. Am J Physiol 250:C807–C811
28. Poggioli J, Sulpice JC, Vassort G (1986) Inositol phosphate production following alpha$_1$-adrenergic, muscarinic or electrical stimulation in isolated rat hearts. FEBS Lett 206:292–298
29. Renard D, Poggioli J (1987) Does the inositol tris/tetrakisphosphate pathway exist in rat heart? FEBS Lett 217:117–123
30. Rüegg JC, Pfitzer G, Eubler D, Zeugner C (1984) Effect on contractility of skinned fibres from mammalian heart and smooth muscle by a new benzimidazole derivative, 4,5-dihydro-6-(2-(4-

methodyphenyl)-1H-benzimidazol-5-yl)-5-methyl-3(2H)-pyridazinone. Drug Res 34:1736–1738

31. Scherrer NM, Ferguson JE (1985) Inositol 1,4,5-trisphosphate is not effective in releasing calcium from skeletal sarcoplasmic reticulum microsomes. Biochem Biophys Res Commun 128:1064–1070

32. Schmitz W, Scholz H, Scholz J, Steinfarth M (1987) Increase in IP$_3$ precedes alpha-adrenoceptor-induced increase in force of contraction in cardicac muscle. Eur J Pharmacol 140:109–111

33. Schmitz W, Scholz H, Scholz J, Steinfath M, Lohse M, Puurunen J, Schwabe U (1987) Pertussis toxin does not inhibit the alpha$_1$-adrenoceptor-mediated effect on inositol phosphate production in the heart. Eur J Pharmacol 134:377–378

34. Schmitz W, Scholz H, Erdmann E (1987) Effects of α- and β-adrenergic agonists, phosphodiesterase inhibitors and adenosine on isolated human heart muscle preparations. TIPS 8:447–450

35. Scholz H (1980) Effects of beta- and alpha-adrenoceptor activators and adrenergic transmitter releasing agents on the mechanical activity in the heart. In: Szekeres L (ed) Adrenergic activators and inhibitors, Handbook of experimental pharmacology; Springer, Heidelberg, Vol 54/I, 651–733

35a. Scholz J, Schaefer B, Schmitz W, Scholz H, Steinfath M, Lohse M, Schwabe U, Puurunen J (1988) Alpha$_1$-adrenoceptor-mediated positive inotropic effect and inositol trisphosphate increase in mammalian heart. J Pharmacol Exp Ther 245:327–335

36. Schümann HJ, Wagner J, Knorr A, Reidemeister JC, Sadony V, Schramm G (1978) Demonstration in human atrial preparations of α-adrenoceptors mediating positive inotropic effects. Naunyn Schmiedebergs Arch Pharmacol 302:333–336

37. Streb H, Irvine RF, Berridge MJ, Schulz I: Release of Ca^{2+} from a nonmitochondrial intracellular store in pancreatic cells by inositol-1,4,5-trisphosphate. Nature 306:67–69

38. Thieleczek R, Heilmeyer Jr LMG (1986) Inositol 1,4,5-trisphosphate enhances Ca^{2+}-sensitivity of the contractile mechanism of chemically skinned rabbit skeletal muscle fibres. Biochem Biophys Res Commun 135:662–669

39. Vergara J, Tsien RY, Delay M (1985) Inositol 1,4,5-trisphosphate: A possible chemical link in excitation-contraction coupling in muscle. Proc Natl Acad Sci 82:6352–6356

40. Volpe P, Salviati G, Di Virgilio F, Pozzan T (1985) Inositol 1,4,5-trisphosphate induces calcium release from sarcoplasmic reticulum of skeletal muscle. Nature 316:347–349

41. Woodcock EA, Schmank White LB, Smith A, Mc Leod JK (1987) Stimulation of phosphatidylinositol metabolism in the isolated, perfused rat heart. Circ Res 61:625–631

Author's address:

Priv.-Doz. Dr. Wilhelm Schmitz, Abteilung Allgemeine Pharmakologie, Universitäts-Krankenhaus Eppendorf, Martinistraße 52, D-2000 Hamburg 20, FRG

Cytosolic sodium concentration regulates contractility of cardiac muscle

S.-S. Sheu

Department of Pharmacology, University of Rochester,
School of Medicine and Dentistry, Rochester, New York, USA

Summary

The present chapter provides experimental evidence to show that intracellular Na^+ concentration regulates cardiac contractility effectively by altering intracellular Ca^{2+} concentration via the Na-Ca exchange. This steep coupling between the Na^+ and Ca^{2+} electrochemical gradients implies that a change in intracellular Na^+ concentration is accompanied by a concomitant change in intracellular Ca^{2+} concentration (and, therefore, contractility). Under the physiologic conditions, each cardiac action potential alters intracellular Na^+ concentration in a dynamic manner. Therefore, Na-Ca exchange can regulate cardiac contraction from a beat-to-beat basis.

Introduction

In mammalian hearts, the magnitude of contraction depends intimately on the amount of Ca^{2+} release from the sarcoplasmic reticulum [13]. During the plateau phase of the cardiac action potential Ca^{2+} flows into the cells, mainly through voltage-operated Ca^{2+} channels, and triggers release of Ca^{2+} from the sarcoplasmic reticulum. This transient increase in cytosolic Ca^{2+} concentration removes the inhibitory function of the troponin-tropomyosin complex through Ca^{2+} binding to the troponin C. Consequently, actin and myosin are able to form crossbridges that result in the generation of contractile force.

It is generally agreed that the sarcoplasmic reticulum has high affinity Ca^{2+} transport proteins for sequestering Ca^{2+} in the cytosol. Therefore, the level of cytosolic Ca^{2+} could influence the degree of Ca^{2+} loading in the sarcoplasmic reticulum. The steady state cytosolic Ca^{2+} concentration is reached when the Ca^{2+} entry and Ca^{2+} exit across the plasma membrane are equal.

In cardiac muscle cells, Na-Ca exchange is one of the major Ca^{2+} transporting systems in the plasma membrane that establishes steady state cytosolic Ca^{2+} concentration [16]. Na-Ca exchange uses the energy of the Na^+ electrochemical gradient to transport Ca^{2+} either into or out of the cytosol. At equilibrium state for such a transport system, the following relationship will hold according to thermodynamic principles [11]:

$$\Delta\mu_{Ca} = n\Delta\mu_{Na}, \tag{1}$$

where $\Delta\mu_{Ca}$ is the Ca^{2+} electrochemical gradient, $\Delta\mu_{Na}$ is the Na^+ electrochemical gradient, and n is the stoichiometry of the exchange (the number of Na^+ ions exchanged for each Ca^{2+}). The terms $\Delta\mu_{Ca}$ and $\Delta\mu_{Na}$ are defined as:

$$\Delta\mu_{Ca} = Z_{Ca}F(E_{Ca} - V_m) = RT\ln\frac{[Ca^{2+}]_0}{[Ca^{2+}]_i} - 2FV_m, \tag{2}$$

$$\Delta\mu_{Na} = Z_{Na}F(E_{Na} - V_m) = RT\ln\frac{[Na^+]_0}{[Na^+]_i} - FV_m, \tag{3}$$

where Z_{Ca} and Z_{Na} are the valences of Ca^{2+} and Na^+, respectively, E_{Ca} and E_{Na} are the Ca^{2+} and Na^+ equilibrium potential, respectively, V_m is the membrane potential, and R, T, and F have their usual meanings.

Using Eqs. (2) and (3), Eq. (1) can be expanded to give:

$$[Ca^{2+}]_i = [Ca^{2+}]_0 \left(\frac{[Na^+]_i}{[Na^+]_0}\right)^n \exp\frac{(n-2)V_mF}{RT}. \tag{4}$$

This equation indicates that the cytosolic Na^+ concentration is one of the major determinants for the cytosolic Ca^{2+} concentration. Therefore, cytosolic Na^+ concentration can regulate contractility of cardiac muscle cell.

In this chapter, I will use the following three examples to demonstrate the importance of cytosolic Na^+ concentration in influencing the myocardial contractility: (a) force-frequency relationship; (b) negative inotropic and antiarrhythmic actions by lidocaine; and (c) positive inotropic actions of cardiotonic steroids.

Role of intracellular Na^+ in rate-dependent changes of cardiac contraction

In cardiac muscle cells, the force of contraction is markedly dependent on the rate of beating [1]. This force-frequency relationship is illustrated in Fig. 1 [3]. The upper trace shows

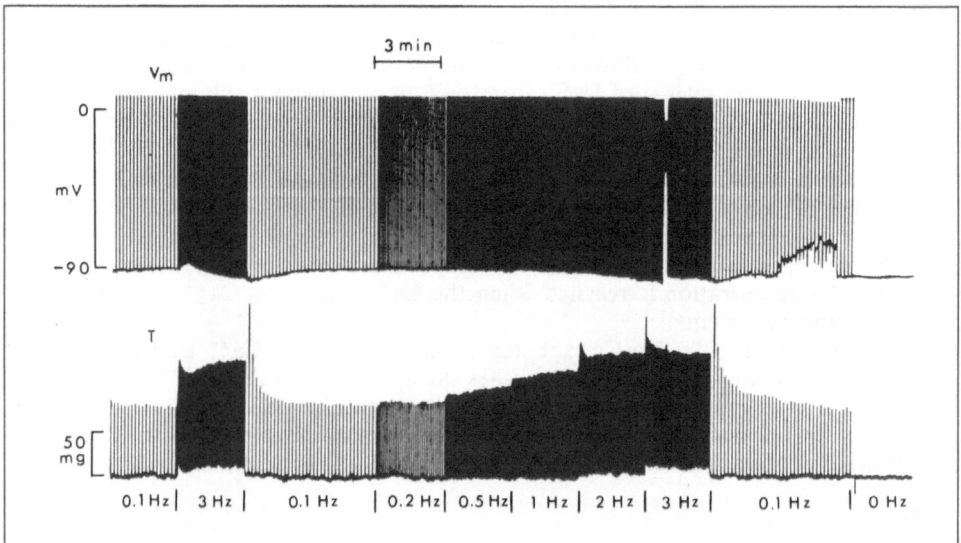

Fig. 1. Recording of membrane potential (V_m) in the upper trace and tension (T) in the lower trace during changes in stimulation rate (Hz) in a sheep ventricular trabecula (from Fig. 3 of [3])

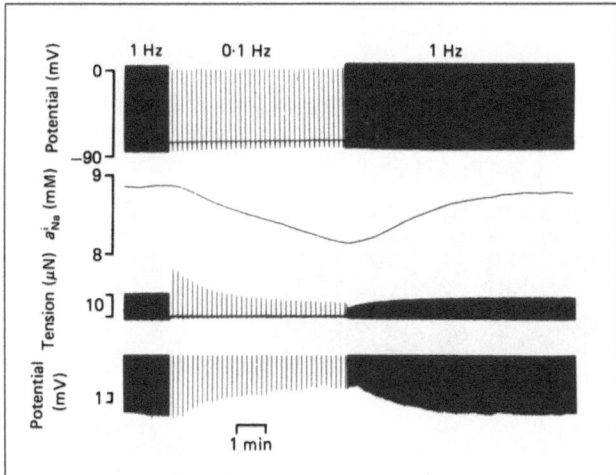

Fig. 2. Effects of stimulation rate on intracellular Na$^+$ activity (a$_{Na}^i$), membrane potential, and twitch tension in sheep cardiac Purkinje fibers. The membrane potential at the bottom trace is an amplification of the lower part of the record shown at the top trace (from Fig. 1 of [9])

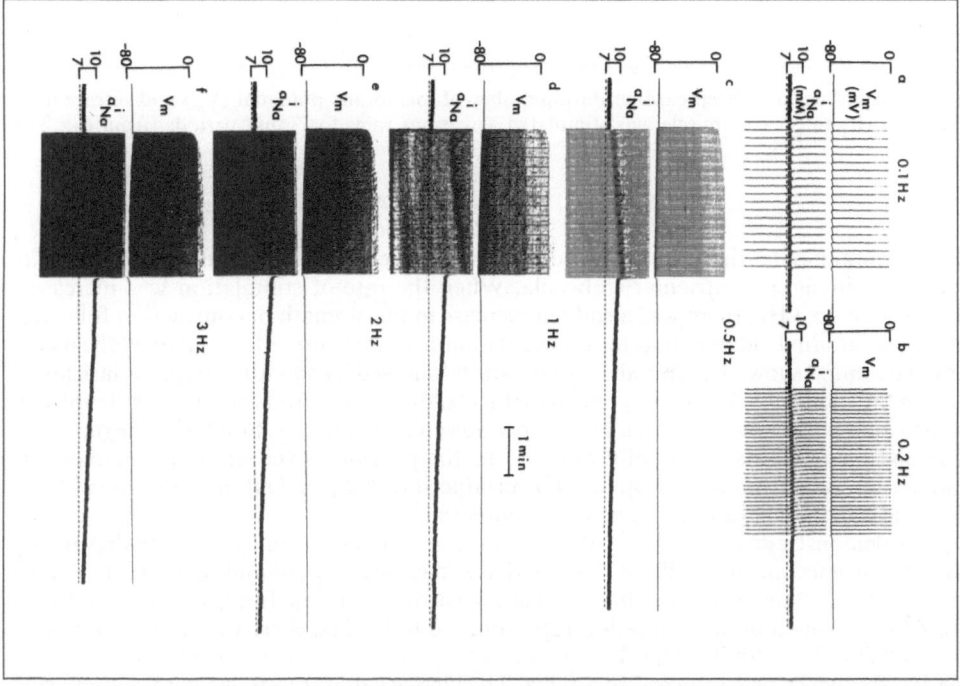

Fig. 3. Recording in a sheep cardiac Purkinje fiber of membrane potential (V$_m$) and intracellular Na$^+$ activity (a$_{Na}^i$). The muscle was stimulated at various rates for 3-min periods (from Fig. 6 of [3])

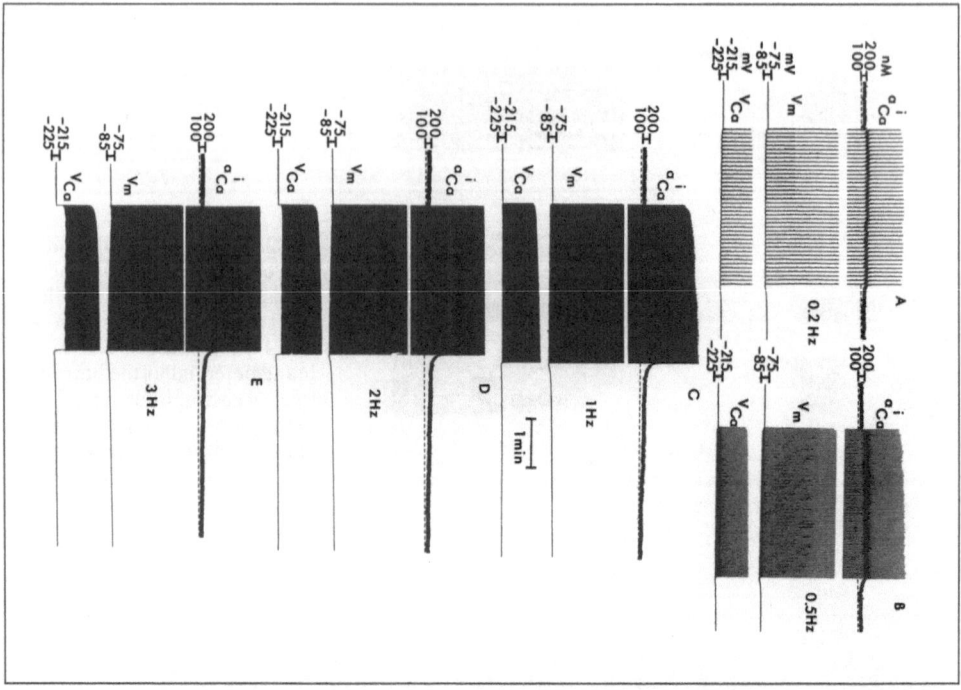

Fig. 4. Recording in a sheep cardiac Purkinje fiber of membrane potential (V_m) and intracellular Ca^{2+} activity (a_{Ca}^i). The muscle was stimulated at various rates for 3-min periods (from Fig. 2 of [8])

the membrane potential recording and the lower trace shows the contractile force measurements in sheep ventricular trabecula. When the rate of stimulation was increased from 0.1 Hz to 3 Hz, there was a sudden increase in the strength of contraction followed by a more gradual increase that took several minutes to develop. Maximal diastolic membrane potential showed an initial depolarization followed by a gradual hyperpolarization during the same time. Following return to 0.1 Hz, there was a transient increase in muscle contraction and a gradual return to the previous contraction size. Similarly, the maximal diastolic potential also gradually returned to the previous level. Stepwise increases in stimulation rate, from 0.1 Hz up to 3 Hz, produced rate-dependent increases in contraction and hyperpolarization of membrane potential.

To demonstrate the role of intracellular Na^+ in regulating heart rate-dependent changes in tension, intracellular Na^+ activity, membrane potential, and tension were measured simultaneously in a shortened sheep cardiac Purkinje fiber, as shown in Fig. 2 [9]. After a reduction in stimulation rate from 1.0 to 0.1 Hz, there was a gradual fall of intracellular Na^+ activity. This decrease in intracellular Na^+ was reversed when the stimulation frequency was returned to 1.0 Hz. The changes in intracellular Na^+ activity can be explained if each action potential is associated with net influx of Na^+ and if the Na^+-K^+ pump slowly accommodates to the new rate of Na^+ influx when the frequency is altered. Note that whereas the changes in intracellular Na^+ activity in response to changes in frequency were monophasic, the tension responses were biphasic. For ex-

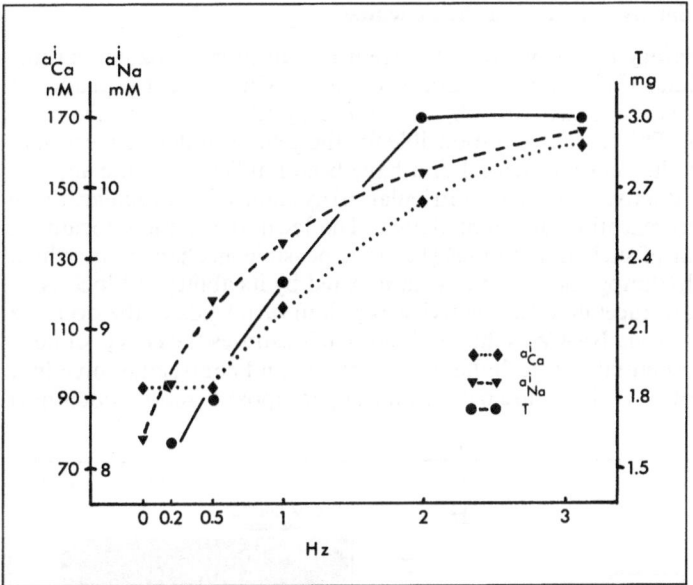

Fig. 5. Relationship between stimulation rate, intracellular Na^+ and Ca^{2+} activities, and contractile force (from Fig. 4 of [8] and Table 1 of [3])

ample, the reduction in frequency was associated with an immediate increase in twitch tension, followed by a negative staircase, whereas the opposite was true when the frequency was increased. The transient increase in twitch tension can be explained by the sarcoplasmic reticulum having more time to fill with Ca^{2+} when the frequency was slowed, so that the next action potential released a substantially increased amount of Ca^{2+} from the sarcoplasmic reticulum. The gradual fall in twitch tension can be explained by a progressive decrease in intracellular Na^+ activity. When intracellular Na^+ activity fell, the intracellular Ca^{2+} concentration also fell as expected from an increased Ca^{2+} efflux and decreased Ca^{2+} influx via Na-Ca exchange. This decline in intracellular Ca^{2+} concentration should decrease the uptake of Ca^{2+} by the sarcoplasmic reticulum and result in a decrease in muscle contraction.

The dependence of intracellular Na^+ and Ca^{2+} activities on stimulus rate is demonstrated in Figs. 3 and 4 [3, 8]. Electrical stimulation was applied to a cardiac Purkinje fiber at rates of 0.1, 0.2, 0.5, 1, 2, and 3 Hz for 3 min. Upon termination of stimulation, there was an elevation of intracellular Na^+ and Ca^{2+} activities. This increase gradually decayed to the control level with a half-time of 80–120 s. These results are consistent with the idea that the rate-dependent changes in intracellular Na^+ activity are associated with changes in intracellular Ca^{2+} activity possibly via a Na-Ca exchange mechanism.

Figure 5 summarizes the relationship between stimulation rate, intracellular Na^+ and Ca^{2+} activity, and contractile force. It is apparent that the force-frequency relationship correlates quite well with the rate-dependent changes in intracellular Na^+ and Ca^{2+} activities.

In summary, these results support the hypothesis that the Na-Ca exchange system plays an important role in the force-frequency relationship.

Negative inotropic and antiarrhythmic actions by lidocaine

An arrhythmogenic oscillatory transient inward current is seen in mammalian cardiac muscle when the intracellular Ca^{2+} concentration is elevated. This current is produced on repolarization and appears to be activated by an oscillatory release of Ca^{2+} from the sarcoplasmic reticulum [6]. The current is responsible for the genesis of delayed after-de-polarization which follow the action potential and have been implicated in the appearance of ectopic pace-maker activity in some ventricular arrhythmias, in particular, those associated with digitalis intoxication or hypokalemia. The local anesthetic lidocaine is often used in the treatment of such arrhythmias [12]. One possible mechanism by which lidocaine reduces this arrhythmogenic current, as implicated by its ability to block Na^+ channels, is by decreasing intracellular Na^+ activity, which in turn leads to the decrease in intracellular Ca^{2+} activity via Na-Ca exchange. Figure 6 illustrates the effects of lidocaine (20 µM) on action potential, intracellular Na^+ activity, and contractile force in a sheep cardiac Purkinje fiber [15]. Figure 6 A shows superimposed action potentials

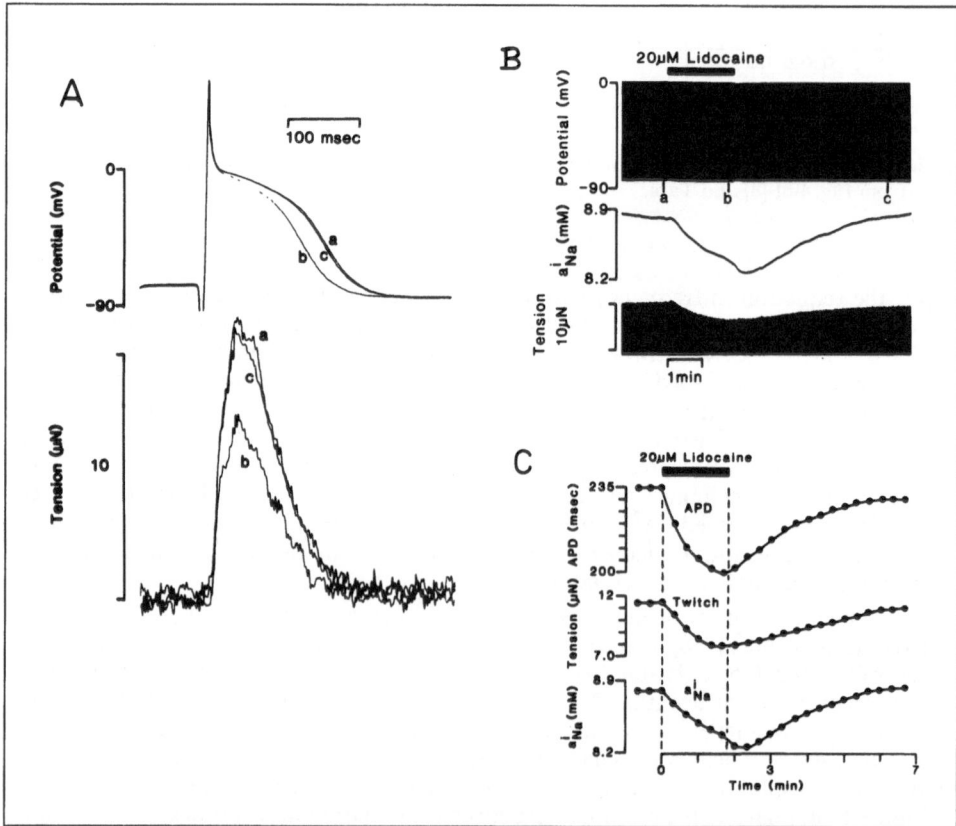

Fig. 6. The action of lidocaine (20 µM) on action potential duration, intracellular Na^+ activity, and tension. Panel A: Superimposed action potentials where curve a is control, curve b is in 20 µM lidocaine, and curve c is recontrol. Corresponding tension records are shown below. Panel B: Time-course of changes in intracellular Na^+ activity and tension. Panel C: Plot showing how action potential duration changed relative to changes of twitch and intracellular Na^+ activity (from Fig. 1 of [15])

(above) and twitches (below) obtained from the control solution (a), after exposure to lidocaine (b), and recontrol solution (c). The time-course of the action of lidocaine to reduce intracellular Na^+ activity and tension is shown in Fig. 6B. The effect of lidocaine on the action potential duration (measured at 75% repolarization), twitch tension, and intracellular Na^+ activity are plotted as a function of time in Fig. 6C. The results show that lidocaine reduced action potential duration, decreased intracellular Na^+ activity, and producd a negative inotropic effect. The reduction of action potential duration by lidocaine was due to blockage of steady state Na^+ "window" current [2]. A reduction in action potential duration can decrease twitch tension, possibly by decreasing Ca^{2+} influx during the plateau of action potential. The decrease in intracellular Na^+ activity can also decrease the twitch tension possibly by decreasing intracellular Ca^{2+} activity via Na-Ca exchange.

As for the antiarrhythmic action by lidocaine, the immediate effect on Na^+ channel activity would decrease the excitability of the heart. Moreover, as described earlier, the decrease of action potential duration and intracellular Na^+ activity would lead to a net decrease in intracellular Ca^{2+} concentration, which in turn leads to a depletion of Ca^{2+} in the sarcoplasmic reticulum. Because the arrhythmogenic transient inward current is due to oscillatory Ca^{2+} release from the sarcoplasmic reticulum, such a depletion of Ca^{2+} in the sarcoplasmic reticulum would diminish this arrhythmogenic activity. The experiment illustrated in Fig. 7 was designed to examine the relationship between intracel-

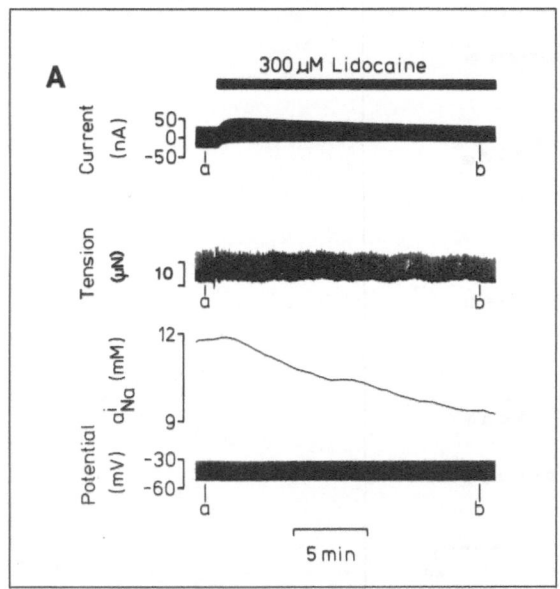

Fig. 7 A

Fig. 7 A–C. The effect of lidocaine on membrane current, tension, and intracellular Na^+ activity. **A** time-course of onset of the effects of lidocaine. Traces show (from top to bottom): membrane current, tension, intracellular Na^+ activity, and membrane potential. Lidocaine (300 µM) was added at the time indicated by a solid bar. The membrane potential was held at −51 mV and 1-s duration pulses were applied to −31 mV at 0.2 Hz. **B** records of current and tension. Traces show (from top to bottom): potential, current, tension. The records were obtained at the time indicated on **A**. **C** comparison of the fast rate of onset of the effects of lidocaine on holding current (I_{HP}) with its slower effect on intracellular Na^+ activity, transient inward current, and after contraction

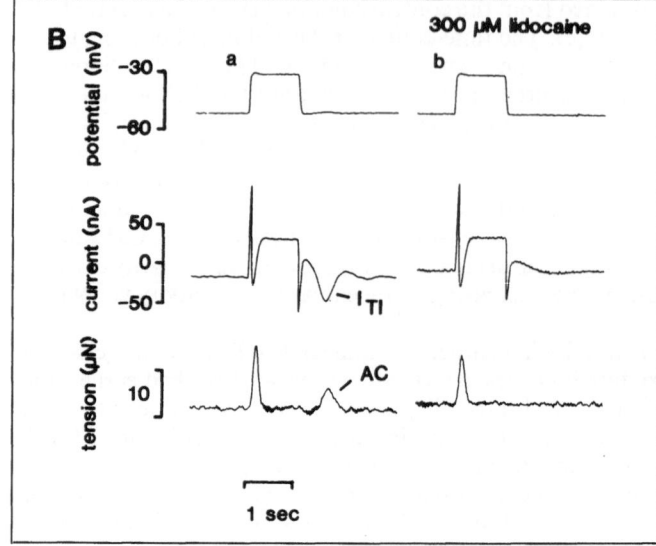

Fig. 7 B

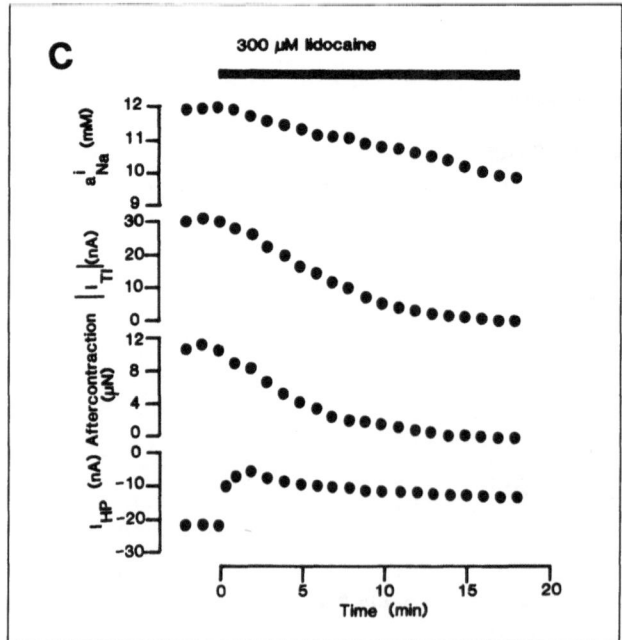

Fig. 7 C

lular Na$^+$ activity and the arrhythmogenic transient inward current [4]. In order to produce a transient inward current in a voltage-clamped sheep cardiac Purkinje fiber, the Na-K pump was partly inhibited by exposing the preparation to a K-free solution containing 0.25 mM Rb. Intracellular Na$^+$ activity increased from 6 mM to around 12 mM under this condition and the transient inward current and after-contraction appeared.

Figure 7 A shows that the subsequent addition of lidocaine (300 μM) produced a reduction of intracellular Na^+ activity. Records of current and tension are shown in Fig. 7 B. As can be seen, the large transient inward current and after-contraction were abolished by lidocaine. The time-course of the effect of lidocaine on intracellular Na^+ activity, transient inward current, and after-contraction is shown in Fig. 7 C. Lidocaine decreased the transient inward current and after-contraction with a time-course that was comparable to the fall of intracellular Na^+ activity. Therefore, it is suggested that the fall of intracellular Na^+ activity is a major factor in the reduction of arrhythmogenic transient inward current.

In sum, these results show that the direct action of lidocaine is the blockage of Na^+ channels; however, this can result in a decrease in cardiac contractility and arrhythmogenicity by decreasing cytosolic Ca^{2+} concentration via Na-Ca exchange.

Positive inotropic actions of cardiotonic steroids

It is now two centuries since Withering documented the effectiveness of digitalis as a therapy for those forms of dropsy we now recognize as congestive heart failure with peripheral edema. The mechanism that is responsible for the therapeutic effectiveness of digitalis in increasing cardiac contractility is still an interesting subject [17]. However, recent studies support the hypothesis of the inhibition of Na-K pump and leads to an alteration in the Na-Ca exchange [5, 7, 10]. There is no question about the fact that cardiac glycosides inhibit the Na-K pump and its biochemical correlate, the Na-K ATPase. Figure 8 shows an example of the positive inotropic effect produced by ouabain (5 μM) in a sheep ventricular trabecula stimulated at 0.2 Hz. Application of 5 μM ouabain (as indicated by the first arrow) caused a gradual increase in contractile force and reached a peak after 15 min of exposure. After reaching the maximal increase in tension, there was a gradual decay of the tension and a development of oscillatory after-contractions (as shown in a single trace at bottom right). This was due to the toxic effects of digitalis resulting in an oscillatory Ca^{2+} release from the sarcoplasmic reticulum when the cells were overloaded with Ca^{2+}. Figure 9 shows an experiment in which the membrane potential, intracellular Na^+ and Ca^{2+} activities were measured in a sheep cardiac Purkinje fiber. Exposure of

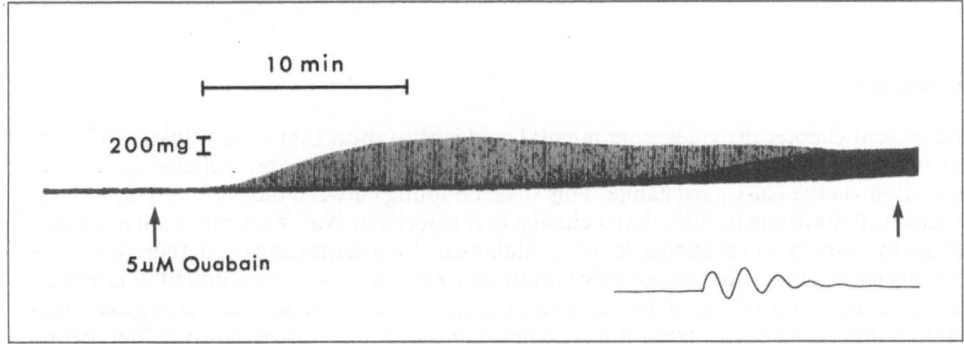

Fig. 8. The effect of ouabain (5 μM) on contractile force of a sheep ventricular trabeculae stimulated at 0.2 Hz. The single trace at bottom right (taken at the time as indicated by the second arrow) shows the oscillatory after-contractions induced by digitalis toxicity. (From Sheu and Fozzard, unpublished results)

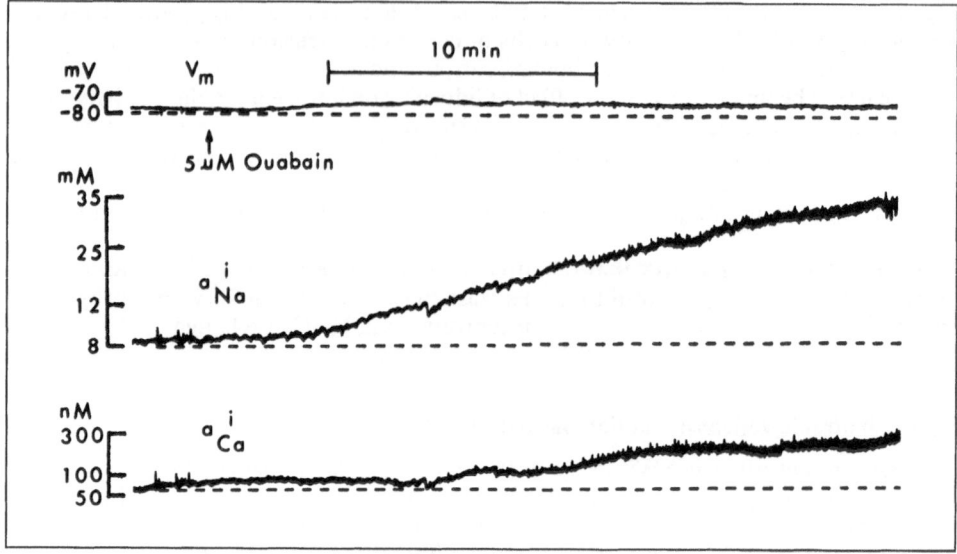

Fig. 9 Simultaneous measurements of membrane potential, intracellular Na^+ activity, and Ca^{2+} activity in a sheep cardiac Purkinje fiber exposed to ouabain (5 μM). (From Sheu and Fozzard, unpublished results, see also Fig. 12 of [14])

the muscle to ouabain (5 μM) for 25 min caused a small depolarization in the membrane potential (5 mV), an increase in intracellular Na^+ activity ($\simeq$4-fold increase), and an increase in intracellular Ca^{2+} activity ($\simeq$3.5-fold increase). These results strongly support the idea that digitalis inhibits Na-K pump, thereby elevating the intracellular Na^+ activity; this leads to an increase in intracellular Ca^{2+} activity via Na-Ca exchange. The increase in the intracellular Ca^{2+} concentration in turn leads to an increase in the Ca^{2+} content of the sarcoplasmic reticulum so that Ca^{2+} release from the sarcoplasmic reticulum and tension are increased when the muscle is subsequently activated.

Conclusions

The present chapter provides experimental evidence to show that intracellular Na^+ concentration regulates cardiac contractility effectively by altering intracellular Ca^{2+} concentration via the Na-Ca exchange. This steep coupling between the Na^+ and Ca^{2+} electrochemical gradients implies that a change in intracellular Na^+ concentration is accompanied by a concomitant change in intracellular Ca^{2+} concentration (and, therefore, contractility) [14]. Under the physiologic conditions, each cardiac action potential alters intracellular Na^+ concentration in a dynamic manner. Therefore, Na-Ca exchange can regulate cardiac contraction from a beat-to-beat basis. Moreover, numerous therapeutic agents such as cardiac inotropic agents and antiarrhythmic agents can modify cardiac contractility via their direct effect on Na^+ transport. Finally, under pathological conditions, intracellular Na^+ concentration could also be altered (e.g., myocardial ischemia) and influence cardiac function adversely.

Acknowledgements. I thank Dr. Daniel Williford, Dr. Jorge Arreola-Gomez, Mr. Allen Christie, and Mr. Robert Dirksen for valuable comments on the manuscript. This work is supported by National Heart, Lung, and Blood Institute Grant HL-33333, American Heart Association Grant-in-Aid (87-1320), and American Heart Association Established Investigatorship Award (85-163). I thank Ms. Wendy Roesch for typing the manuscript.

References

1. Bowditch HP (1871) Über die Eigentümlichkeiten der Reizbarkeit welche die Muskelfasern des Herzens zeigen. Ber Saechs Wiss 23:652–689
2. Carmeliet E, Saikawa T (1982) Shortening of the action potential and reduction of pacemaker activity by lidocaine, quinidine, and procainamide in sheep cardiac Purkinje fibers. An effect on Na or K current. Circ Res 50:257–272
3. Cohen CJ, Fozzard HA, Sheu S-S (1982) Increase in intracellular sodium ion activity during stimulation in mammalian cardiac muscle. Circ Res 50:651–662
4. Eisner DA, Lederer WJ, Sheu S-S (1983) The role of intracellular sodium activity in the antiarrhythmic action of local anaesthetics in sheep Purkinje fibers. J Physiol (Lond) 340:239–257
5. Eisner DA, Lederer WJ, Vaughan-Jones RD (1984) The quantitative relationship between twitch tension and intracellular sodium activity in sheep cardiac Purkinje fibers. J Physiol (Lond) 355:251–266
6. Kass RS, Lederer WJ, Tsien RW, Weingart R (1978) Role of calcium ions in transient inward currents and after contractions induced by strophanthidin in cardiac Purkinje fibers. J Physiol (Lond) 281:187–208
7. Kim D, Barry WH, Smith TW (1984) Kinetics of ouabain binding and changes in cellular sodium content, ^{42}K transport and contractile state during ouabain exposure in cultured chick heart cells. J Pharmacol Exp Ther 231:326–333
8. Lado MG, Sheu S-S, Fozzard HA (1982) Changes in intracellular Ca^{2+} activity with stimulation in sheep cardiac Purkinje strands. Am J Physiol 243(Heart Circ Physiol 12):H133–H137
9. Lederer WJ, Sheu S-S (1983) Heart rate-dependent changes in intracellular sodium activity and twitch tension in sheep cardiac Purkinje fibers. J Physiol (Lond) 345:44P
10. Lee CO, Kang DH, Sokol JH, Lee KS (1980) Relationship between intracellular Na ion activity and tension of sheep cardiac Purkinje fibers exposed to dihydro-ouabain. Biophys J 29:315–330
11. Mullins LJ (1981) Ion transport in heart. Raven Press, New York
12. Rosen MR, Wit AL (1983) Electropharmacology of antiarrhythmic drugs. Am Heart J 106:829–839
13. Rüegg JC (1988) Calcium in muscle activation. Springer-Verlag, Berlin Heidelberg New York
14. Sheu S-S, Fozzard HA (1982) Transmembrane Na^+ and Ca^{2+} electrochemical gradients in cardiac muscle and their relationship to force development. J Gen Physiol 80:325–351
15. Sheu S-S, Lederer WJ (1985) Lidocaine's negative inotropic and antiarrhythmic actions – dependence on shortening of action potential duration and reduction of intracellular sodium activity. Circ Res 57:578–590
16. Sheu S-S, Blaustein MP (1986) Sodium/calcium exchange and regulation of cell calcium and contractility in cardiac muscle, with a note about vascular smooth muscle. In: Fozzard HA, Haber E, Jennings RB, Katz AM, Morgan HE (eds) The heart and cardiovascular system. Raven Press, New York
17. Smith TW, Antman EA, Friedman PL, Blatt CM, Marsh JD (1984) Digitalis glycosides: Mechanisms and manifestation of toxicity. Prog Cardiovasc Dis 26:413–441, 495–523, 27:21–56

Author's address:

Shey-Shing Sheu, Ph. D., Department of Pharmacology, University of Rochester, School of Medicine and Dentistry, 601 Elmwood Avenue, Rochester, New York 14642, USA

Energy requirements of contraction and relaxation: implications for inotropic stimulation of the failing heart

A. M. Katz

Cardiology Division, Department of Medicine, University of Connecticut, USA

Summary

It is likely that the myocardium in the patient with congestive heart failure is unable to provide enough chemical energy to meet its mechanical requirements. If this interpretation is correct, inotropic stimulation, by increasing energy utilization, could contribute to the progressive myocardial cell death that characterizes end-stage cardiac hypertrophy. This deterioration could be delayed by the depressed myocardial contractility in the chronically overloaded heart, which reduces myocardial energy utilization, and delayed by changes in the expression of myosin isoforms that improve cardiac efficiency. An important goal of therapy in congestive heart failure, therefore, may be to reduce energy expenditure by unloading the failing heart and, in some cases, by administration of negative inotropic drugs.

Introduction

The clinical picture in congestive heart failure is generally dominated by signs and symptoms related to a fall in cardiac output, and increased systemic and pulmonary venous pressures. Fatigue, dyspnea, and anasarca most commonly cause the patient to seek treatment, and therapy is usually tailored to minimize these symptoms. Less apparent are important changes in the composition of the cells of the hypertrophied and failing myocardium, which also play an important role in the pathogenesis of the abnormal cardiac function seen in these patients. As these cellular changes are hard to quantify and their progression difficult to follow, therapy is generally directed to the circulatory and renal responses to the impaired cardiac function [13]. Yet recent data indicate that the poor longterm prognosis in most patients with congestive heart failure is probably determined mainly by deterioration of the myocardium, which culminates in myocardial cell death.

Energetics of the hypertrophied and failing heart

In most patients with heart failure, whether due to hemodynamic overloading or loss of functional myocardial tissue, the work of the active myocardial cells is increased. Although hypertrophy distributes the increased load among a greater number of sarcomeres, the chronic increase in the rate of energy expenditure makes it likely that the cells of the hypertrophied and failing heart are in an energy-starved state [22]. This reflects a number of changes in the cells of the hypertrophied and failing heart which influence their ability to provide for their energetic needs.

Increased intercapillary distance

Over 50 years ago it was shown that the number of capillaries supplying the hypertrophied heart does not increase in proportion to the increased mass of myocardium, suggesting that these hearts are inadequately perfused [40]. More recent morphometric studies of the hypertrophied myocardium have shown that the number of transverse capillary profiles per mm^2 is decreased, indicating that intracapillary distance is increased [2]. By increasing the diffusion distance for substrates and metabolites, most important of which is oxygen, this capillary deficit can contribute to a relative deficit in energy supply for contraction.

Altered contents of myofibrils and mitochondria

With the development of established myocardial hypertrophy, the fraction of cell volume occupied by myofibrils is increased, whereas the mass of mitochondria decreases [2, 30, 33]. The resulting disproportionate increase in the volume fraction of the cardiac myocyte occupied by energy-consuming myofibrils, relative to the volume of mitochondria which regenerate ATP, could also contribute to a lack of chemical energy in the failing heart.

Decrease in high energy phosphate compounds

More direct evidence that the cells of the hypertrophied and failing heart are in an energy-starved state has been obtained in studies of animal models of heart failure, which show decreased myocardial high energy phosphate contents following pressure overloading of the left [10] and right [32] ventricles. Endomyocardial biopsies taken from the hearts of patients with congestive heart failure have shown decreased levels of ATP and phosphocreatine [36], and a correlation has been found between decreased ATP content and impaired contraction and relaxation [3].

Natural history of the myocardium in the response to heart failure

The likelihood that hypertrophied, failing hearts are energy depleted is of considerable importance to our understanding of pathogenesis and management of patients with congestive heart failure. Most important of the clinical implications of such an energy deficit is the possibility that it hastens the deterioration of the chronically overloaded myocardium and so may lead to the death of the patient.

Almost a century ago, Osler [29] recognized that the clinical state in patients with congestive heart failure could be divided into three stages. A period of "development", during which the myocardium hypertrophies to meet a chronically increased load, is followed by a period of "full compensation" in which hypertrophy has allowed the heart to meet its new increased workload. Osler, however, recognized that the hypertrophied heart was not normal, but that it tended to undergo spontaneous and progressive deterioration, which he called the phase of "broken compensation".

Meerson [25, 26], who examined the mechanism for these processes in an animal model of systemic hypertension produced by aortic banding, provided a clear description of the response of the heart to chronic hemodynamic overloading. He characterized the three phases described by Osler as a "shortterm stage of damage", a "longterm stage of

Table 1. Three stages in the response to a sudden hemodynamic overlaod

Stage 1: (Days) Transient breakdown	
Circulatory:	Acute heart failure; pulmonary congestion, low output
Cardiac:	Acute left ventricular dilatation, early hypertrophy
Myocardial:	Increased content of mitochondria relative to myofibrils
Stage 2: (Weeks) Stable hyperfunction	
Circulatory:	Improved pulmonary congestion and cardiac output
Cardiac:	Established hypertrophy
Myocardial:	Increased content of myofibrils relative to mitochondria
Stage 3: (Months) Exhaustion and progressive cardiosclerosis	
Circulatory:	Progressive left ventricular failure
Cardiac:	Further hypertrophy with progressive fibrosis
Myocardial:	Cell death

Based on studies of Meerson [28].

relatively constant hyperfunction", and a "longterm stage of progressive cardiosclerosis and exhaustion" (Table 1). Understanding the mechanisms that give rise to this third stage in the myocardial response to a chronic overload is of greatest significance in planning therapy for the patient with congestive heart failure at this stage, which appears to be progressive, probably represents a vicious cycle that ends with the death of myocardial cells and their replacement with fibrous tissue. In light of evidence that the failing heart is energy-starved, the ultimate deterioration of the heart in patients with congestive heart failure probably arises, at least in part, from inadequate provision of chemical energy to meet the abnormal, and sustained, increase in cardiac work.

Relaxation and contraction are both impaired in the failing heart

The fact that relaxation, like contraction, is impaired by a deficit in chemical energy within the myocardial cell [22] may explain observations that the hemodynamic abnormalities in patients with heart failure arise from impaired relaxation in the overloaded myocardium [11]. While both contraction and relaxation are active processes, energy is used quite differently during the systolic and diastolic phases of the cardiac cycle.

Energy expenditure during systole

As energy must be expended to pump blood under pressure into the aorta and pulmonary artery, generation of mechanical work by the contractile proteins requires that the chemical energy of ATP is released in order for the walls of the heart to shorten and develop tension. However, while systole is an active process, activation is not. Muscular contraction is initiated when Ca^{2+} enters the cytosol of the heart through channels in the sarcolemmal and sarcoplasmic reticulum membranes. Because cytosolic Ca^{2+} concentration in the resting heart is much lower than that in the extracellular space and sarcoplasmic reticulum, where the activator Ca^{2+} is stored, activation is effected by the very rapid passive diffusion of Ca^{2+} into the cytosol [21]. Thus, while energy is utilized by the contractile proteins during contraction, systole is initiated by downhill Ca^{2+} fluxes that do not require the direct expenditure of energy.

Energy expenditure during diastole

The heart relaxes when activator Ca^{2+} is transported out of the cytosol back into the sarcoplasmic reticulum and extracellular space. As this Ca^{2+} leaves the cell against a strong electrochemical gradient, relaxation requires energy in order to move Ca^{2+} uphill, out of the cell [20, 38]. The fact that energy must be expended by the heart during both systole and diastole means that energy starvation can cause both contraction and relaxation abnormalities in patients with heart failure.

Changes in the composition of the hypertrophied and failing heart

It is now clear that subtle changes occur in the proteins in the hypertrophied and failing heart, and that alterations in protein structure can give rise to chronic changes in cardiac function in the patient with congestive heart failure. A regulatory mechanism based on plasticity in the structures of individual myocardial proteins was initially suggested by findings that the ATPase activity of cardiac myosin could change in response to pathophysiological abnormalities like chronic hemodynamic overloading, aging, and endocrinopathies [1, 16]. The importance of changing molecular properties of highly purified myocardial proteins raised the possibility that functionally significant alterations of protein structure in the heart participated in the adaptation of the myocardium to longterm circulatory changes such as chronic hemodynamic overloading [18].

Variability of gene expression in the heart

Rapid developments in molecular biology and growing knowledge of the mechanisms responsible for the remarkable variability of gene expression in the heart are beginning to illuminate the nature of the changes in cardiac proteins that are responsible for abnormal myocardial cell function in patients with heart failure. The best understood of these changes are those involving myosin, which exhibits a number of important structural and biochemical properties. Each myosin molecule is made up of heavy and light chains, the heavy chain being responsible for the intrinsic rate of energy liberation by myosin both in vitro (ATPase activity) and in vivo (muscle shortening velocity). Myosin heavy and light chains represent families of isoforms that are encoded by different gene families, and expression of these genes differs among different muscles and at different times in a single muscle during ontogeny [9, 37]. Variability in the expression of the genes that code for different isoforms of the peptide chains of myosin [5] and such mechanisms as alternate gene splicing [4] are now recognized to play an important role in the adaptation of the heart to a chronic overload.

Response of the myocardium to chronic overloading

Longstanding hemodynamic overloading of the myocardium leads to changes in the expression of genes that encode cardiac myosin [15, 23, 24, 34, 35, 39], the sodium pump [6], and creatine phosphokinase [14, 27]. Functional studies also suggest that the calcium pump of the sarcoplasmic reticulum may change in response to altered physiological state [31]. These changes appear to play an important role in the adaptation of the heart to a chronically increased hemodynamic load.

The ability of the heart to adjust its composition in response to a chronic overload has been thoroughly studied in the case of myosin. In the rat ventricle, the presence of the V1 (or alpha) myosin heavy chain determines a high rate of myosin ATPase activity and rapid velocity of shortening, while the V3 (or beta) myosin heavy chain determines a low myosin ATPase activity and slow shortening velocity. Both myosin ATPase activity and the muscle shortening velocity that this enzymatic activity determines, can be depressed in the failing heart. Chronic pressure overloading of the rat ventricle increases the expression of the gene that encodes the V3 isoform, thereby increasing the proportion of this "slow" myosin heavy chain in the ventricle. Replacement of "fast" V1 myosin heavy chains with "slow" V3 heavy chains increases mechanical efficiency, although at the expense of a slowing of cross-bridge cycling [12]. Similar changes in myocardial composition have been observed in the human heart, where a direct relationship has been observed between the proportion of "slow" myosin in the atria of patients with left atrial enlargement and left atrial size [28]. While this adaptation to a chronic overload impairs myocardial contractility, it is likely that the resulting energy-sparing effect could prolong survival in a patient with congestive heart failure [17, 19].

Clinical implications

A deficit of chemical energy in the hypertrophied or failing heart would have important implications for the patient with congestive heart failure. Most important of these is that inotropic stimulation, by increasing energy utilization in an energy-starved heart, could exacerbate the detrimental effects of a deficit in chemical energy. It is reasonable to postulate, although not yet established, that energy starvation contributes to the progressive myocardial cell death that characterizes end-stage cardiac hypertrophy. If such a causal relationship exists, then reducing energy expenditure by unloading the hypertrophied and failing heart might explain the ability of some vasodilators to prolong life in patients with heart failure [7]. The ability of converting enzyme inhibitors to prolong survival in patients with severe congestive heart failure [8] might also arise in part from the actions of these drugs to inhibit the influence of neurotransmitters and hormones to stimulate the myocardium, thereby preserving myocardial cell viability. While still unproven, these considerations should be taken into account in tailoring therapy for the patient with congestive heart failure.

References

1. Alpert NR, Gordon MS (1962) Myofibrillar adenosine triphosphatase activity in congestive heart failure. Am J Physiol 202:940–946
2. Anversa P, Olivetti G, Melissari M, Loud AV (1980) Stereological measurement of cellular and subcellular hypertrophy and hyperplasia in the papillary muscle of adult rat. J Mol Cell Cardiol 12:781–795
3. Bashore TM, Magorien DJ, Letterio J, Shaffer P, Unverferth DV (1987) Histologic and biochemical correlates of left ventricular chamber dynamics in man. JACC 9:734–742
4. Breitbart RE, Andreadis A, Nadal-Ginard B (1987) Alternative splicing: a ubiquitous mechanism for the generation of multiple protein isoforms from single genes. Ann Rev Biochem 56:467–495
5. Bugaisky L, Zak R (1986) Biological mechanisms of hypertrophy. In: Fozzard H, Haber E, Katz A, Jennings R, Morgan HE (eds) The heart and cardiovascular system. Raven Press, New York, pp 1491–1506

6. CharlemagneD, Maixen J-M, Preteseille M, Lelievre LG (1986) Ouabain binding sites and (Na^+, K^+)-ATPase activity in rat cardiac hypertrophy: expression of neonatal forms. J Biol Chem 261:185–189
7. Cohn JN, Archibald DG, Ziesche S et al. (1986) Effect of vasodilator therapy on mortality on chronic congestive heart failure. Results of a Veterans Administration cooperative study (V-HeFT). New Eng J Med 314:1547–1552
8. CONSENSUS Trial Study Group (1987) Effects of enalapril on mortality in severe congestive heart failure: results of the Cooperative North Scandinavian Enalapril Survival Study (CONSENSUS). New Eng J Med 316:1429–1435
9. Emerson CP Jr, Berstein SI (1987) Molecular genetics of myosin. Ann Rev Biochem 56:695–726
10. Furchgott RF, Lee KS (1961) High energy phosphates and the force of contraction of cardiac muscle. Circulation 24:416–428
11. Grossman W, Lorell B (1988) "Diastolic Relaxation of the Heart." Martinus Nijoff Pub, Boston
12. Hamrell BB, Alpert NR (1986) Cellular basis of the mechanical properties of hypertrophied myocardium. In: Fozzard H, Haber E, Katz A, Jennings R, Morgan HE (eds) The heart and cardiovascular system. Raven Press, New York, pp 1507–1524
13. Harris P (1983) Evolution and the cardiac patient. Cardiovasc Res 17:313–319, 373–378, 437–445
14. Ingwall JS, Kramer MF, Fifer MA, Lorell BH, Shemin R, Grossman W, Allen PD (1985) The creatine kinase system in normal and diseased human myocardium. New Eng J Med 313:1050–1054
15. Izumo S, Lompre A-M, Matsuoka R, Koren G, Schwartz K, Nadal-Ginard B, Mahdavi V (1987) Myosin heavy chain messenger RNA and protein isoform transitions during cardiac hypertrophy. J Clin Invest 79:970–977
16. Katz AM (1970) Contractile proteins of the heart. Physiol Rev 50:58–163
17. Katz AM (1973) Biochemical "defect" in the hypertrophied and failing heart. Deleterious or compensatory? Circ 47:1076–1079
18. Katz AM (1976) Tonic and phasic mechanisms in the regulation of myocardial contractility. Basic Res Cardiol 71:447–455, 1976
19. Katz AM (1978) A new inotropic drug: its promise and a caution. N Eng J Med 299:1409–1410
20. Katz AM (1984) Calcium fluxes across the sarcoplasmic reticulum. In: Opie LH (ed) Calcium antagonists and cardiovascular disease. Persp Cardiovasc Res 9:53–66
21. Katz AM (1986) Potential deleterious effects of inotropic agents in the therapy of chronic heart failure. Circ 73 (Suppl III):III-184–III-188
22. Katz AM (1988) Cellular mechanisms in congestive heart failure. Am J Cardiol (in press)
23. Litten RZ, Martin BJ, Low RB, Alpert NR (1982) Altered myosin isozyme pattern from pressure-overloaded and thyrotoxic hypertrophied rabbit hearts. Circ Res 50:856–864
24. Lompre AM, Schwartz K, D'Albis A, Lacombe G, Van Thiem N, Swynghedauw B (1975) Myosin isoenzyme redistribution in chronic heart overload. Nature 282:105–107
25. Meerson FZ (1961) On the mechanism of compensatory hyperfunction and insufficiency of the heart. Cor et Vasa 3:161–177
26. Meerson FZ (1969) The myocardium in hyperfunction, hypertrophy and heart failure. Circ Res 25 (Suppl II):II-1–II-163
27. Meerson FZ, Javick MP (1982) Isozyme pattern and activity of myocardial creatine phosphokinase under heart adaptation to chronic overload. Basic Res Cardiol 77:349–358
28. Mercadier JJ, DeLaBastoe D, Menasche P, N'Guyen Van Cao A, Bouveret P, Lorente P, Piwnica A, Slama R, Schwartz K (1987) Alpha-myosin heavy chain isoform and atrial size in patients with various types of mitral valve dysfunction: a quantitative study. JACC 9:1024–1030
29. Osler W (1892) The principles and practice of medicine. Appleton & Co, p 634
30. Page E, McCalister LP (1973) Quantitative electron microscopic description of heart muscle cells. Application to normal, hypertrophied and thyroxin-stimulated hearts. Am J Cardiol 31:172–181

31. Penpargkul S, Repke DI, Katz AM, Scheuer J, Effect of physical training on calcium transport by rat cardiac sarcoplasmic reticulum. Circ Res 40:134–138, 1977
32. Pool PE, Spann JF, Buccino RA, Sonnenblick EH, Braunwald E (1967) Myocardial high energy phosphate stores in cardiac hypertrophy and heart failure. Circ Res 21:365–373
33. Rabinowitz M (1973) Protein synthesis and turnover in normal and hypertrophied heart. Am J Cardiol 31:202–210
34. Rupp H (1981) The adaptive changes in the isoenzyme pattern of myosin from hypertrophied rat myocardium as a result of pressure overload and physical training. Basic Res Cardiol 76: 79–88
35. Scheuer J, Malhotra A, Hirsch C, Capasso J, Schaible TF (1982) Physiologic cardiac hypertrophy corrects contractile protein abnormalities associated with pathologic hypertrophy in rats. J Clin Invest 70:1300–1305
36. Swain JL, Sabina RL, Peyton RB et al. (1982) Derangements in myocardial purine and pyrimidine nucleotide metabolism in patients with coronary artery disease and left ventricular hypertrophy. Proc Nat Acad Sci USA 79:655–659
37. Swynghedauw B (1986) Developmental and functional adaptation of contractile proteins in cardiac and skeletal muscles. Physiol Rev 66:710–771
38. Tada M, Yamamoto T, Tonomura Y (1978) Molecular mechanisms of active calcium transport by sarcoplasmic reticulum. Physiol Rev 58:1–79
39. Tusuchimochi H, Kuro-o M, Takaku F, Yoshida K, Kawana M, Kimata S-I, Yazaki Y (1986) Expression of myosin isozymes during the developmental stage and their redistribution induced by pressure overload. Jap Circ J 50:1044–1052
40. Wearn JT (1939–40) Morphological and functional alterations of the coronary circulation. Harvey Lect 35:243–270, 1939–1940

Author's address:

Arnold M. Katz, M.D., Cardiology Division, Department of Medicine, University of Connecticut, Farmington, CT, 06032, USA

Genetic and non-genetic control of myocardial calcium

N. R. Alpert, E. M. Blanchard, L. A. Mulieri, R. Nagai, A. Zarain-Herberg, M. Periasamy

Department of Physiology and Biophysics, University of Vermont College of Medicine, Burlington, Vermont, USA

Summary

Some aspects of the genetic and non-genetic control of the amount and rate of calcium cycled during steady-state activation of papillary muscles from right ventricular rabbit myocardium are presented. Genetic reorganization of the intracellular structure of the myocardium is achieved by producing right ventricular pressure overload and thyrotoxic hypertrophy. The mechanical performance of the pressure overload heart is slowed while time to peak tension is increased. These changes are associated with an increase in myothermal economy. In thyrotoxic hypertrophy the rate of mechanical performance is increased while time to peak tension is decreased. These alterations are associated with a decrease in myothermal economy. Tension-independent heat is used as an index of calcium cycling. In pressure overload hearts the amount and rate of calcium cycling is decreased. In contrast in thyrotoxic hypertrophy the amount of calcium cycled is unchanged while the rate is increased. In the pressure overload hearts there is a decrease in sarcoplasmic reticular (SR) Ca^{++} ATPase, whereas in the thyrotoxic preparations the message is increased. The change in the rate of calcium uptake in pressure overload and thyrotoxic hearts is correlated with a change in the amount of SR Ca^{++} ATPase mRNA. Calcium cycling was also altered by non-genetic inotropic intervention. Isoproterenol (1 µM) increases the amount of calcium cycled during each contraction relaxation cycle and the rate at which it is removed. These alterations are associated with an increase in force and a foreshortened twitch. Incubating the papillary muscle in high calcium (11 mM) also increases the force and the amount of calcium released into the cytosol. Unter these circumstances the rate of uptake is not significantly increased and, accordingly, the isometric twitch is not foreshortened. In the presence of verapamil (14 µM) the peak twitch force is decreased and the isometric myogram is foreshortened. These changes are associated with a decrease in the amount of calcium released during activation and the rate at which it is removed.

Introduction

Periodic alteration in the levels of free cytosolic calcium play a central role in myocardial excitation contraction coupling. Contraction and relaxation of heart muscle involves cyclic changes of cytosolic calcium from low resting levels ($< 10^{-7}$ M) to activating levels (10^{-6} M) and then the return to the original resting levels. The time-course of the free calcium transient is determined by the complex interaction of 1) trigger calcium crossing the sarcolemma (SL), 2) the quantity and rate of calcium release from the sarcoplasmic reticulum, 3) the calcium buffering systems, in particular the binding to the troponin system, 4) the quantity and rate of calcium uptake by the sarcoplasmic reticulum, and 5) the transport of calcium across the SL from the cytosol. The major portion of the calcium

Supported in part by the following USPHS grants: PHS 28001-07, HL 39303-02.

involved in the changes that occur during excitation, contraction, and relaxation involve the sarcoplasmic reticular release and uptake. This paper addresses the nature of the genetic and non-genetic control of the amount and rate of calcium cycled during steady state activation of papillary muscles from right ventricular rabbit myocardium.

Our approach to the problem involves the use of a new myothermal method for assessing the amount and rate of calcium released and removed during each contraction relaxation cycle [4, 6]. The rationale for the new myothermal method involves the partitioning of heat liberated by isometrically contracting papillary muscle preparations. In steady state, paced isometric contraction, activity-related heat is liberated from muscle in two phases, an initial rapid phase associated with mechanical contraction and relaxation and a slower phase associated with recovery processes. The initial heat originates from high-energy phosphate hydrolysis during crossbridge cycling (tension-dependent heat, TDH) and high energy phosphate hydrolysis during calcium cycling (tension-independent heat, TIH). When the muscle is incubated in a medium that eliminates crossbridge cycling, a triggerable heat output remains that reflects the energy cost of calcium cycling. The time-course and extent of this output is used to calculate the amount and rate of calcium cycling.

Chronic alterations in the hemodynamic state, as well as acute pharmacologic intervention produce distinct changes in isometric peak twitch force (P_{tw}), rate of force development (dp/dt), time to peak twitch tension (TPT) and tension-time integral ($\int Pdt$) at optimal muscle length. In part, these alterations are a reflection of changes in the quantity and rate of calcium cycled following activation. The chronic increase in hemodynamic demand on the heart results in myocardial hypertrophy and restructuring of the intracellular components. Both of these changes are under genetic control and affect calcium cycling. Acute pharmacologic intervention also affects calcium cycling. In this review, we plan to illustrate the genetic effects on calcium cycling by examining the tension-independent heat in normal, pressure overload and thyrotoxic hearts and relate changes in these to alterations in mRNA associated with the synthesis of sarcoplasmic reticular calcium ATPase (the SR calcium pump). A spectrum of pharmacologic interventions are achieved by examining the effects of isoproterenol, high external calcium and verapamil on tension-independent heat and isometric force.

Methods and rationale

Animal and papillary muscle preparations

The animal and muscle preparations have been previously described [2, 14]. Hearts were obtained from male rabbits (2.0 kg) after sacrifice by cervical dislocation. The hearts were rapidly removed from the thorax and washed in 37 °C, oxygenated Krebs-Ringer solution. The right ventricle was opened, a long thin papillary muscle identified, tied at the tendonous and base ends with 4.0 silk ligatures, dissected free of the heart wall and mounted on the thermopile. The tendonous end was attached to a stationary glass hook at the bottom of the thermopile while the base end was attached to the capacitance force transducer above the thermophile [8] by a thin glass rod. The muscle was mounted over the centrally located hot junctions of the thermophile. It was brought into close thermal contact by means of a tether that forces the tendonous end of the muscle against the thermopile. Platinum wires, embedded in the ligatures at each end of the muscle, are used to deliver square wave stimuli (0.2 Hz, 2 ms duration, 10% above threshold). The entire apparatus was enclosed in a muscle chamber and submerged in a constant temperature water bath (21 °C). The muscle was incubated while being stimulated in normal Krebs-Ringer solution for 2 h. It was then stretched in small increments until L_0 (muscle length for maximum active tension) was reached. Heat measurements were made after draining the Krebs-Ringer solution from the chamber. Right ventricular pressure overload hypertrophy was induced by banding the pulmonary artery with a spiral monel metal clip that reduced the diameter 67% [9]. Thyroxine-induced hypertrophy was produced by intramuscular injection of 0.2 mg/kg of L-thyroxine with the injection omitted if the animal's weight loss exceeded 20% [5].

Myothermal measurements

The time-course of changes in muscle temperature was measured using ultra-thin vacuum deposited bismuth and antimony piles sandwiched between two thin sheets of mica [14]. The configuration was similar to that used by Hill [11]. The measuring junctions in the central zone were in thermal contact with the muscle. Reference junctions on both sides of the measuring junctions were in thermal contact laterally with the brass jaws of the thermopile frame clamping the mica.

Analysis of myothermal measurements

The time-course of heat evolution from heart muscle is an index of the biochemical activities taking place in that muscle. The heat evolved can be partitioned into four components. The quiescent muscle produces resting or basal heat H_B that is associated with the maintenance of ionic gradients, protein synthesis and degradation as well as futile crossbridge cycling. When the muscle is activated (isometric contraction, 0.2 Hz stimulus frequency), the activity-related heat output (T_A) is liberated at a rate approximately three-times that of the basal rate and appears in an initial rapid and secondary slower phase. The initial heat (I) is associated with high energy phosphate hydrolyzed by crossbridge cycling and excitation contraction coupling phenomena (calcium cycling). The secondary slower phase (R, recovery heat) is a mono exponential process and goes on throughout the period between stimuli. When it is stripped from the total activity-related heat it is approximately 55% of the total activity-related heat in normal or pressure overload hearts (0.55 T_A) and 41% in the thyrotoxic hearts (0.41 T_A). Recovery heat is associated with the resynthesis of high energy phosphate compounds by the mitochondria. The initial heat is made up of the tension-dependent heat (TDH) and tension-independent heat (TIH). The former is a reflection of crossbridge cycling while the latter is related to calcium cycling. This report is focussed on measurements of tension-independent heat. Tension-independent heat is separated from initial heat by incubating the muscle in a 2,3-butanedione monoxime solution that eliminates force while leaving the excitation contraction coupling phenomena unaltered [4, 6]. The triggerable heat output under these conditions reflects the amount and rate of calcium cycling during each contraction relaxation cycle.

Measurement of sarcoplasmic reticulum Ca^{++} ATPase mRNA

Slot blot analysis was carried out as previously described using cardiac/slow twitch SR calcium ATPase cDNA probes [17].

Incubating solutions and inotropic interventions

The papillary muscle, dissected from the right ventricle was incubated in Krebs-Ringer solution containing the following in mM: Na^+, 152; K^+, 3.6; Cl^-, 135; HCO_3^-, 25; Mg^{++}, 0.6; $H_2PO_4^-$, 1.3; SO_4^{2-}, 0.6; Ca^{2+}, 2.5; glucose, 5.6. Solutions were continuously bubbled with 95% O_2–5% CO_2 during dissection and equilibration. When termal and force measurements are made the muscle is incubated in a moist chamber and allowed to drain. The inotropic agents isoproterenol, calcium and verapil were added to the Krebs solution to give final concentrations of 1 μM, 11 mM, and 14 μM, respectively.

Results and discussion

General description and mechanical measurements

The control and experimental animals exhibited normal appearance. There were no obvious external signs of heart failure. The right ventricular systolic and diastolic pressures for the control group were 16.2 and 1.2 mm Hg, respectively. These pressures were increased in the pressure overload and thyrotoxic preparations (Table 1). The heart weight of the right ventricular free wall from control preparations was 1.17 ± 0.05 g. In the pressure overload and thyrotoxic preparations this increased to 1.70 ± 0.13 g and 1.31 ± 0.05 g, respectively. In right ventricular papillary muscles from normal hearts, the peak isometric twitch force was 5.90 g/mm² while the time to peak tension was 627 ms and the tension time integral was 6.12 g s/mm² (Table 2). The peak twitch force was unchanged in the pressure overload hearts and depressed in the thyrotoxic preparations (Table 2). Time to peak tension was increased in the pressure overload and decreased in

Tabel 1. Right ventricular systolic and diastolic pressures (mmHg) in control (C), pressure overload (P), and thyrotoxic preparations (T)

Preparations	Systolic pressure	Diastolic pressure
C	16.5 ± 1.2	1.2 ± 0.1
P	31.9 ± 3.4	2.2 ± 0.2
T	46.4 ± 4.7	4.1 ± 1.1

Table 2. Mechanical parameters of papillary muscles from control (C), pressure overload (P), and thyrotoxic hearts (T). P_{TW} is the peak isometric twitch tension (g/mm²), TPT is time to peak tension (ms), and ∫Pdt is the isometric tension time integral $(g \cdot s/mm^2)$

Preparation	P_{TW}	TPT	∫Pdt
C	5.90 ± 0.25	627 ± 20	6.12 ± 0.57
P	5.11 ± 0.47	816 ± 21	5.64 ± 0.57
T	4.55 ± 0.76	382 ± 36	4.83 ± 0.86

Table 3. Heat production in control (C), pressure overload (P), and thryotoxic (T) hearts. Initial heat (I) and tension-independent heat (TIH) in mcal/g for isometrically contracting papillary muscles paced at 0.2 Hz incubated at 21 °C. The numbers represent the mean and standard error of the mean

Preparation	I	TIH
C	1.7±0.1	0.25±0.1
P	1.2±0.1	0.13±0.1
T	1.7±0.2	0.17±0.1

the thyrotoxic hearts (Table 2). The isometric tension-time integral was unchanged in the pressure overload and decreased in the thyrotoxic hearts (Table 2).

The genetic regulation of calcium cycling

Hypertrophy was produced using two very different types of stress. In pressure over-loaded hearts, the contractile performance is slower with time to peak tension prolonged. This is consistent with a decrease in the rate of calcium removal. In myocardial hypertrophy secondary to thyrotoxicosis, the contractile performance is faster with time to peak tension foreshortened. These observations are consistant with an increased rate of calcium uptake. In each of these preparations the heart responds to stress by increasing in size and altering the expression of the intracellular constituents. These changes are under genetic control. The goal of these experiments was to assess the nature of these changes and to understand the underlying mechanisms responsible for them. Heat measurements are used to evaluate how the restructured heart handles calcium. Initial heat is liberated as a result of the activity of the contractile machinery and excitation contraction coupling phenomena. The initial heat liberated in control hearts was 1.7 ± 0.1 mcal/g (Table 3). This was reduced in pressure overload and unchanged in thyrotoxic hearts (Table 3). The initial heat is composed of the tension-dependent heat (crossbridge cycling) and tension-independent heat (excitation contraction coupling phenomena, primarily calcium cycling). When the papillary muscle is incubated in a BDM-manitol solution [4, 6] force is eliminated while a triggerable heat output remains. This is the tension-independent heat. In control hearts the tension-independent heat is 0.25 ± 0.1 mcal/g (Table 3). In pressure overload hearts the tension-independent heat is reduced while in thyrotoxic hearts it is not significantly different from control values (Table 3).

Calculation of the calcium cycled from the tension independent heat

The tension-independent heat is a result of pump activity in the sarcolemma, mitochondria, and sarcoplasmic reticulum [4]. Seventy-five percent of the heat is liberated by the sarcoplasmic reticulum calcium ATPase pump (Fig. 1). The amount of calcium cycled by the sarcoplasmic reticulum during each contraction relaxation cycle of the heart can be calculated from the tension independent heat using the following relationship:

$$\text{SR-Ca} = (\text{TIH}) \, (1/E) \, (R) \, (K) , \tag{1}$$

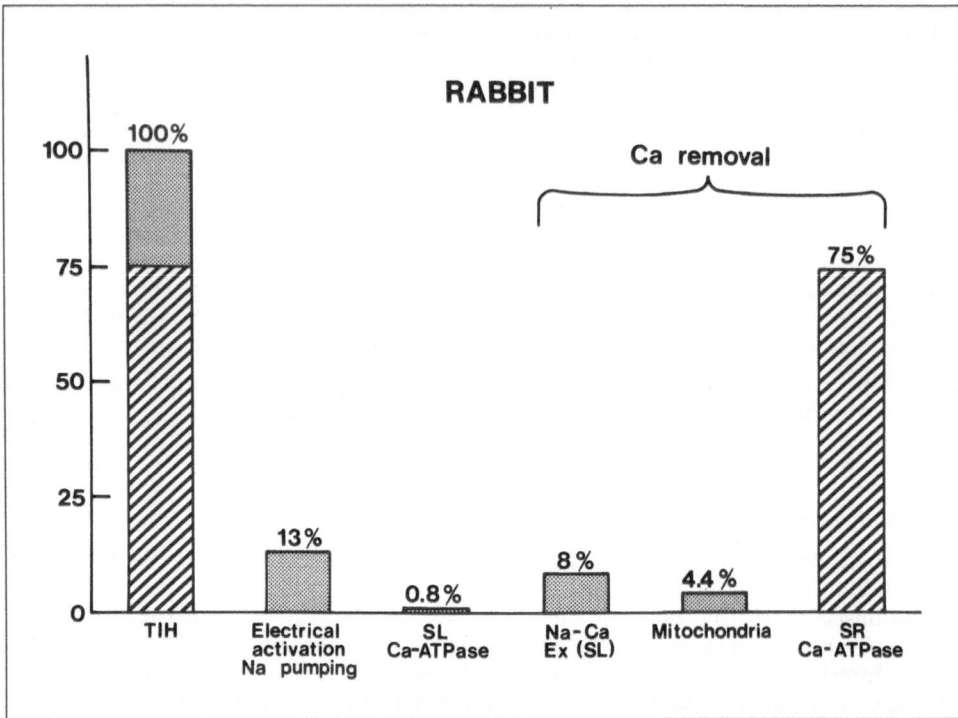

Fig. 1. The percent of the tension-independent heat associated with the various excitation contraction coupling phenomena [4]. The stippled portion (25% of the total) is unrelated to the SR calcium ATPase; it consists of heat liberation associated with membrane depolarization and sodium pumping, sarcolemmal calcium ATPase, sodium calcium exchange, and mitochondrial transport. The cross-hatched portion is that part of the tension-independent heat directly related to SR calcium cycling (75% of total)

where SR-Ca is the sarcoplasmic reticular calcium cycled per gram wet weight of muscle per beat; TIH is the tension-independent heat; (1/E) is the reciprocal enthalpy of creatine phosphate hydrolysis; R is the ratio of calcium cylced for every ATP hydrolyzed by the SR calcium ATPase pump (2 Ca/ATP); and K is the portion of the tension-independent heat associated with SR calcium cycling (0.75) (Fig. 1). The calcium cycled per beat in nmol Ca/g wet weight was 45, 23, and 31 for control, pressure overload, and thyrotoxic hearts, respectively.

The relation among the rate of calcium uptake,
time to peak tension and SR Ca^{++} mRNA

The rate of calcium uptake as a function of the total calcium released can be assessed from the tension-independent heat measurements. When the rate of uptake is plotted as a function of time following the stimulus, there is a substantial difference among the three preparations, with the thyrotoxic hearts removing calcium the fastest and the pressure over-

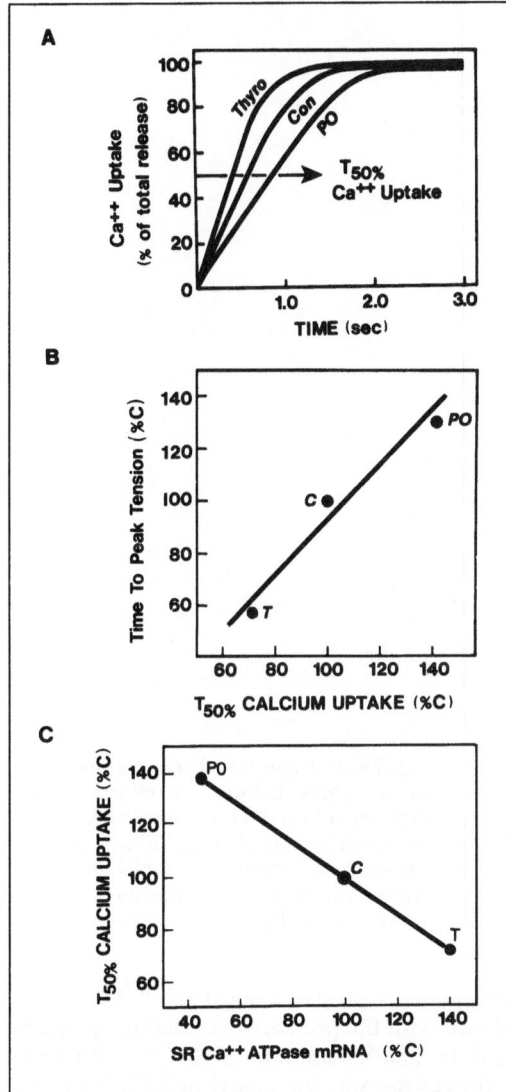

Fig. 2 A–C. The relation among rate of calcium uptake, time to peak tension, and SR Ca^{++} ATPase mRNA in control (Con, C) pressure overload (PO), and thyrotoxic (Thyro, T) hearts (expressed as percent of control). **A** The time course of calcium uptake as a function of total calcium released: this rate of uptake is assessed from tension-independent heat measurements as described in Eq. (1). An index of the rate of uptake is the time to 50% uptake (see arrow); **B** the relation between time to 50% uptake and time to peak tension; **C** the relation between time to 50% uptake and SR Ca^{++} ATPase mRNA

load hearts removing it the slowest (Fig. 2 A). This difference is readily quantitated by measuring the time for removal of 50% of the calcium released during activation (arrow, Fig. 2 A). There is a linear relationship between the time to 50% calcium uptake and the time to peak tension (Fig. 2 B). The change in the rate of calcium removal might result from an alteration in the type of calcium pump (fast vs slow) or the number of pumps expressed as a result of the pressure overload or thyrotoxic stress. It was determined that there was only one type of SR Ca^{++} ATPase in heart muscle, namely the cardiac/slow twitch type [17]. Measurement of SR Ca^{++} ATPase mRNA decreased to $40 \pm 5\%$ of control in the pressure overload hearts and increased to $140 \pm 18\%$ in the thyrotoxic prepa-

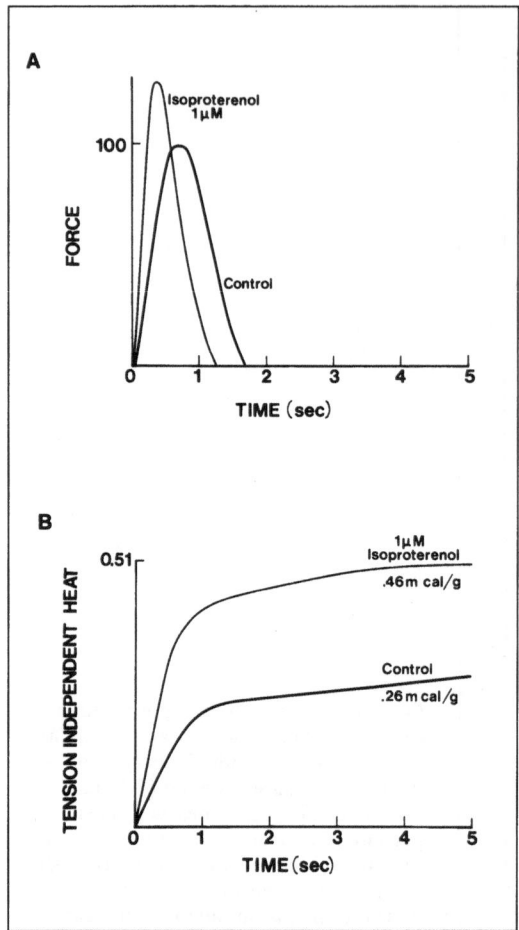

Fig. 3 A, B. Right ventricular papillary muscle isometric force (**A**) and tension-independent heat (**B**) for control and isoproterenol-treated hearts. Force is expressed as a percent of control values, while tension independent heat is expressed in mcal/g

rations [17]. When the time to 50% calcium uptake is plotted against the %SR Ca^{++} ATPase mRNA, a linear relationship is obtained with the pressure overload hearts having the longest time to 50% calcium uptake and the smallest amount of SR Ca^{++} ATPase mRNA, while in the thyrotoxic preparation the reverse was the case (Fig. 3 C). It would appear that in response to stress the gene controls the expression of the SR pumps by reducing the specific message in the pressure overload hearts and increasing it in the thyrotoxic hearts. The physiological consequences of the changes in SR Ca^{++} ATPase mRNA is seen in the altered mechanics (time to peak tension) and altered rate of tension independent heat liberation. Both of these changes appear to be directly related to the change in the rate at which calcium is removed from the cytosol following activation.

Non-genetic alterations in calcium cycling

It is well known that calcium cycling can be altered by a number of non-genetic interventions. We have chosen to examine the effects of isoproterenol, high calcium, and verapa-

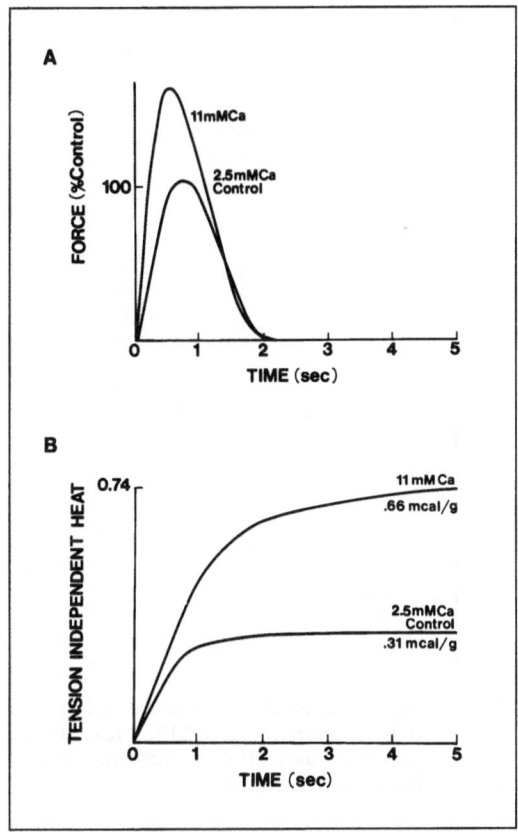

Fig. 4 A, B. Right ventricular papillary isometric force (**A**) and tension-independent heat (**B**) for control muscle and muscle incubated in high calcium Krebs-Ringer solution

mil on the force-developed and tension-independent heat libration. Isoproterenol produces a 25% increase in force with the rate of force development and relaxation increased (Fig. 3 A). The tension-independent heat liberation is significantly increased in the hearts incubated in isoproterenol, as is the rate of calcium removal. The increase in the total calcium cycled and its rate of uptake result in the increased force and the foreshortening of the twitch time. Papillary muscle incubated in high calcium also exhibits an increase in isometric force development (60%) (Fig. 4 A). In contrast to the inotropic effects of isoproterenol the twitch is not foreshortened to any great extent. In the high calcium incubating medium the tension-independent heat is markedly increased but the rate of uptake (expressed as a percent of the total release) does not seem to be altered (Fig. 4 B). When the papillary muscles are incubated in verapamil there is a marked depression in force development along with the foreshortening of the twitch (Fig. 5 A). Tension-independent heat measurements in these preparations indicate a decrease in the amount of calcium cycled (Fig. 5 B). It would appear that the cytosolic calcium required for force development is reduced below threshold levels earlier in the presence of verapamil than in normal Krebs-Ringer.

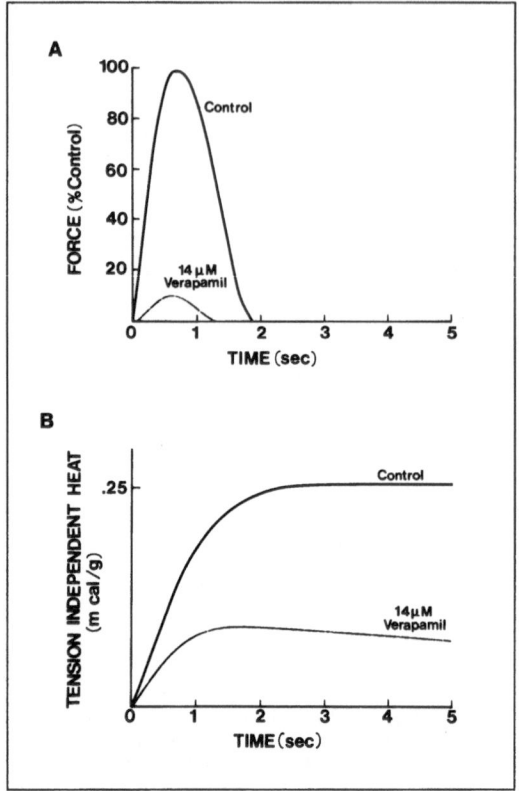

Fig. 5 A, B. Right ventricular papillary muscle isometric force (**A**) and tension-independent heat (**B**) for control and verapamil-treated muscles

Discussion and conclusions

The response of the heart to stress involves an increase in mass and a restructuring of the intracellular components. The increase in mass as well as the change in expression is under genetic control. Pressure overload and thyrotoxic stress were chosen because they represent the extremes of reorganization of the hypertrophied heart. In the former the heart develops force more slowly, has an increased time to peak tension and exhibits an increase in the economy of force development [2]. Conversely, thyrotoxic hearts exhibit the opposite changes [3]. The mechanical and myothermal changes in the hypertrophied hearts are attributable to alterations in myosin crossbridge head cycling and calcium cycling. The increase in muscle mass is associated with an overall increase in total mRNA [7, 15, 16]. With regard to the myosin in pressure overload there is an increase in β myosin mRNA while in the thyrotoxic hearts there is an increase in α myosin mRNA. Thus, there is an increase in overall mass of the heart, the amount of myosin synthesized and a switch in myosin isoenzymes. The β myosin is associated with a low ATPase activity, slow rate of force development and high thermal economy. The α myosin is associated with a high ATPase activity, a fast rate of force development and a low thermal economy. Thus, genetic control for the myosin involves an increase in the rate of synthesis and isoenzyme switching which are directly related to the total and type of myosin mRNA. It was surprising to find that the genetic control for calcium cycling is entirely different. First, there is only one type of sarcoplasmic reticular calcium ATPase pump in contrast to the isoenzyme found for myosin; this is the cardiac/slow twitch type [17]. Regulation occurs by decreasing (pressure overload) or increasing (thyrotoxicosis) the specific SR Ca^{++} ATPase mRNA while the mRNA for the heart as a whole is elevated. The genetic control exerted bears a direct relation to the rate of calcium uptake and the time to peak tension and provides a tentative explanation for these phenomena. It does not provide insight into the regulation of the amount of calcium cycled in each of the preparations.

We have also shown that calcium cycling can be altered with interventions that do not affect the genetic apparatus. Isoproterenol results in an increase in the amount of calcium cycled, along with an increase in the rate at which it is removed from the cytosol. These changes may in part be related to the phosphorylation of phospholamban [18]. An increase in the calcium bathing the muscle also results in an increase in the amount of calcium cycled but under these conditions the rate of uptake is not increased. The mechanical consequence of this difference is clearly seen in the isometric myogram for each treatment with both resulting in an increase in force but with isoproterenol producing a foreshortened twitch in contrast to that seen with high calcium. Verapamil results in a foreshortened twitch with lower force development. These changes also appear to be related to the amount and rate of calcium cycling.

From the experiments described above it is apparent that the amount and rate of calcium cycling can be altered by the genetic control of intracellular components as well as by non-genetic means. Although the conclusions presented seem well supported, there are some reservations. The mRNA was obtained from the free wall of the right ventricle while the measurements of tension-independent heat and force were made on right ventricular papillary muscles. Although it is reasonable to believe that the free wall and papillary muscle cells are very similar, one needs to consider the possibility that there may be some difference between the two. When this question was addressed with respect to myosin expression, some heterogeneity in different regions of the heart was observed, although in general there was substantial regional similarity of myosin expression [12].

Attention must also be directed at the use of tension-independent heat as an index of the amount and rate of calcium cycled during each contraction-relaxation cycle. The major assumption involves assigning 75% of that heat to sarcoplasmic reticular calcium movement. It is uncertain whether this percent is identical for all preparations (C, P, T). Furthermore, tension-independent heat is an indirect measurement of calcium cycling. In contrast the aequorin light measurements provide a direct measure of the cytosolic calcium concentration [1, 13]. Here the limitation resides in the method used to quantify and calibrate the results along with the fact that the measurements reflect calcium concentration and not the quantity and rate of cycling. The interaction of the pumps and various buffers must be considered before the aequorin method can be used to quantify the amount of calcium cycled during a twitch. Future studies will have to address these reservations.

References

1. Allen DG, Kentish JC (1985) The cellular basis of the length-tension relation in cardiac muscle. J Mol Cell Cardiol 17:821–840
2. Alpert NR, Mulieri LA (1982) Increased myothermal economy of isometric force generation in compensated cardiac hypertrophy induced by pulmonary artery constriction in the rabbit. Circ Res 50:491–500
3. Alpert NR, Mulieri LA (1984) Hypertrophic adaptation of the heart to stress: A myothermal analysis. In: Zak R (ed) Growth of the heart in health and disease. Raven Press, New York, pp 363–379

4. Alpert NR, Blanchard EM, Mulieri LA (1989) Tension Independent Heat in Rabbit Papillary Muscle. J Physiol (in press)
5. Banarjee SK, Flink IL, Morkin E (1976) Enzymatic properties of natural and N-ethylmaleimide modified cardiac myosin from normal and thyrotoxic rabbits. Circ Res 34:319–326
6. Blanchard EM, Mulieri LA, Alpert NR (1987) The effects of acute and chronic inotropic interventions on tension independent heat of rabbit papillary muscle. Basic Res Cardiol 82: Suppl 2, 127–135
7. Everrett AW, Clark WA, Chizzonite RA, Zak R (1983) Changes in synthesis rates of alpha and beta myosin heavy chains in rabbit heart after treatment with thyroid hormone. J Biol Chem 258:2421–2425
8. Hamrell BB, Panannen R, Trono J, Alpert NR (1975) A stable, sensitive, low compliance capacitance force transducer. J Appl Physiol 38:190–193
9. Hamrell BB, Alpert NR (1979) The mechanical characteristics of hypertrophied rabbit muscle cardiac muscle in the absence of congestive heart failure. Circ Res 40:20–25
11. Hill AV (1939) Recovery heat in muscle. Proc Roy Soc Lond (Biol) 127:297–307
12. Litten RZ, Martin BJ, Buchtal RH, Nagai R, Low RB, Alpert NR (1985) Heterogeneity of myosin isoenzyme content of rabbit heart. Circ Res 57:406–414
13. Morgan JP, Blinks JR (1982) Intracellular Ca^{++} transients in the cat papillary muscle. Can J Physiol Pharmacol 60:524–528
14. Mulieri LA, Luhr G, Trefry J, Alpert NR (1977) Metal-film thermopiles for use with rabbit right ventricular papillary muscles. Am J Physiol 233:C146–C156
15. Nagai R, Pritzl N, Low RB, Stirewalt WS, Zak R, Alpert NR, Litten RZ (1987) Circ Res 60:692–699
16. Nagai R, Low RB, Stirewalt WS, Alpert NR, Litten RZ (1988) Efficiency and capacity of protein synthesis are increased in pressure overload cardiac hypertrophy. Am J Physiol 255:H325–H328
17. Nagai R, Zarain-Herzberg A, Brandl CJ, Fujii J, Tada M, MacLennan DH, Alpert NR, Periasamy M (1989) Regulated expression of myocardial Ca^{++} SR ATPase and Phospholamban in response to pressure overload and thyroid hormone. Proc National Academy of Sciences (in press)
18. Tada M, Kirchberger MA, Katz AM (1975) Phosphorylation of a 22000-dalton component of the cardiac sarcoplasmic reticulum by adenosine 3′:5′ monophosphate dependent protein kinase. J Biol Chem 250:2640–2647

Author's address:

Norman R. Alpert, Department of Physiology and Biophysics, University of Vermont College of Medicine, Burlington, Vermont 05405, USA

II. Pharmacology of positive inotropic substances

Regulation of force and intracellular calcium transients by cyclic AMP generated by forskolin, MDL 17,043 and isoprenaline, and its modulation by muscarinic receptor agents: a novel mechanism for accentuated antagonism

M. Endoh

Department of Pharmacology, Yamagata University School of Medicine, Japan

Summary

The relation of changes in intracellular calcium transients and force of isometric contractions in response to an elevation or reduction of cyclic AMP levels was investigated in isolated dog ventricular trabeculae and rabbit papillary muscles, in which multiple superficial cells have been microinjected with the calcium sensitive bioluminescent protein aequorin. Forskolin, MDL 17,043 and isoprenaline elevated the tissue cyclic AMP level, increased consistently the peak aequorin signals and force, and abbreviated the duration of both signals in a concentration-dependent manner. When the effect of isoprenaline was compared with that of alteration of extracellular calcium concentration ($[Ca^{2+}]_0$), the increase in force by isoprenaline was associated with higher peak aequorin signals than that by alteration of $[Ca^{2+}]_0$ for a given increase in force, indicating the decrease in calcium sensitivity of myofibrils by cyclic AMP generated by β-adrenoceptor stimulation. Carbachol, which did not affect significantly the basal force and cyclic AMP levels, lowered the cyclic AMP levels elevated previously by forskolin, MDL 17,043 or isoprenaline in the isolated dog ventricular trabeculae. It antagonized the increase in peak aequorin signals and force caused by these agents in a concentration-dependent manner. When carbachol had been administered prior to isoprenaline and the concentration-response curve for isoprenaline was determined in the presence of carbachol, the relation of force peak aequorin signals was not modified by carbachol in the rabbit papillary muscle. Carbachol, when administered during induction of the positive inotropic action by forskolin, MDL 17,043 and isoprenaline, decreased the force more than peak aequorin signals in a concentration-dependent manner in the dog ventricular trabeculae. Therefore, the relation of force to peak aequorin signals was shifted downwards during the carbachol-induced inhibition, indicating a further decrease of calcium sensitivity of myofibrils by carbachol. This effect of carbachol appears to be specific to the cyclic AMP-mediated positive inotropic action, since the α-adrenoceptor-mediated (cyclic AMP-independent) action was unaffected by carbachol. This mechanism may play an important role for "accentuated antagonism" in the mammalian ventricular myocardium.

Introduction

Noradrenaline released from sympathetic nerve terminals by excitation triggers a series of intracellular biochemical processes involving cyclic AMP by binding to β-adrenoceptors located on the outer surface of myocardial cell membrane, resulting in modulation

This paper is supported partly by Grant-in-Aid for Scientific Research on Priority Areas No. 62624004 and No. 63641002 from the Ministry of Education, Science and Culture, Japan and National Institutes of Health grant HL-12186.

of contractility. Physiological regulation of myocardial contractility in vivo is largely achieved by this cyclic AMP-mediated signal transduction process. The cyclic AMP-mediated phosphorylation of functional proteins of various intracellular organelles is considered to be the essential subcellular mechanism. The Ca^{2+} influx via voltage-dependent Ca^{2+} channels in sarcolemma is increased by an increase in open probability of channels induced by cyclic AMP-dependent phosphorylation of Ca^{2+} channel proteins. The amount of Ca^{2+} released from sarcoplasmic reticulum (SR) by excitation, which constitutes the peak Ca^{2+} transients in physiological contractions is increased, because more Ca^{2+} are taken up into SR by cyclic AMP-dependent phosphorylation of phospholamban in SR membrane. These changes in Ca^{2+} mobilization are considered to be essential for the cyclic AMP-dependent positive inotropic effect. The characteristic change in myocardial contractility caused via cyclic AMP-dependent process is an acceleration of relaxation. The increased rate of Ca^{2+} uptake into SR, and decrease in Ca^{2+} sensitivity of myofibrils being exerted in relation to cyclic AMP-dependent phosphorylation of troponin I may contribute to the acceleration of relaxation (for reviews see [9, 27, 28, 35, 36, 38, 39, 42]).

Inhibitory signals through excitation of the vagus nerve are transmitted to individual myocardial cells via the subcellular process triggered by binding of acetylcholine to muscarinic receptors. The mechanism involved in this inhibitory signal transduction is much more intricate than that of stimulatory regulation. The inhibitory effect on ventricular contractility is exerted mostly through the cyclic AMP-dependent process. It is known that muscarinic receptor activation inhibits the ventricular contraction only when the force of contraction has been increased by sympathetic nerve excitation [20], i.e., during cyclic AMP accumulation [7, 41], which has been termed "accentuated antagonism" [30]. The primary mechanism involved in this inhibition may be lowering of cyclic AMP levels resulting from inhibition of adenylate cyclase through the inhibitory GTP-binding protein, G_i [19, 37, 40]. In addition, activation of potassium channels is likewise promoted through muscarinic receptors via the GTP-binding protein, G_k [2, 29, 32]. The latter mechanism operates preferentially in atrial and nodal cells, while the former, in both atrial and ventricular myocardium [10].

Stimulatory as well as inhibitory signals are finally transformed to facilitation or reduction of intracellular free calcium ion ($[Ca^{2+}]_i$) mobilization [11, 12], which is directly related to Ca^{2+} binding to troponin C resulting in generation of force. Sensitivity of myofilaments to intracellular Ca^{2+} may be also modified [42]. In the present study it was examined how the modulation of calcium mobilization and calcium sensitivity of myofibrils was involved in the regulation of myocardial contractility elicited by accumulation of cyclic AMP. For this purpose, isolated dog ventricular trabeculae and rabbit papillary muscles, multiple superficial cells of which had been microinjected with the Ca^{2+} sensitive bioluminescent protein aequorin were used for the experiments. The subcellular mechanisms involved in inhibition of these cyclic AMP-mediated effects by muscarinic receptor activation were also focused. It is postulated that the decrease of calcium sensitivity of myofibrils elicited by muscarinic receptor activation during the interation may play an important role as subcellular mechanism of "accentuated antagonism".

Methods and materials

Hearts were excised from mongrel dogs of either sex (8–16 kg), anesthetized with sodium pentobarbital (30 mg/kg i.v.). Free-running trabeculae or thin papillary muscles (< 1 mm diam.) were dissected from the wall of the right ventricle. Papillary muscles were also isolated from rabbit right ventricle.

Determination of force of isometric contractions and cyclic nucleotide levels

Details of experimental procedures have been described elsewhere [7, 14]. Briefly, ventricular trabeculae or papillary muscles were mounted in 20 ml organ baths containing bicarbonate-buffered Krebs-Henseleit solution (with 0.057 mM ascorbic acid and 0.027 mM disodium EDTA). The solution was bubbled with 95% O_2–5% CO_2 at 37 °C (pH 7.4). The muscles were stimulated through platinum field electrodes at 0.5 (dog ventricular trabeculae) or 1.0 (rabbit papillary muscle) Hz with square pulses of 5 ms duration and voltage about 20% above threshold. During an equilibration period of 1 h, length of the muscle was adjusted to give the maximum contractile force. Force of contraction of muscles was recorded on a thermal pen-writing oscillograph by means of strain-gauge transducers. ($\pm$)-Pindolol (3×10^{-8} M) was added to the bath to eliminate the possibility that noradrenaline released from the tissue might modify the direct effect of the drugs being tested. For determination of cyclic nucleotide levels, the muscles were removed from the bath at various times after administration of the drug, immediately frozen in liquid nitrogen, weighed, and stored overnight at -30 °C. Each muscle sample was frozen with 0.5% trichloroacetic acid in a Teflon capsule precooled in liquid nitrogen. The frozen sample was then homogenized mechanically by shaking for 30 s in a Mikro-Dismembrator (B. Brown, Melsungen, FRG). After adding 10 µl of 1 N HCl, aliquots of 100 µl of the supernatant were extracted five times with 1 ml of water-saturated ether, heated at 80 °C for 3 min to evaporate the residual ether, lyophilized overnight, and resuspended in 100 µl of distilled water. The cyclic nucleotide contents were determined with a sensitive radioimmunoassay method [21].

Determination of aequorin signals

Isolated muscles were mounted horizontally in an organ bath constructed for aequorin injection. The experimental procedures have also been described in detail previously [12]. The aequorin solution was injected by the application of gas pressure to the aequorin-loaded fine tipped micropipettes (resistance of 35 to 50 MΩ in 150 mM KCl) at a temperature of 32 °C. The potential was monitored to confirm the cell penetration into the quiescent myocardial cell preparation. After aequorin-injection sufficient to detect the intracellular calcium transients, the muscle was transferred to an apparatus composed of a vertically mounted ellipsoidal reflector in which the muscle preparation was positioned at the upper focal point and the photocathode of the photomultiplier (EMI 9635A) at the lower. The muscle was mounted inside a glass tube that extends axially into the top of the ellipsoidal reflector from a water-jacketed organ bath, being electrically driven at 0.5 or 1.0 Hz at 37.5 °C essentially in the same experimental condition in which cyclic nucleotide contents were determined. In order to obtain a satisfactory signal-to-noise ratio the aequorin signals of 64 successive contractions were averaged. Drugs were administered in a cumulative manner in a volume of 0.1 ml into the 50 ml organ bath. In the experiments with forskolin and MDL 17,043, ($\pm$)-bupranolol (3×10^{-7} M) was allowed to act for more than 20 min before administration of these agents.

Drugs and chemicals used

($-$)-Isoprenaline hydrochloride (Sigma Chemical Co., St. Louis, Missouri, USA); ($\pm$)-pindolol base (Sandoz AG, Basel, Switzerland); forskolin (Calbiochem-Behring Corp., La Jolla, California, USA, or Nippon Kayaku Co. Ltd., Tokyo, Japan); MDL 17,043

(Merrel Dow Research Center, Cinncinnati, Ohio, USA); ($\pm$)-bupranolol (Sanol, Mon-
heim, FRG). The anti-cyclic AMP and anti-cyclic GMP antisera were generously sup-
plied by Yamasa Shoyu Co., Choshi, Japan). Forskolin was dissolved in 50% methanol
in a concentration of 4 mM as stock solution and diluted with 0.9% NaCl; MDL 17,043,
in 0.12 ml of 1.0 N sodium hydroxide for each 10 mg of the compound.

Experimental values are presented as means $\pm$ SEM. The statistical significance was
estimated by Student's t-test; P values smaller than 0.05 were considered to be significant.

Results

Effects of forskolin

In isolated dog right ventricular trabeculae (in the presence of 3×10^{-8} M ($\pm$)-pindolol),
forskolin (10^{-6} M) elevated the cyclic AMP level and force substantially in parallel in a
time-dependent manner. The force and cyclic AMP levels reached a steady state about
10 min after the administration (Table 1). The cyclic GMP level was not affected significantly by forskolin. The increase in force and cyclic AMP levels was dependent on the
concentration of forskolin (Table 2).

Table 1. Time-course of changes in force of contraction and cyclic nucleotide levels after
administration of 10^{-6} M forskolin in the isolated dog right ventricular trabeculae

Time (min)	Contractile force (mN or %)	Cyclic AMP (pmol/mg w.w.)	Cyclic GMP (fmol/mg w.w.)
0 (control)	4.6±0.34 mN (19)	0.84±0.05 (20)	19.4±4.4 (14)
1	109.5±2.1*% (19)	0.89±0.08* (6)	21.6±6.8 (7)
3	139.9±6.8*% (19)	0.93±0.07* (6)	17.7±4.0 (5)
5	155.8±8.1*% (19)	0.99±0.07* (6)	16.1±2.0 (5)
10	173.6±9.0*% (19)	1.27±0.07* (10)	10.7±2.0 (11)
20	178.5±9.9*% (13)	1.40±0.10* (10)	14.3±2.9 (9)

* <0.05 vs the corresponding control values. Numbers in parentheses represent numbers of muscle
preparations.

Table 2. Concentration-dependent effects of forskolin on the force of contraction and cyclic
nucleotide levels determined at 20 min after the administration in the isolated dog ventricular
trabeculae

Forskolin (M)	Force of contraction (mN or %)	Cyclic AMP (pmol/mg w.w.)	Cyclic GMP (fmol/mg w.w.)
0	3.04± 0.43 mN (16)	0.79±0.04 (14)	35.9± 5.2 (6)
3×10^{-7}	130.9 ± 9.0*% (4)	0.92±0.08* (9)	56.9±15.5 (9)
10^{-6}	176.9 ±15.5*% (8)	1.35±0.08* (7)	–
3×10^{-6}	154.6 ±17.4*% (4)	1.40±0.09* (8)	59.1±13.1 (8)

* P <0.05 vs the corresponding control values. Numbers in parentheses represent numbers of muscle
preparations.

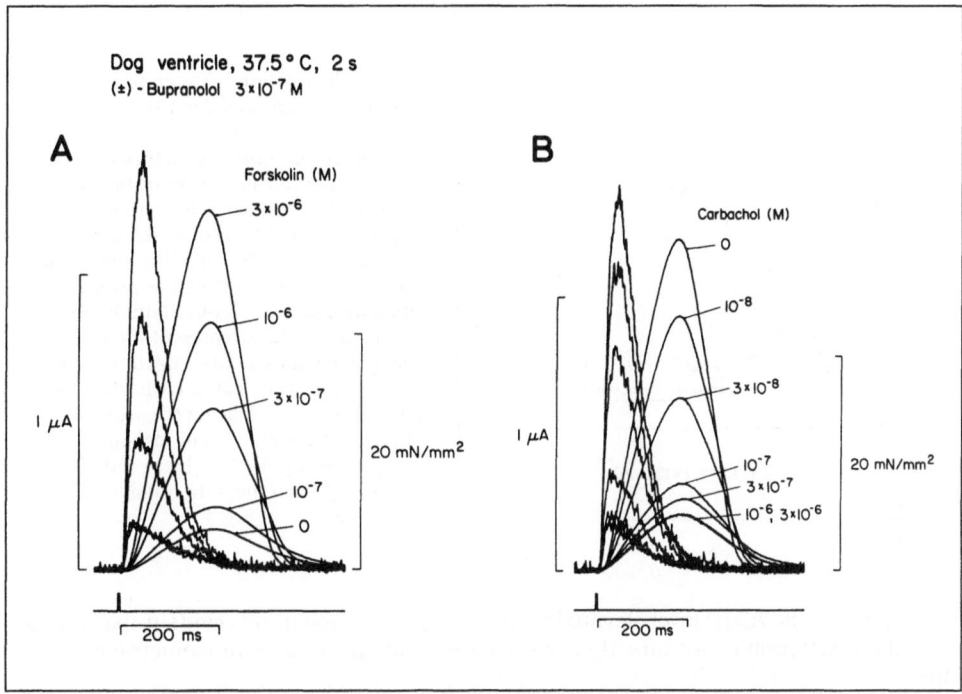

Fig. 1 A, B. Concentration-dependent effects of forskolin (**A**), and carbachol (**B**) administered during induction of the positive inotropic action 3×10^{-6} M forskolin on aequorin signals (noisy records) and isometric contractions recorded from a dog ventricular trabecula (electrically driven at 0.5 Hz at 37.5 °C). Both drugs were administered in a cumulative manner to the same muscle in the presence of 3×10^{-7} M ($\pm$)-bupranolol. Aequorin signals and isometric contractions are signal-averaged records of 128 successive contractions superimposed [from [15] with permission of the publisher]

Carbachol (3×10^{-6} M) added to the organ bath at 7 min after the administration of 10^{-6} M forskolin reversed the positive inotropic effect of forskolin. At 10 min of forskolin administration the cyclic AMP level in the presence of 3×10^{-6} M carbachol (1.01 ± 0.04 pmol/mg w.w.; n = 6) was significantly ($P < 0.05$) lower than the level without carbachol (1.40 ± 0.09 pmol/mg w.w.; n = 8). The increase in force by 10^{-6} M forskolin (at 10 min of administration) in the presence of carbachol (3×10^{-6} M) was $90.9 \pm 6.1\%$ (of the baseline force [n = 6]), which was significantly ($P < 0.05$) lower than the value with forskolin alone ($186.6 \pm 19.3\%$; n = 6).

Influence of carbachol (3×10^{-6} M) on the effect of forskolin was investigated also at 20 min after the administration of forskolin (10^{-6} M). The positive inotropic effect of forskolin was significantly ($P < 0.05$) decreased by carbachol added at 17 min after the forskolin administration (197.3 ± 14.6, and $114.9 \pm 5.0\%$ with and without carbachol, respectively [n = 7 each]). The cyclic AMP levels at 20 min after the administration of forskolin in the absence or presence of carbachol (being allowed to act for 3 min before the determination) were not significantly different from each other (1.35 ± 0.09 [n = 5] or 1.37 ± 0.88 [n = 7] pmol/mg w.w. without or with carbachol, respectively). Although the change in force was dissociated from that of the cyclic AMP level by carbachol at 20 min after the administration of forskolin, the reason for the absence of carbachol's effect in

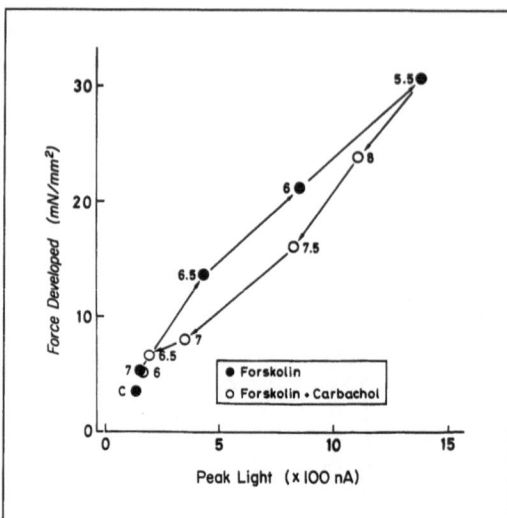

Fig. 2. The relation between peak aequorin signals and force of isometric contractions during development of positive inotropic action of forskolin (closed circles), and during concentration-dependent inhibition elicited by carbachol (open circles) of forskolin-induced response (3×10^{-6} M) in a dog ventricular trabecula. C: baseline values prior to administration of forskolin. Numbers attached to symbols represent the negative logarithmic molar concentration of forskolin (closed circles) or carbachol (open circles). Arrows indicate the sequence of changes in the relation in response to the compounds applied. Values plotted were calculated from the data obtained from actual records in Fig. 1

lowering the cyclic AMP level elevated by forskolin is unknown. It is evident that the tissue cyclic AMP level is not directly correlated with the peak force of isometric contractions.

Forskolin increased the intracellular calcium transient in a concentration-dependent manner as shown in Fig. 1 A. Force increased substantially in parallel to the increase in peak aequorin signals. The duration of contraction was markedly abbreviated (Fig. 1 A). Carbachol administered in the presence of forskolin antagonized the increases in peak aequorin signals and force of contraction elevated previously by forskolin in a concentration-dependent manner as shown in Fig. 1 B.

It was examined whether or not the relationship between peak aequorin signals and force is modified during the concentration-dependent development of effects of forskolin, and during the graded inhibition of forskolin-induced effects by carbachol. For this purpose, the relationship of force to peak aequorin signals was plotted as shown in Fig. 2. The relation during development of the forskolin-induced effects, and that during the carbachol-induced inhibition were not superimposable. This implies that the calcium sensitivity of contractile proteins was modified in a different manner during the development of forskolin-induced effects and during the forskolin-carbachol interaction. The curve during the carbachol-induced inhibition located downwards compared to that during the development of the forskolin-induced effects, i.e., for a given level of the peak aequorin signals the associated force was lower during the application of carbachol, indicating that the calcium sensitivity of myofibrils was decreased during the carbachol-induced inhibition of forskolin's effects.

Effects of MDL 17,043 (enoximone)

MDL 17,043 has been shown to be a positive inotropic agent acting by inhibition of the low K_m cyclic AMP specific phosphodiesterase (PDE) and thereby accumulating cyclic AMP (14,26). After the administration of 2×10^{-4} M MDL 17,043, the cyclic AMP level

Table 3. Time-course of changes in force of contraction and cyclic nucleotide levels after administration of 2×10^{-4} M MDL 17,043 in the isolated dog right ventricular trabeculae

Time (min)	Contractile force (mN or %)	Cyclic AMP (pmol/mg w.w.)	Cyclic GMP (fmol/mg w.w.)
0 (control)	2.8± 0.51 mN (15)	0.90±0.04 (22)	25.4± 5.1 (7)
0.5	124.1± 6.9*% (15)	1.06±0.09 (7)	20.8± 6.6 (6)
1	132.6± 8.8*% (15)	1.07±0.13 (4)	16.3± 2.7 (4)
2.5	148.7± 9.9*% (15)	1.24±0.13 (5)	23.2± 8.0 (2)
5	165.1±16.4*% (19)	1.36±0.10* (7)	–
10	172.6±15.9*% (19)	1.16±0.09* (10)	23.5±13.6 (2)
20	180.5±37.4*% (5)	1.42±0.07* (3)	–

* $P < 0.05$ vs the corresponding control values. Numbers in parentheses represent numbers of muscle preparations.

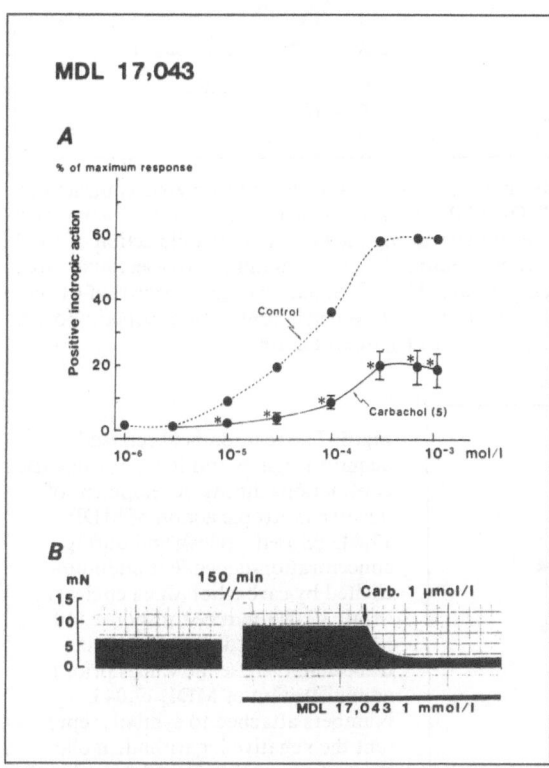

Fig. 3 A, B. Differential inhibitory effect of carbachol being dependent on the order of application on the positive inotropic action of MDL 17,043 in isolated dog ventricular trabeculae (electrically driven at 0.5 Hz at 37 °C). Experiments were performed in the presence of a β-adrenoceptor antagonist, 3×10^{-8} M pindolol. **A** Carbachol (10^{-6} M) had been administered prior to MDL 17,043, and the concentration-response relationship for MDL 17,043 was determined in the presence of carbachol when force achieved a steady level. **B** MDL 17,043 (10^{-3} M) had been administered prior to carbachol. Carbachol (10^{-6} M) was administered during the positive inotropic action of MDL 17,043 (reproduced from [14] with permission of the publisher)

and force were increased in parallel in a time-dependent manner (Table 3). Both the force and cyclic AMP levels reached a steady level in 5 min after the administration.

The concentration-response curve for the positive inotropic effect of MDL 17,043 was shifted to the right and downwards in the presence of 10^{-6} M carbachol (Fig. 3 A). When carbachol (10^{-6} M) was administered during the positive inotropic action of MDL

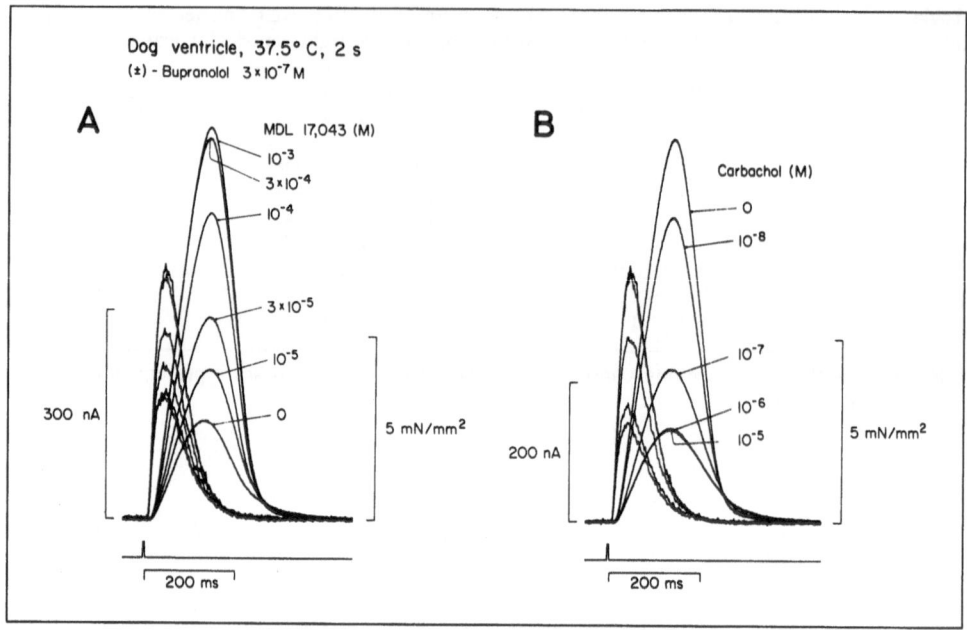

Fig. 4 A, B. Effects of MDL 17,043 and carbachol on aequorin signals and isometric contractions of an isolated dog ventricular trabeculae. A MDL 17,043 was administered in a cumulative manner. B Carbachol was administered in a cumulative manner during the positive inotropic action of 10^{-3} M MDL 17,043. A and B are from the same muscle (length 7.5 mm; cross-sectional area 0.75 mm²). Stimulation frequency, 0.5 Hz; temperature, 37.5 °C. Signal-averaged records of 64 successive contractions with aequorin signals (noisy records) and isometric contractions superimposed. Experiment was performed in the presence of 3×10^{-7} M ($\pm$)-bupranolol

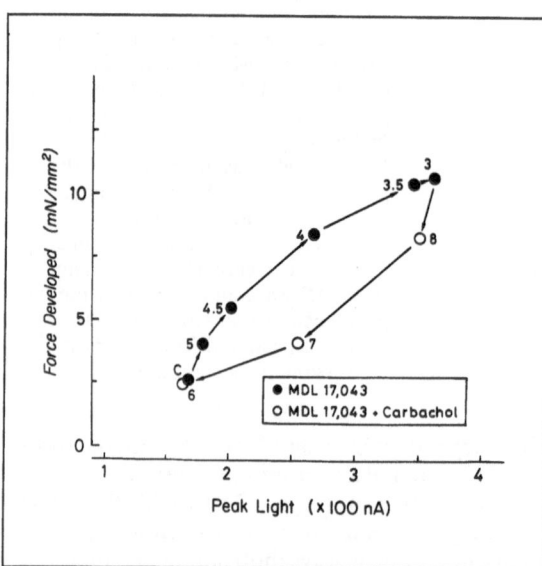

Fig. 5. The relation between peak aequorin signals and force of isometric contractions during development of positive inotropic action of MDL 17,043 (closed circles), and during concentration-dependent inhibition elicited by carbachol (open circles) of MDL 17,043-induced response (10^{-3} M) in an isolated dog ventricular trabecula. C: baseline values prior to administration of MDL 17,043. Numbers attached to symbols represent the negative logarithmic molar concentration of MDL 17,043 (closed circles) or carbachol (open circles). Arrows indicate the sequence of changes in the relation in response to the compounds applied. Values plotted were calculated from the data obtained from actual records shown in Fig. 4

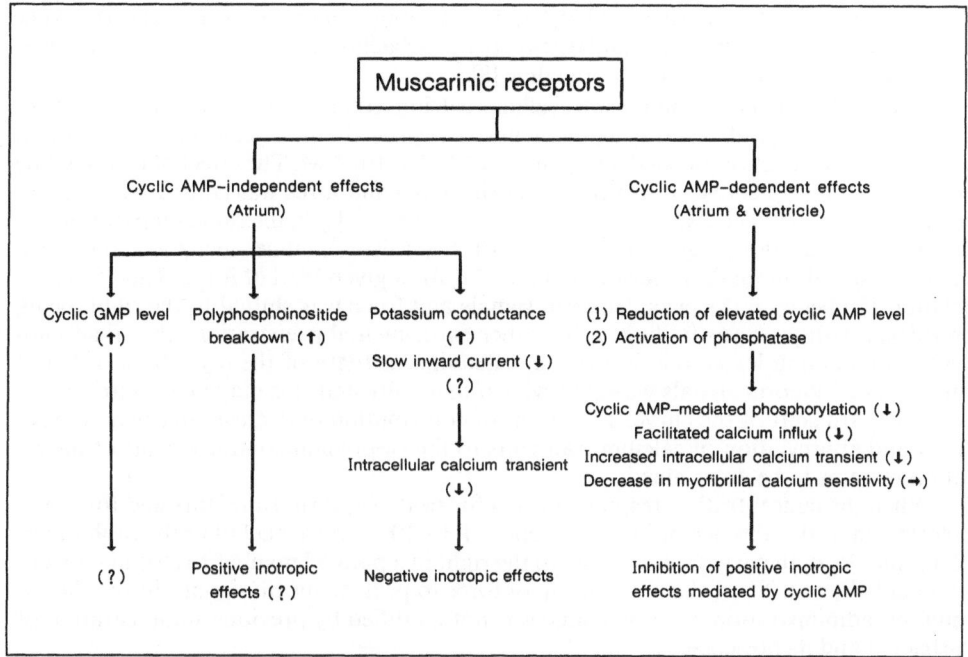

Fig. 6. Schematic representation of the intracellular signal transduction process subsequent to muscarinic receptor activation in mammalian myocardium. Arrows in parentheses indicate increase (upward) or decrease (downward) of the respective parameters

17,043, carbachol inhibited the positive inotropic effect of MDL 17,043 more effectively. The force was decreased to a level even lower than the baseline force (Fig. 3 B). Thus, Fig. 3 shows that carbachol was more effective in antagonizing the positive inotropic action of MDL 17,043, when it was administered during MDL 17,043-induced action than when it was administered prior to MDL 17,043.

The cyclic AMP level elevated by MDL 17,043 $(2 \times 10^{-4}$ M) was decreased to a level close to control $(10^{-6}$ M), when the force was antagonized by carbachol [14].

MDL 17.043 increased the intracellular calcium transient in a concentration-dependent manner up to 10^{-3} M in parallel to the increase in force of isometric contractions. Carbachol added in the presence of 10^{-3} M MDL 17,043 suppressed the force more effectively than the peak aequorin signals in a concentration-dependent manner (Fig. 4 B). Figure 5 shows the relationship of force to peak aequorin signals during the graded induction of the effects of MDL 17,043, and during inhibition of the MDL 17,043-induced effects by carbachol. It is evident that the curve was shifted downwards by carbachol, indicating that the calcium sensitivity of myofibrils decreased by carbachol administered in the presence of MDL 17,043.

Effects of isoprenaline

The positive inotropic effect of isoprenaline in the rabbit papillary muscle and canine ventricular trabeculae is associated with the elevation of cyclic AMP levels as has been shown in cardiac muscle of other species [7–9, 24]. In the canine ventricular trabeculae the cyclic AMP level was reduced rapidly and markedly by carbachol, when carbachol antagonized the positive inotropic effect of isoprenaline [8].

Effects of isoprenaline on the intracellular calcium transient and isometric force have been investigated in detail [12]. Isoprenaline increased the peak calcium transient and force in a concentration-dependent manner at 10^{-9} to 10^{-6} M. The effect of isoprenaline on the relationship between the peak aequorin signal and force was compared with that of elevation of extracellular calcium concentration ($[Ca^{2+}]_0$) in the same preparation. At higher concentrations (generally above 10^{-8} M) isoprenaline increased the peak aequorin signals more than elevation of $[Ca^{2+}]_0$ for a given level of force. Therefore, the relationship between the peak aequorin signals and force was shifted to the right, being consistent with previous findings using other experimental preparations that β-adrenoceptor stimulation leads to a decrease in calcium sensitivity of the myofibrils [12]. The duration of aequorin signals was abbreviated markedly and this may also contribute to the rightward shift of the curve. The extent of contribution of decrease in calcium sensitivity and abbreviation of calcium transients to the isoprenaline-induced shift of the relation remains to be determined.

When the concentration-response curves for peak aequorin transients and force was determined in the absence or in the presence of 3×10^{-6} M carbachol in the rabbit papillary muscle, both curves were shifted to the right in a parallel manner to a similar extent (data not shown). Thus, the relationship of force to peak aequorin signals during the cumulative administration of isoprenaline was not modified by previous administration of carbachol and its presence.

In contrast, when carbachol was administered to the isolated dog ventricular trabecula during induction of the positive inotropic action by isoprenaline, the relation of force to peak aequorin signals was shifted markedly downwards as observed with forskolin and MDL 17,043 (data not shown).

Discussion

Cyclic AMP-mediated changes in intracellular calcium transients

Accumulation of cyclic AMP in the myocardial cells results in an activation of protein kinase A which plays a key role in the system by causing the phosphorylation and subsequent conformational changes of functional proteins in sarcolemma, SR membrane, and myofibrils. All three positive inotropic agents used in the present study accumulated cyclic AMP in the dog and rabbit ventricular myocardium: forskolin by direct activation of catalytic subunit of adenylate cyclase; MDL 17,043 by suppressing the breakdown of cyclic AMP by inhibition of FIII fraction of PDE; isoprenaline through coupling via G_s to activation of catalytic subunit. The present findings on the effects of these agents on the intracellular calcium transients are essentially in accordance with the current view on the mechanism of action mediated by cyclic AMP in the following respects: 1) The peak aequorin signals which represent the peak calcium transients were elevated in a concentration-dependent manner by these agents with concomitant abbreviation of duration of the transients. These changes in calcium transients may be produced by facilitation of Ca^{2+} mobilization at the level of SR and sarcolemma, i.e., an increased release of Ca^{2+} from SR associated with phospholamban phosphorylation, and an increased probability of calcium channel opening related to the phosphorylation of sarcolemmal calcium channel proteins; 2) The shift of the relationship of peak force of contraction to the peak calcium transients to the direction indicating a decrease in calcium sensitivity of myofibrils, which may be partly ascribed to the lowered calcium binding affinity of troponin C associated with phosphorylation of troponin I (for reviews, see [6, 27, 35, 39]). The abbreviation of calcium transients may partly contribute to the shift, since the equilibrium is not achieved between peak calcium transients and force during the physiological twitch contraction of cardiac muscle [43, 44].

Novel mechanism of "accentuated antagonism"

Calcium ions play a key role as the final intracellular mediator of signal transduction in the contractile responses of all types of muscles. Therefore, it is reasonable to suppose that the inhibitory signal transmitted via muscarinic receptors is finally transformed to a reduction of the intracellular calcium ion concentration or modification of calcium sensitivity of contractile proteins in myocardial cells. Winegrad et al. [22, 23] showed that the calcium sensitivity of myofibrils of rat ventricular myocardium that had been rendered hyperpermeable was increased via a cyclic GMP-dependent process that favored dephosphorylation of troponin I. Although muscarinic receptor agonists elevate the cyclic GMP level in mammalian atrial muscle, it remains unsettled whether or not cyclic GMP modulates contractility through changes in calcium sensitivity in the intact cardiac muscle [11].

The muscarinic inhibition of cyclic AMP-mediated increase in force induced by isoprenaline was associated with parallel changes in peak calcium transients when carbachol was administered prior to isoprenaline, and the effect of isoprenaline was elicited in its presence in the rabbit papillary muscle. Therefore, the relation between force and peak calcium transients was not modified appreciably by muscarinic stimulation [11]. Thus, in this condition the reduction of cyclic AMP production caused by muscarinic inhibition via G_i appears to be reflected directly to the suppression of the positive inotropic effect of β-stimulation caused by muscarinic agonists.

In contrast, when carbachol was administered during induction of the positive inotropic action via accumulation of cyclic AMP by forskolin, MDL 17,043 or isoprenaline in the dog ventricular trabeculae, the reduction of force by muscarinic receptor activation was associated with relatively less decrease in peak calcium transients. The difference produced by this and previous experimental protocol may be as follows. In the latter situation the functional proteins would have been fully phosphorylated (by EC_{80-100} of cyclic AMP accumulating agonists) prior to administration of carbachol. If dephosphorylation of functional proteins is responsible for muscarinic inhibition of cyclic AMP-mediated inotropic effects, and the carbachol-induced dephosphorylation proceeds simultaneously in all phosphorylated functional proteins, there would have been no dissociation of peak force from peak calcium transients. In discord with this postulate, cyclic AMP-mediated decrease in calcium sensitivity of myofibrils (which has been observed to be elicited by isoprenaline alone) appears to be more pronounced during carbachol-induced inhibition. This observation implies two potential subcellular mechanisms for "accentuated antagonism". First, if the decrease of calcium sensitivity is related to phosphorylation of troponin I, the phosphorylation of this protein may be rather resistant to carbachol-induced dephosphorylation compared to calcium channel proteins or phospholamban, and this may deserve further biochemical study.

Second, this mechanism may be favorable for "accentuated antagonism" in that a small reduction of intracellular calcium transients is amplified to a pronounced reduction of force, i.e., the "accentuated antagonism" in a real sense in the adrenergic-cholinergic interaction *in situ*. Another possibility is that carbachol may have a potential intrinsic inhibitory activity which is potentiated by accumulation of cyclic AMP. However, considering that this effect of carbachol was elicited only when carbachol was added in the presence of cyclic AMP accumulation, and not vice versa, one cannot accept the latter postulate as the case.

A question arises whether the dissociation of force from intracellular calcium transients may not be specific to muscarinic inhibition, but a general phenomenon such as hysteresis was observed whenever the isometric force previously elevated was lowered via various inhibitory interventions. It is known that the force is reduced to a level lower than the baseline prior to application of β-stimulation, when β-adrenoceptor activation was discontinued by washout or by using β-antagonists [34]. Although the mechanisms underlying these observations have not yet been elucidated, a similar process has been observed in the muscarinic inhibition in the present study. On the other hand, it was recently found that the dissociation of force from calcium transients in an opposite direction also occurs. In the aequorin-injected cat and ferret papillary muscles, a cardiotonic agent, pimobendan increased both the force and peak light signals. In these experiments, peak calcium transients were lowered rapidly during washout of pimobendan, while the force remained at higher levels [25]. Thus, the relation of force to peak calcium transients may be regulated by modulation of calcium sensitivity in both directions at the level of myofibrils, and this may be reflected in the changes in the relation of the aequorin-injected intact myocardial cell preparations. Finally, it has be kept in mind that the potential variation of mode of regulation of myocardial contractility (through autonomic receptors, and Ca^{2+} mobilization and sensitivity) among mammalian species may be involved in modulatory mechanism at subcellular level.

Subcellular mechanisms of muscarinic regulation
of myocardial contraction cyclic AMP-dependent regulation

It is generally accepted that cyclic AMP plays an important role for the inhibitory action of muscarinic receptor agonists on mammalian ventricular myocardium. It appears that different sites of cyclic AMP-mediated processes are regulated by muscarinic receptor activation, and that the significance of the modified process depends much on the species of animals. In dog ventricular myocardium, carbachol inhibited the cyclic AMP-mediated positive inotropic actions of isoproterenol, histamine, glucagon, papaverine, theophylline, and lowered the cyclic AMP level elevated by these agents, while left unchanged both the cyclic AMP-independent positive inotropic actions of g-strophanthin and elevation of $[Ca^{2+}]_0$ [7, 8]. α-Adrenoceptor-mediated effect was unaffected by carbachol in the rabbit papillary muscle [13]. These findings imply that carbachol antagonizes specifically the positive inotropic effect mediated by cyclic AMP, by lowering the cyclic AMP levels that had been elevated previously by these agents. Although the primary mechanism may be inhibition of adenylate cyclase through activation of G_i by carbachol, it is not considered to be the sole mechanism for "accentuated antagonism". In other species, especially in the guinea pig heart, muscarinic receptor agonists did not change or only slightly lowered the cyclic AMP level, while the cyclic AMP-mediated positive inotropic effect was strongly suppressed [41]. Therefore, it has been postulated that the muscarinic inhibition may involve some other mechanisms such as intracellular antagonism of cyclic AMP by cyclic GMP [41] or activation of phosphatase by the muscarinic receptor stimulation resulting in dephosphorylation of the previously phosphorylated functional proteins with little change in cyclic AMP metabolism (Watanabe A, personal communications). The following observations also support the view described above. An extensive study by Scholz et al. [1, 5] addressed the question of whether or not changes in cyclic AMP and/or cyclic GMP metabolism are involved in the muscarinic- and adenosine-induced antiadrenergic effect (that is postulated to be elicited through a mechanism identical to that of muscarinic receptor activation) in the isolated guinea pig atrial and papillary muscle preparations. They observed neihter lowering of cyclic AMP levels nor elevation of cyclic GMP levels during the inhibition of cyclic AMP-mediated mechanical responses. On the other hand, Linden et al. [31] postulated that the muscarinic inhibitory action is entirely ascribed to its effect to lower the cyclic AMP level in the rat myocardium.

Cyclic AMP-independent regulation

Other pathways for signal transduction through muscarinic receptors do not involve changes in cyclic AMP metabolism in the myocardial cells. These are operative largely in the regulation of atrial, SA and AV nodal functions, and are less important in mammalian ventricular myocardium. The potassium conductance of the atrial cell membrane is increased through G_k [2, 28, 31], resulting in an abbreviation of action potential and negative chronotropic and inotropic actions in atria. The functional role of generation of cyclic GMP [17, 18] and acceleration of phosphatidylinositol turnover [3, 4] via muscarinic receptor activation has not yet been established.

The dual inhibitory regulation of myocardial function via activation of myocardial muscarinic receptors from respect of cyclic AMP-dependent and cyclic AMP-independent processes is illustrated in Fig. 6. The inhibitory signals triggered by the agonist binding to muscarinic receptors are transmitted to the subsequent biochemical, electrophysiological, and mechanical changes through activation of the GTP-binding proteins, G_i and G_k, which couple the receptor activation to the catalytic subunit of adenylate cyclase in the action mediated through the cyclic AMP-dependent mechanism, or to potassium channels in that mediated by cyclic AMP-independent processes preferentially exerted in atrial and nodal cells.

Cyclic AMP-induced decrease in calcium sensitivity of myofibrils, which is unveiled or enhanced by muscarinic receptor activation, may play an important role for "accentuated antagonism" of cyclic AMP-mediated positive inotropic effect by muscarinic receptor activation in mammalian ventricular myocardium.

Acknowledgements. The present study was supported partly by Grant-in-Aid for Scientific Research on Priority Area No. 62624004 and No. 63641002 from the Ministry of Education, Sciences, and Culture, Japan and the National Institute of Health, USA, grant HL-12186.

References

1. Böhm M, Brückner R, Hackbarth I, Haubitz B, Linhart R, Meyer W, Schmidt B, Schmitz W, Scholz H (1984) Adenosine inhibition of catecholamine-induced increase in force of contraction in guinea-pig atrial and ventricular heart preparations. Evidence against a cyclic AMP- and cyclic GMP-dependent effect. J Pharmacol Exp Ther 230:483–492
2. Brown AM, Birnbaumer L (1988) Direct G protein gating of ion channels. Am J Physiol 254:H401–H410
3. Brown JH, Buxton IL, Brunton LL (1985) α_1-Adrenergic and muscarinic cholinergic stimulation of phosphoinositide hydrolysis in adult rat cardiomyocytes. Circ Res 57:532–537

4. Brown JH, Masters SB (1984) Does phosphoinositide hydrolysis mediate "inhibitory" as well as "excitatory" muscarinic responses? Trends Pharmacol Sci 5:417–419
5. Brückner R, Fenner A, Meyer W, Nobis TM, Schmitz W, Scholz H (1985) Cardiac effects of adenosine and adenosine analogs in guinea-pig atrial and ventricular preparations: evidence against a role of cyclic AMP and cyclic GMP. J Pharmacol Exp Ther 234:766–774
6. Demaille JG, Jean-Francois Pechére's Group (1983) The control of contractility by protein phosphorylation. Adv Cyclic Nucleotide Protein Res 15:337–371
7. Endoh M (1979) Correlation of cyclic AMP and cyclic GMP levels with changes in contractile force of dog ventricular myocardium during cholinergic antagonism of positive inotropic actions of histamine, glucagon, theophylline and papaverine. Japan J Pharmacol 29:855–864
8. Endoh M (1980) The time course of changes in cyclic nucleotide levels during cholinergic inhibition of positive inotropic actions of isoprenaline and theophylline in the isolated canine ventricular myocardium. Naunyn-Schmiedeberg's Arch Pharmacol 312:175–182
9. Endoh M (1986) Regulation of myocardial contractility via adrenoceptors: differential mechanisms of α- and β-adrenoceptor-mediated actions. In: Grobecker H, Philippu A, Starke K (eds) New aspects of the role of adrenoceptors in the cardiovascular system. Springer, Berlin Heidelberg New York, pp 78–105
10. Endoh M (1987) Dual inhibition of myocardial function through muscarinic and adenosine receptors in the mammalian heart. J Appl Cardiol 2:213–230
11. Endoh M, Blinks JR (1984) Effects of endogenous neurotransmitters on calcium transients in mammalian atrial muscle. In: Fleming WW, Langer SZ, Graefe KH, Weiner N (eds) Neuronal and extraneuronal events in autonomic pharmacology. Raven Press, New York, pp 221–230
12. Endoh M, Blinks JR (1988) Actions of sympathomimetic amines on the Ca^{2+} transients and contractions of rabbit myocardium: reciprocal changes in myofibrillar responsiveness to Ca^{2+} mediated through α- and β-adrenoceptors. Circ Res 62:247–265
13. Endoh M, Motomura S (1979) Differentiation by cholinergic stimulation of positive inotropic actions mediated via α- and β-adrenoceptors in the rabbit heart. Life Sci 25:759–768
14. Endoh M, Yanagisawa T, Morita T, Taira N (1985) Differential effects of sulmazole (AR-L 115BS) on contractile force and cyclic AMP levels in canine ventricular muscle: comparison with MDL 17,043. J Pharmacol Exp Ther 234:267–273
15. Endoh M, Yanagisawa T, Taira N, Blinks JR (1986) Effects of new inotropic agents on cyclic nucleotide metabolism and calcium transients in canine ventricular muscle. Circulation 73 (suppl III):III-117–III-133
16. Fleming JW, Watanabe AM (1986) Biochemical mechanisms of parasympathetic regulation of cardiac function. In: Fozzard HA, Haber E, Jennings RB, Katz AM, Morgan HE (eds) The heart and cardiovascular system, vol 2. Raven Press, New York, pp 1679–1688
17. George WJ, Polson JB, O'Toole AG, Goldberg ND (1970) Elevation of guanosine 3',5'-cyclic phosphate in rat heart after perfusion with acetylcholine. Proc Natl Acad Sci USA 66:398–403
18. George WJ, Wilkerson RD, Kadowitz PJ (1973) Influence of acetylcholine on contractile force and cyclic nucleotide levels in the isolated perfused rat heart. J Pharmacol Exp Ther 184:228–235
19. Gilman AG (1984) G proteins and dual control of adenylate cyclase. Cell 36:577–579
20. Hollenberg M, Carriere S, Barger AC (1965) Biphasic action of acetylcholine on ventricular myocardium. Circ Res 16:527–536
21. Honma M, Satoh T, Takezawa J, Ui M (1977) An ultrasensitive method for the simultaneous determination of cyclic AMP and cyclic GMP in small-volume samples from blood and tissue. Biochem Med 18:257–273
22. Horowits R, Winegrad S (1981) Cholinergic regulation of troponin phosphorylation in cardiac muscle. Biophys J 33:85a
23. Horowits R, Winegrad S (1983) Cholinergic regulation of calcium sensitivity in cardiac muscle. J Mol Cell Cardiol 15:277–280
24. Inui J, Brodde OE, Schumann HJ (1982) Influence of acetylcholine on the positive inotropic effect evoked by α- or β-adrenoceptor stimulation in the rabbit heart. Naunyn-Schmiedeberg's Arch Pharmacol 320:152–159
25. Kappler JH, Lee NKM (1988) Pimobendan increases myofibrillar responsiveness to Ca^{++} in intact mammalian myocardium. Abstract in Fed Proc (APS/ASPET Fall Meeting at Montreal)

26. Kariya T, Wille LJ, Dage RC (1982) Biochemical studies on the mechanism of cardiotonic activity of MDL 17,043. J Cardiovasc Pharmacol 4:509–514
27. Katz AM (1983) Cyclic adenosine monophosphate effects on the myocardium: a man who blows hot and cold with one breath. J Am Coll Cardiol 2:143–149
28. Kranias EG (1986) Protein phosphorylation and the cardiac sarcoplasmic reticulum. In: Solaro RJ (ed) Protein phosphorylation in heart muscle. CRC Press, Boca Raton, pp 105–128
29. Kurachi Y, Nakajima T, Sugimoto T (1986) On the mechanism of activation of muscarinic K^+ channels by adenosine in isolated atrial cells: involvement of GTP-binding proteins. Pfluegers Arch 407:264–274
30. Levy MN (1971) Sympathetic-parasympathetic interactions in the heart. Circ Res 29:437–445
31. Linden J, Hollen CE, Patel A (1985) The mechanism by which adenosine and cholinergic agents reduce contractility in rat myocardium. Correlation with cyclic adenosine monophosphate and receptor densities. Circ Res 56:728–735
32. Pfaffinger PJ, Martin JM, Hunter DD, Nathanson NM, Hille B (1985) GTP-binding proteins couple cardiac muscarinic receptors to a K channel. Nature 317:536–538
33. Scholz H (1980) Effects of beta- and alpha-adrenoceptor activators and adrenergic transmitter releasing agents on the mechanical activity of the heart. In: Szekeres L (ed) Adrenergic activators and inhibitors. Springer, Berlin Heidelberg New York, pp 651–733
34. Skomedal T, Schiander IG, Osnes JB (1988) Both an *alpha* adrenoceptor mediated and a *beta* adrenoceptor mediated component contribute to the final inotropic response to norepinephrine in rat heart. J Pharmacol Exp Ther (in press)
35. Solaro RJ (1986) Protein phosphorylation and the cardiac myofilaments. In: Solaro RJ (ed) Protein phosphorylation in heart muscle. CRC Press, Boca Raton, pp 129–156
36. Solaro RJ, Disalvo J, Paul RJ (1983) Coordination of metabolism and contractility by phosphorylation in cardiac, skeletal, and smooth muscle. Fed Proc 42:7–71
37. Spiegel AM, Gierschik P, Levine MA, Downs RW Jr (1985) Clinical implications of guanine nucleotide-binding proteins as receptor-effector couplers. New Engl J Med 312:26–33
38. Tada M, Katz AM (1982) Phosphorylation of the sarcoplasmic reticulum and sarcolemma. Annu Rev Physiol 44:401–423
39. Tsien RW (1977) Cyclic AMP and contractile activity in heart. Adv Cyclic Nucleotide Res 8:368–420
40. Ui M (1984) Islet-activating protein, pertussis toxin: a probe for functions of the inhibitory guanine regulatory component of adenylate cyclase. Trends Pharmacol Sci 5:277–279
41. Watanabe AM, Besch HR Jr (1975) Interaction between cyclic adenosine monophosphate and cyclic guanosine monophosphate in guinea pig ventricular myocardium. Circ Res 37:309–317
42. Winegrad S (1984) Regulation of cardiac contractile proteins: correlation between physiology and biochemistry. Circ Res 55:565–574
43. Yue DT (1987) Intracellular $[Ca^{2+}]$ related to rate of force development in twitch contraction of heart. Am J Physiol 252:H760–H770
44. Yue DT, Marban E, Wier WG (1986) Relationship between force and intracellular $[Ca^{2+}]$ in tetanized mammalian heart muscle. J Gen Physiol 87:223–242

Author's address:

Masao Endoh, M.D. Department of Pharmacology, Yamagata University School of Medicine, 2-2-2 Iida-nishi, 990-23 Yamagata, Japan

Studies on the mechanism of action of the bipyridine milrinone on the heart

A. E. Farah, C. J. Frangakis

Dept. of Molecular Pharmacology, Glaxo Research Laboratories,
Research Triangle Park, North Carolina, USA

Summary

Milrinone is a positive inotropic and vasodilator agent when tested in experimental animals and in human heart-failure patients. It is generally believed that milrinone acts by inhibiting phosphodiesterase IV, thus increasing cyclic AMP, $[Ca^{++}]_i$ and cardiac contractile force and relaxation.

Maximal force produced by milrinone is greater when single-dose response curves are compared to cumulative dose-response curves. In vitro, milrinone produces a tachyphylaxis, the extent of which is both dose- and time-dependent. Recovery of tachyphylaxis is both dose- and time-dependent and is not influenced by inhibitors of protein or RNA synthesis. There is a specific cross-tachyphylaxis between milrinone and amrinone, theophylline, papaverine, and Bay K8644. This tachyphylaxis may explain the low maximal contractile response of the cumulative dose-response observed in isolated tissues.

Milrinone increased cyclic AMP in dog and guinea pig cardiac muscle. As previously shown by Endoh et al. [17], milrinone in low doses produced a biphasic effect on cyclic AMP. The early increase (first 60–70 s) in cyclic AMP shows a good correlation with contractile force changes. If cyclic AMP is determined at maximal contractile force this correlation was poor.

Here we also present instances where the increase in cyclic AMP after milrinone (determined at maximal effect) does not correlate with the contractile response.

The cross-tachyphylaxis of milrinone with Bay K8644 suggests that milrinone has an action on the sarcolemmal Ca^{++} channels. Bay K8644 suppresses the positive inotropic effect of catecholamines by 50%, but not the cyclic AMP response. The inotropic effect of milrinone, in contrast to norepinephrine is highly sensitive to $[Ca^{++}]_o$, stimulation rate, and $[K^+]_o$. In this respect milrinone behaves more like Bay K8644. We postulate that the main inotropic action of milrinone is due to a sarcolemmal effect. The early cyclic AMP production described could be in the sarcolemmal compartment and this may explain some of the similarities of milrinone's actions with those of Bay K8644.

The tachyphylaxis observed with the inotropic effect of milrinone does not extend to the decreases in relaxation time. This and other findings to be discussed suggest that the positive inotropic and reduction in relaxation time by milrinone depend on different mechanisms, possibly through differential compartmentalization of cyclic AMP.

Introduction

Amrinone (Inocor) has been approved in the USA and other countries for intravenous treatment of severe heart-failure cases. On oral medication, this drug produced a series of side actions which necessitated the reduction of dosage to ineffective levels. The second generation bipyridine, milrinone (Corotrope), was synthesized by Lesher and Phillione. Its relation to amrinone is shown in Fig. 1 and some of its cardiovascular phar-

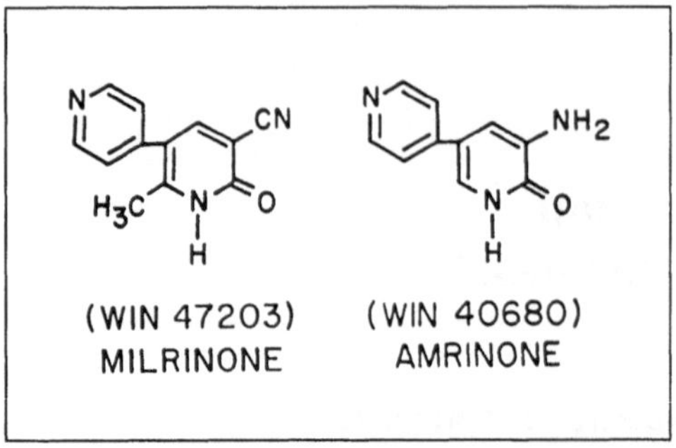

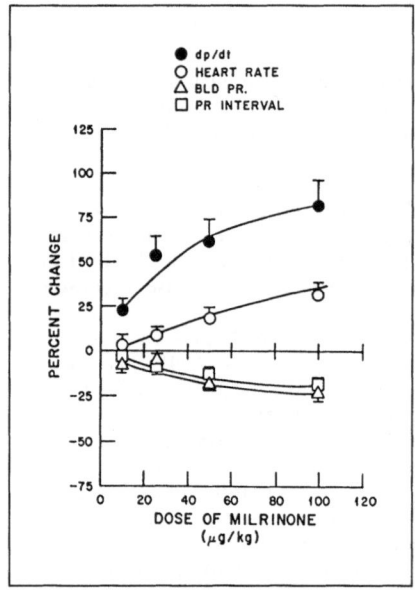

Fig. 1. Structural formulae of amrinone and milrinone

Fig. 2. Effects of milrinone in the intact pentobarbital anesthetized female dogs. Values represent the averages $\pm$ SEM of six animals, and are represented as percent change from the control pre-medicated values

macology was described by Alousi et al. [1–3] and Pastelin et al. [55] (Figs. 2 and 3). Milrinone is now in advanced clinical trials, and available evidence indicates that it is a useful therapeutic agent in the treatment of chronic heart failure, both in dogs [34] and in man [5, 10, 13, 30, 37, 44]. Both the intravenous and oral administrations of milrinone are effective and this drug has fewer side effects than amrinone when given chronically. In normal individuals serum half-life of milrinone is about 60 min with a duration of action of 4–6 h with a 10 mg dose, both orally and intravenously. Effective plasma concentrations are 50–300 ng per ml [38, 61]. In heart-failure cases the half-life is prolonged, probably due to the reduction of the renal excretion of milrinone [38].

The studies of Alousi et al. [1, 2] and Pastelin et al. [55] have shown that milrinone is a positive inotropic agent with vasodilatory properties (Fig. 2). It is a phosphodiesterase inhibitor with no effect on the sodium potassium dependent ATPase [1, 2]. However Mylotte et al. [52] have demonstrated a stimulation of Ca^{++}-ATPase in rabbit myo-

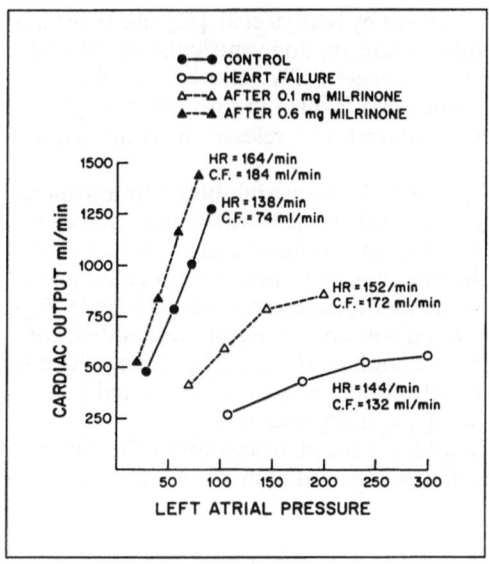

Fig. 3. Effects of milrinone on the dog heart-lung preparation during pentobarbital-induced heart failure. The preload was increased by increasing the height of the inflow vessel by three, 5-cm-steps. During the control period the increase in preload produced an increase in cardiac output with minimal changes in left atrial pressure. Heart failure increased the left atrial pressure and reduced cardiac output. When the preload was increased the left atrial pressure increased by 200 mm of water and cardiac output increased by about 250 ml/min. Administration of 0.1 mg of milrinone improved the response of the heart to the increase in preload and an additional 0.5 mg of milrinone restored cardiac function to control levels. Note the increase in heart rate and coronary flow (CF) produced by milrinone. Results are the averages of six individual preparations

cardial membranes by milrinone. The importance of this observation is not clear since the closely related amrinone does not have this action.

The effects of milrinone on heart muscle are not inhibited by either alpha-or beta-adrenergic blocking agents or by antihistamines or anti-seratonergic agents. It is about 30–50-times as effective as amrinone when tested in isolated preparations [1, 2]. Milrinone could overcome acute heart failure produced by pentobarbital, (see Fig. 3) propranolol, dihydropyridine, Ca^{++} blockers, and spontaneous heart failure [55]. However, the acute heart failure in the dog heart-lung preparation produced by papaverine was resistant to milrinone, although it still responded to epinephrine [55]. In the normal heart-lung preparation, milrinone increased coronary blood flow, oxygen consumption, and heart action, but had no significant effect on the calculated external efficiency of the normal heart. In the failing heart milrinone did not increase or even decrease the oxygen consumption but markedly increased the work performance of these failing hearts [31]. Similar observations have been made in human patients with congestive heart failure [7, 48]. The overall clinical and experimental findings show that amrinone and milrinone are positive inotropic and vasodilator agents and seem to have clinical utility for the treatment of severe cases of heart failure.

How do these drugs produce the increase in contractile force of the heart? As previously stated amrinone inhibits cardiac phosphodiesterase, and seems to have a specificity for cyclic AMP phosphodiesterase IV as first shown by Kariya et al. [32] and Weishaar et al. [71]. In our laboratories we obtained similar results on dog ventricular cyclic AMP phosphodiesterase IV (unpublished observations). It is generally believed that phosphodiesterase inhibition results in an increase in cyclic AMP, which in turn will phosphorylate sarcolemmal and sarcoplasmic proteins essential for Ca^{++} release and translocation across these membranes [70].

The earliest observations on crude cardiac phosphodiesterase inhibition by amrinone were those of Honerjager et a. [27] and Endoh et al. [16]. These authors also showed increases in cyclic AMP following addition of high concentrations of amrinone (> 1 mM). Work published by Alousi and Johnson [2] showed that amrinone did not cause an increase in cyclic AMP when doses used were in the therapeutic range, (1–10 μg/ml). Significant increases in cyclic AMP were observed when doses of 100 μg/ml were added, confirming the findings of Honerjager et al. [27] and Endoh et al. [16]. Milrinone, which is about 30-times as effective as amrinone, is also a phosphodiesterase inhibitor and in high concentrations increased cyclic AMP in guinea pig papillary muscles [3].

These discrepancies thus prompted us to investigate the pharmacology of milrinone, especially as it related to its possible action via the production of cyclic AMP.

Tachyphylaxis to milrinone

Figure 4 shows a dose-response curve of milrinone obtained with isolated dog right ventricular trabeculae bathed in Krebs-Henseleit nutrient solution at 37 °C. Under these conditions we observed a tachyphylaxis when a second dose of milrinone was added [20]. This tachyphylaxis was not observed in intact dogs, confirming the findings of Sys and Brutsaert [68]. We used only one dose of milrinone for each muscle strip and in Fig. 4, we see a flattening out of the dose-response curve between 1.0 and 5.0 μg/ml of milrinone. Concentrations above 5 μg/ml produced a second increase in contractile force. We applied a third-order polynomial analysis to this curve and confirmed that the observed flattening out of the curve was statistically significant [20]. These data suggested that we may be dealing with more than one mechanism of action of milrinone. The lower curve in Fig. 4 shows the increase in contractile force produced by a second or tachyphylactic dose of milrinone. The third curve is the cumulative dose-response curve and its maximum corresponds to that of the first portion of the single-dose dose-response curve. Further evidence for a biphasic effect of milrinone was demonstrated in ferret cardiac muscle kept at 26 °C [42]. Milrinone produced a biphasic contraction, one phase being sensitive to Ca^{++} channel blockers, the other to ryanodine.

Effect of milrinone on sarcolemmal Ca^{++} channels

To further substantiate a biphasic dose-response curve we utilized the Ca^{++} channel blockers verapamil and ruthenium red (see Fig. 5). The latter agent is a red dye, which according to Frank et al. [23] and Luft [39], localizes in the sarcolemma of normal cardiac cells. As can be seen from Fig. 5, both verapamil and ruthenium red inhibited the effects on contractile force produced by milrinone in doses below 2.5 μg/ml, but had no significant effect on milrinone doses above 5 μg/ml. A similar differential effect of verapamil and ruthenium red was also demonstrated when studying the effects of milrinone on the action potential of dog ventricular trabeculae. Verapamil reduced contractile force and phase 2 of the action potential changes produced by a dose of 0.25 μg, but had no sig-

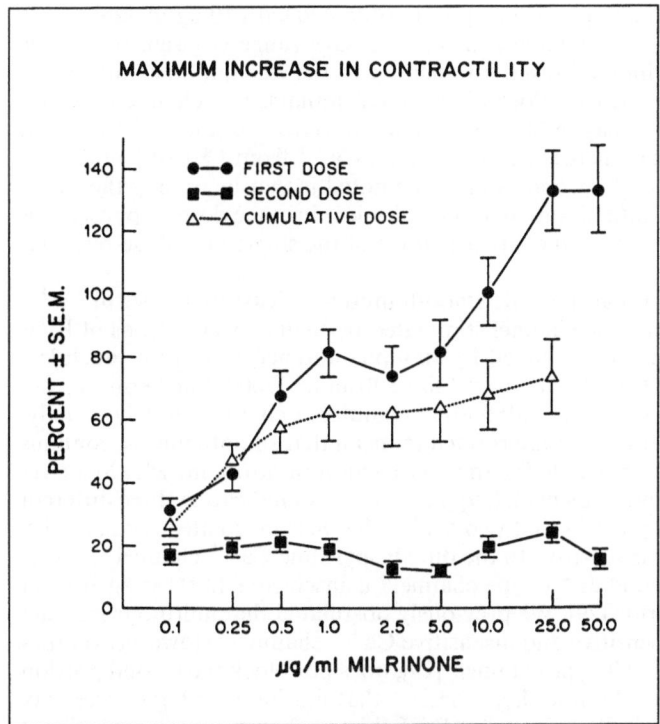

Fig. 4. Dose response curves of milrinone in isolated dog trabeculae. ●----● = Single dose response curve; one dose applied to each trabeculum. ■----■ = Tachyphylaxis; same dose applied after washout of initial dose. △----△ = Cumulative dose response curve; note the flattened portion of the single-dose response curve between 1.0 and 5 μg of milrinone per ml of bathing fluid. Results are the averages of six to eight individual determinations ± SEM

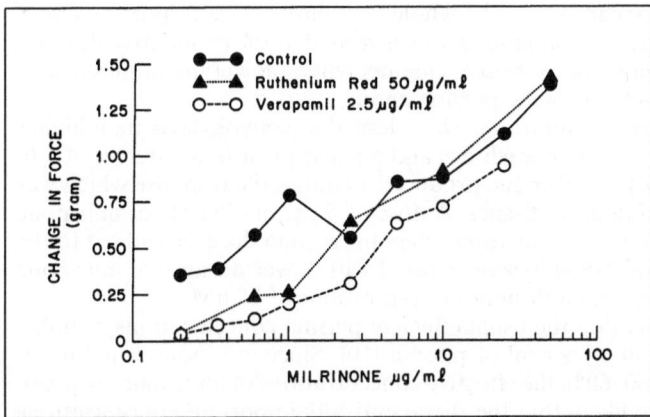

Fig. 5. Effects of verapamil and ruthenium red on the single-dose response curve of milrinone in dog trabeculae; conditions as in Fig. 4. Results are the averages of six to eight individual muscle preparations

nificant effect on the contractile or phase 2 responses obtained with 10 µg/ml [20]. These data thus show that milrinone, in the lower effective dosage range was sensitive, while higher concentrations of milrinone showed a resistance to the calcium blocking agents. If the effects of verapamil are due to a blockade of sarcolemmal Ca^{++} channels then increasing the extracellular Ca^{++} may reduce or eliminate the effects of the Ca^{++} blocker. Our data show that the effects of increasing extracellular Ca^{++} from 1.8 to 4.5 mM eliminated the effects of verapamil. A second point to be noted is that increasing the extracellular Ca^{++} from 1.8 to 4.5 mM eliminated the biphasic milrinone dose response, thus suggesting that Ca^{++} plays a role in the biphasic form of the single-dose dose-response curve of milrinone [20].

Meisheri et al. [47] showed that in aortic smooth muscle at least two types of Ca^{++} channels can be demonstrated – the channel stimulated by high concentrations of K^+ is sensitive while the Ca^{++} channel stimulated by an alpha adrenergic agent is much less sensitive to the inhibitory effects of the verapamil-like substance D600. Furthermore, amrinone inhibits both the K^+ and alpha adrenergic stimulated channels nearly equally. These data of Meisheri et al. have been interpreted to mean that smooth muscle contains at least two Ca^{++} channels responsible for the constriction of aortic muscle. More recently Tsien's group [53, 54] and Tang et al. [69] published data indicating three different Ca^{++} channels in cultured nerve cells and two Ca^{++} channels in isolated cardiac cells. In heart cells the "L channel" is sensitive to the dihydropyridine Ca^{++} channel agonists and antagonists, while the second or "T type channel" is insensitive to these agents but sensitive to the diuretic amiloride [29]. We previously postulated that milrinone may act on both the dihydropyridine sensitive and insensitive Ca^{++} channels. However, verapamil which blocks both the L and T type channels [29], does not block the second portion of the milrinone dose-response. Another hypothesis is that milrinone in high doses may activate the Na^+-Ca^{++} exchange mechanism [4]. If this mechanism were operative it probably would be blocked by amiloride which blocks the T channels [69], as well as the Ca^{++}-H^+ and Ca^{++}-Na^+ exchange system [4, 56] – however, data on amiloride's and milironine's effects are not available.

The in vitro cumulative dose-response curve of milrinone in the presence of 1.8 mM Ca^{++} is monophasic and reaches a maximum increase of about 60% over basal force (see Fig. 4). This cut-off of contractile force may be related to the tachyphylaxis that has been demonstrated in the in vitro preparations. The whole cumulative dose response curve is inhibited by 2.5 µg/ml (5.5 µM) of verapamil. This suggests that the cumulative dose-response curve represents the verapamil-sensitive channel while the verapamil-insensitive channel is eliminated by the tachyphylactic phenomenon.

In our experience small doses of milrinone show less of a tachyphylaxis than higher dosages. Thus, following five repetitive washings and a subsequent time interval of 1 h, addition of 0.25 µg/ml (1.2 µM) of milrinone produced a contractile response which was 20–30% lower than those seen in naive tissues. A dose of 5 µg/ml (25 µM) of milrinone showed a 60–80% reduction of contractile force when the second dose was added to the bath. Furthermore, recovery of tachyphylaxis is rapid with lower dosages of milrinone (0.5 µM) and may take 5 to 6 h with milrinone concentrations of 25 µM.

In humans and in the intact dog the usual effective plasma concentrations of milrinone are between 0.05 and 0.40 µg per ml of plasma [10]. Since milrinone is bound to serum proteins to an extent of 50–60% the effective concentrations of milrinone are probably even lower. It is therefore likely that the therapeutically important concentrations of milrinone are verapamil sensitive and probably would not show any measurable degree of tachyphylaxis when tested in vivo [66, 67].

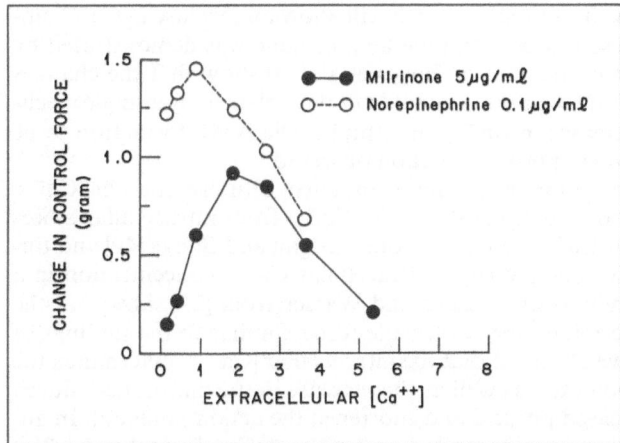

Fig. 6. Effects of extracellular calcium concentration on the positive inotropic effect of milrinone (5 µg/ml) and norepinephrine (0.1 µg/ml) in dog trabeculae. Note the sensitivity of milrinone's effects to extracellular $[Ca^{++}]$. Each of the dose-response curves is the average of six individual muscle preparations

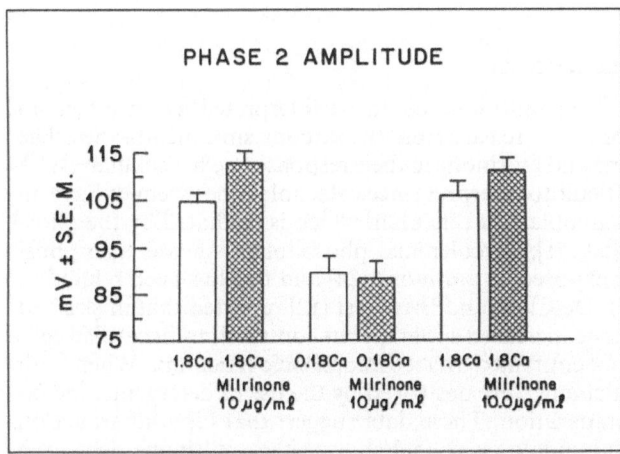

Fig. 7. Effect of extracellular calcium on the increase of the phase 2 of the intracellular action potential produced by milrinone in dog trabeculae. Results are the average of six individual determinations ± SEM

Evidence that milrinone may act on a Ca^{++} entry via the sarcolemma can also be seen from Fig. 6. The positive inotropic effect of milrinone is highly sensitive, while the effects of norepinephrine are relatively insensitive to $[Ca^{++}]_0$ (Fig. 6). Figure 7 shows that the effect of milrinone on the phase 2 of the action potential was sensitive to the $[Ca^{++}]_0$. These data thus support the hypothesis that a major effect of milrinone is the activation of Ca^{++} channels, either directly or indirectly, which will allow an increase in Ca^{++} uptake by cardiac cells. This increase in Ca^{++} uptake by milrinone was demonstrated by Frangakis et al. [22]. Furthermore, the data of Hayes et al. [24] show that the changes in protein phosphorylation induced by amrinone could not be explained by a single mechanism of action but required increases in Ca^{++} entry and cyclic AMP formation to be responsible for the pattern of protein phosphorylation observed.

If entry of Ca^{++} via the sarcolemma is an important effect of milrinone, this Ca^{++} should, according to Fabiato and Fabiato [18], release Ca^{++} from intracellular stores and thus increase the intracellular Ca^{++} concentration. Morgan and Blinks [49] and Endoh et al. [17] reported that milrinone increases intracellular Ca^{++} concentration in a dose-dependent manner. The findings of Fozzard and Wasserstrom [21] show that the resting Ca^{++} concentration determined the contractile force. Similar to the findings of Fozzard and Wasserstrom [21], we observed that the intracellular $[Ca^{++}]$ determined the height of phase 2 of the action potential as well as the increase in the milrinone-induced increase in force. Milrinone increased phase 2 and shortened the action potential. In another study, Frangakis et al. [22] have demonstrated an increase $^{45}Ca^{++}$ uptake of rabbit myocytes when exposed to low concentrations of milrinone. In addition, electrophysiological observations of Malecot et al. [43], Sutko et al. [65], and Canniff et al. [9] have shown that amrinone and milrinone increased the inward Ca^{++} current. All these observations thus support the idea that Ca^{++} entry plays a role in the action of the inotropic bipyridines.

Effect of milrinone on sarcoplasmic reticulum

The effects of extracellular Ca^{++} on milrinone could be interpreted as an effect on Ca^{++} entry via the sarcolemma or Ca^{++} release from the sarcoplasmic membranes. The differences between norepinephrine and milrinone in their response to extracellular Ca^{++} (Figs. 6 and 7) are striking but difficult to interpret since catecholamines seem to have an effect on calcium uptake by the sarcoplasmic reticulum which is mediated by the phosphorylation of phospholamban [50, 51]. Sarcolemmal phospholamban was also phosphorylated when heart cells were exposed to amrinone [24] and this has been related to calcium metabolism of heart cells. De Clerk and Brutsaert [12] reported that in skinned cardiac cells amrinone and milrinone increased spontaneous contractions in cardiac cells that were devoid of sarcolemma but contained intact sarcoplasmic reticulum. When both sarcolemma and sarcoplasmic reticulum were destroyed by the use of detergents, the bipyridines had no effect on this preparation. These data suggest that bipyridines act on an intracellular calcium store. Rapundalo et al. [56] showed that milrinone decreased minimally the magnitude of Ca^{++} uptake by sarcoplasmic reticulum of dog heart muscle. This could result in the increase of $[Ca^{++}]_i$ and could explain the results of De Clerk and Brutsaer [12].

It is of interest to note that the frog heart is insensitive to the bipyridine (unpublished results). Similarly, the neonatal dog heart and adult Purkinje tissue of the dog [8, 9, 57] show no positive or even negative inotropic effects following the addition of the bipyridines. All these preparations: frog, Purkinje tissue, and neonatal dog heart have a poorly developed T tubular system and a poorly functioning sarcoplasmic reticulum. This could possibly explain the lack of response to milrinone in these tissues. On the other hand the negative inotropic or poor positive inotropic response of the rat heart to milrinone cannot be explained on this basis since the rat heart has a highly active sarcoplasmic reticulum [63]. Also, rat myocardial contractility is relatively insensitive to Ca^{++}

channel blockers and highly sensitive to ryanodine, an inhibitor of sarcoplasmic Ca^{++} release [62–65]. Radpundalo et al. [56] found that rat heart becomes more sensitive to milrinone when extracellular Ca^{++} is reduced thus indicating that the rat heart is already contracting maximally. If this is the case, then increasing the extracellular Ca^{++} above control levels (1.8 mM) should reduce the response of milrinone in dog trabeculae. That this is the case can be seen in Fig. 6. However, the reduction of contractile response with high extracellular $[Ca^{++}]$ is a phenomenon also seen with cardioactive glycoside, catecholamines, as well as with milrinone. This could be explained on the basis that at high $[Ca^{++}]$ the contractile force is already maximal and cannot be exceeded and therefore, it may have nothing to do with Ca^{++} loading of the sarcoplasmic reticulum. However, when the contractile effects of high Ca^{++} on the trabeculae were reduced by 50–60% with the addition of pentobarbital, milrinone's inotropic effects were still very weak, indicating that contractile force was not the limiting factor under these conditions [19]. This suggests that Ca^{++} saturation of either the Ca^{++} channels or the sarcoplasmic reticulum reduced milrinone's effects. Moreover, we recently substantiated this hypothesis by the observation that addition of verapamil in the presence of high $[Ca^{++}]_0$ increased the response to milrinone [19].

The data with ruthenium red suggest that the effects produced by small doses of mil- and they concluded that ryanodine inhibits cardiac contractile force by blocking release of Ca^{++} from the sarcoplasmic reticulum. Ryanodine in a concentration of 0.05 µg/ml (0.1 µM) blocked the effects of high and low doses of milrinone by a noncompetitive mechanism (unpublished observations). However, this again cannot differentiate between a sarcoplasmic reticulum or sarcolemmal site of action, since a block of calcium release from the sarcoplasmic reticulum would reduce the effectiveness of the increase in Ca^{++} entry produced by milrinone. In a more recent study Sutko et al. [65] showed that ryanodine had no influence on an inward Ca^{++} current and thus its effects were most likely due to the inhibition of Ca^{++} release from an intracellular calcium store.

The data with ruthenium red suggest that the effects produced by small doses of milrinone are blocked by a mechanism localized to the sarcolemma. This could be due to a block of Ca^{++} entry or a block of Ca^{++} release from the sarcolemma as postulated by Lullmann et al. [40, 41]. The electrophysiological data show that phase 2 of the action potential of dog muscle is sensitive to both Ca^{++} [14] and milrinone [9]. At low Ca^{++} concentrations milrinone has markedly reduced effects on both phase 2 of the action potential and contractile force (see Figs. 6 and 7).

All the data presented could be explained by a direct or indirect action of the bipyridines on calcium channels or on calcium release from an intracellular calcium store. It is therefore difficult to differentiate between the two mechanisms based on the data at hand. It is quite possible however that both mechanisms may be involved in the action of milrinone and could explain the biphasic dose response [20, 42].

Relation between positive inotropic effect and cyclic AMP

The second aspect to be discussed is the mechanism whereby Ca^{++} entry or Ca^{++} release in the cardiac cell is increased by milrinone. As previously stated, the general hypothesis is that milrinone and amrinone are phosphodiesterase inhibitors, that they increase cyclic AMP, and this in turn increases the phosphorylation of a protein or a group of proteins controlling the entry of Ca^{++} via calcium channels. The methylxanthines are general phosphodiesterase inhibitors that do not have specificity for any of the three types of isozymes found in heart muscle. The relation between phosphodiesterase inhibition and positive inotropic effects of the methylxanthines has been questioned by Tsien [70], McNeil et al. [45, 46], Mushlin et al. [51], and others. It is thus useful to re-examine the relationship between phosphodiesterase inhibition by amrinone and milrinone and their positive inotropic effect. The data of Honerjager et al. [27] and Endoh et al. [16] show a correlation between phosphodiesterse inhibition and amrinone's positive inotropic effect. Kariya and Dage [32], Hayes et al. [24], and Weishaar et al. [72] showed that amrinone and milrinone are specific inhibitors of phosphodiesterase IV which hydrolyze cyclic AMP with a K_m of 1 μM and does not require calmodulin or Ca^{++} for its action. However, we previously observed that a number of phosphodiesterase inhibitors which were structurally related to the known bipyridines (amrinone and milrinone) had either negative or no inotropic effects [unpublished observations).

Furthermore, in comparing the inotropic response of dog, rabbit, guinea pig, and rat cardiac tissue to milrinone's ability to inhibit cardiac phosphodiesterase activity, it was observed that the rat isozyme was equally sensitive to milrinone's inhibition, although the rat heart showed a poor inotropic response as compared to the other species. A comparison of right ventricular Purkinje fibers and trabeculae showed a similar sensitivity to inhibition of phosphodiesterase, while Purkinje fiber of the dog responded by a slight negative inotropic response to milrinone. Similar discrepancies were observed by Kobeylarz et al. [36] when the responses of various species to imazodan were studied. Thus, in contrast to the canine dog heart tissue, guinea-pig and rat heart phosphodiesterase were insensitive to the inhibitor imazodan. More recently, Weishaar et al. [72] attempted to explain these discrepancies. These authors postulate that the isozyme related to inotropism is associated with the membranous fraction of cardiac homogenates. Thus, the tissues which were insensitive to inotropic stimulation by a select number of the phosphodiesterase inhibitors were those that contained little phosphodiesterase IV (also know as fraction III) in the membranous material. This could explain the discrepancies observed with milrinone in the various species and it would be important to see a confirmation of these findings with milrinone and other phosphodiesterase inotropes. According to the results published by Weishaar et al. [72] guinea-pig hearts had no membrane bound phosphodiesterase III and thus should be insensitive to milrinone. However, data of Alousi et al. [3] and our own data indicated that the guinea pig heart is quite sensitive to milrinone. It is unclear whether the subclasses of phosphodiesterase proposed by Weishaar et al. [72] really explain the differential responses of all of the phosphodiesterase inhibitors.

During the earlier stages of our studies no good correlation could be found between the inotropic response and the increase of cardiac cyclic AMP by milrinone. However, Endo et al. [17] showed that with small doses of milrinone there was an early increase in cyclic AMP followed by a decline which reached control values at about the time maximal inotropic effects were observed. This decline was followed by a secondary rise which peaked long after maximal contractile force had been attained (see Fig. 8 A). When a large dose of milrinone was added the increase in cyclic AMP was monophasic. We confirmed these findings (Fig. 8 A, B).

Since there was a biphasic cyclic AMP response to small doses of milrinone, we determined the cyclic AMP and contractile changes, 60–70 s after the addition of milrinone and our results are shown in Fig. 9. It is clear that the increase in cyclic AMP determined during the first 60–70 s correlates with the inotropic effects of milrinone, while the cyclic AMP determined at maximal contractile response does not show a good correlation. We observed a number of instances where the increase in cyclic AMP did not correlate with the contractile response to milrinone (see Table 1). Thus, the rat heart, which responds poorly to milrinone, shows a normal increase in cyclic AMP when measured at maximal contractile response. In a similar manner dog Purkinje tissue, which does not show a posi-

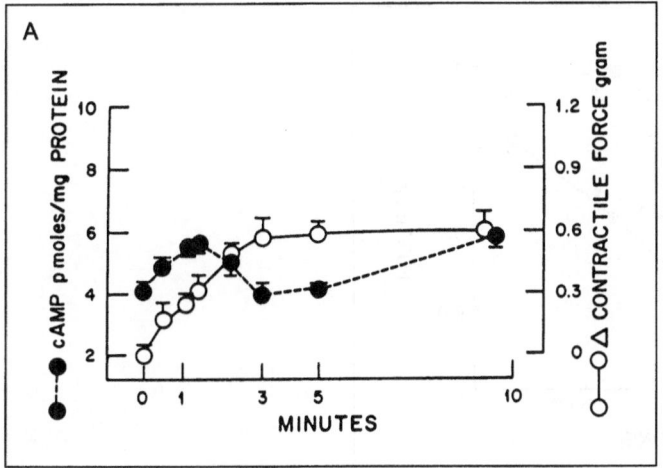

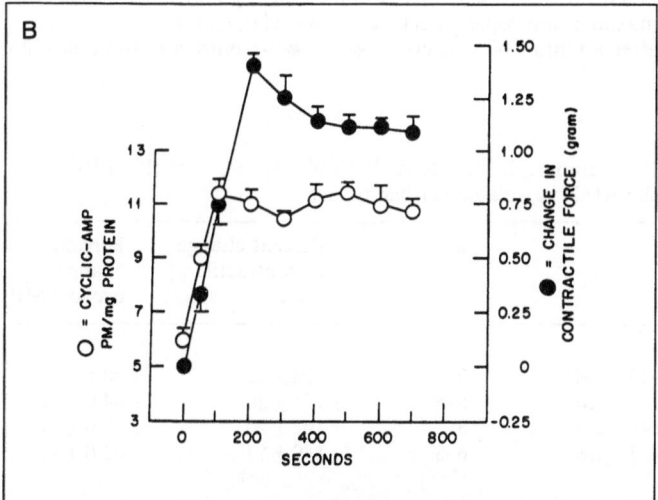

Fig. 8 A, B. Time-dependent changes on the contractile force and cyclic AMP content of dog ventricular trabeculae. **A** Effects of 1.25 µM milrinone. Each point represents the average obtained from eight tissues ± SEM. **B** Effects of 125 µM milrinone. Each point represents the average obtained from six trabeculae ± SEM. Milrinone added at zero time

tive inotropic response to milrinone, shows an increase in cyclic AMP. As previously described the tachyphylaxis produced by a large dose of milrinone reduced the response of a second dose of milrinone following washout of the first dose. The perfused guinea pig heart was rinsed in these experiments and the second dose of milrinone produced no increase in contractile force but increased cyclic AMP similar to the first dose of milrinone (Fig. 10). It is clear that an increase in cyclic AMP does not always signify an increase in cardiac contractile force.

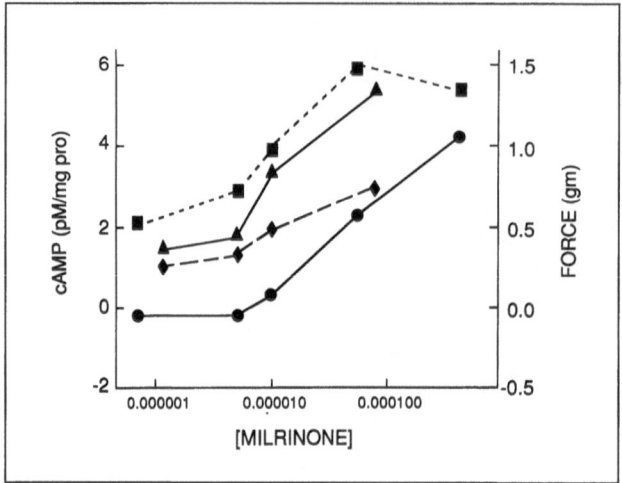

Fig. 9. Effect of milrinone on contractile force and cyclic AMP content of dog ventricular trabeculae. ■----■ = cyclic AMP during maximal inotropic effect; ●----● = Maximal contractile force; ▲----▲ = Cyclic AMP 60–70 s after addition of milrinone; ◆----◆ = Contractile force 60–70 s after addition of milrinone

Table 1. Effect of milrinone (MIL), norepinephrine (NOREPI) and isoproterenol (ISOPROT) on cardiac contractile force and cyclic AMP in isolated cardiac tissues

Tissue	Drug			n	Percent change in contractile force	Percent change cyclic AMP
Canine						
Purkinje	Mil	140	µM	7	− 24 ± 12	+ 50.5 ± 11
Purkinje	Norepi	3	µM	5–8	+ 176 ± 20	+ 64.0 ± 11
Trabeculae	Mil	140	µM	8	+ 167 ± 20	+ 49.6 ± 7
Trabeculae	Norepi	3	µM	6–8	+ 139 ± 17	+ 83.0 ± 17
Perfused rat hearts						
	Mil	500	µM	6	− 8 ± 6	+ 540 ± 80
	Isoprot	0.1	µM	5	+ 173 ± 19	+ 316 ± 47
Perfused guinea pig hearts						
First dose	Mil	140	µM	6	+ 71 ± 9	+ 132 ± 20
Third dose	Mil	140	µM	6	+ 5 ± 2	+ 144 ± 18
	Isoprot	0.1	µM	6	+ 110 ± 16	+ 148 ± 22

More recently we observed a cross-tachyphylaxis between milrinone and the positive inotropic agent Bay K8644 [60], a dihydropyridine related to the calcium channel blockers nifedipine and nitrendipine. This tachyphylaxis had some specificity since it markedly reduced the effects of milrinone but produced only a 43% reduction in norepinephrine or isoproterenol responses and had no significant effect on the Ca^{++} response

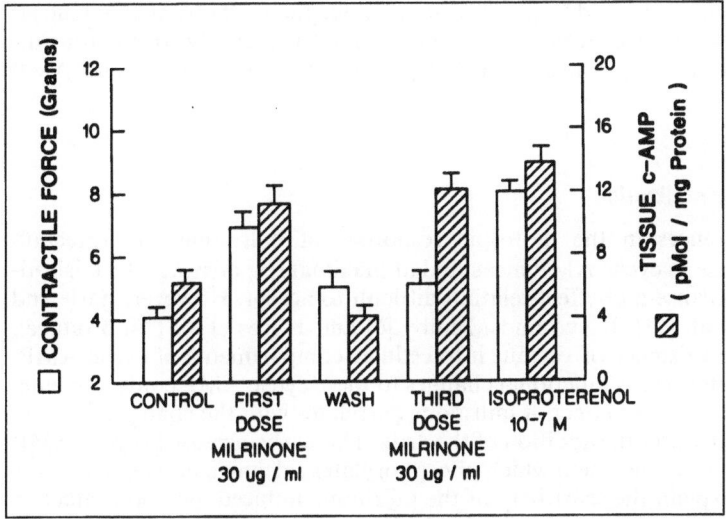

Fig. 10. Effect of tachyphylaxis to milrinone on contractile force changes and cyclic AMP content of perfused guinea pig hearts. The first group of hearts was perfused with 30 µg/ml of milrinone; maximal effects were obtained and quick-frozen for cyclic AMP determination. Group two was allow to recover (10–15 min) after treatment with the same dose of milrinone; AMP levels were determined when control contractility was attained. Hearts in a third group were challenged three times with milrinone and then quick-frozen for assay. Hearts in group four were initially treated with milrinone and then challenged with isoproterenol (0.3 µM). Each column is the average of eight hearts ± SEM

Table 2. Effects of Bay K 8644 on contractile force and cyclic AMP changes produced by milrinone in isolated perfused guinea-pig hearts

	n	Cardiac force		Cyclic AMP pmoles/mg prot.
		Baseline	Percent change from baseline	
Control	10	3.4±0.12	–	6.66±0.59
Isoproterenol 0.25 µg	6	3.0±0.34	233 ±40*	17.68±1.9*
Milrinone 30 µg	8	3.1±0.18	156 ± 9*	12.8 +1.7*
Bay K 8644 0.7 µg	8	3.0±0.21*	160 ±40*	4.79±0.73
Bay K 8644 (two doses)	8	2.7±0.3*	11 ± 4.1*	nd
Milrinone (after Bay K)	8	3.4±0.3	3.4± 1.8	15.97±1.6*
Isoproterenol (after Bay K)	6	3.5±0.4	133 ± 9.4*	17.91±1.57*

Hearts were quick-frozen in liquid nitrogen at time of peak tension (or after 5 min). Values represent the average ± SEM obtained for the contractile force and cyclic AMP content of six to 10 isolated hearts.
* Indicates significance at $p < 0.05$ from control tissues.

(see Table 2). Furthermore, the addition of milrinone after the washout of Bay K8644, increased cyclic AMP without increasing contractile force significantly. In the four instances described, catecholamine increased both contractile force and cyclic AMP (Table 2).

Evidence for compartmentalization

There are clear differences in the inotropic responses of milrinone and catecholamines. Thus an increase in cyclic AMP measured at maximal response, can be misleading and would make a cause-and-effect relation difficult to accept. However, Earle and Steiner [15], Corbin et al. [11], Beavo and Mumby [6] and Hayes et al. [25] produced strong evidence for the existence of various intracellular compartments of cyclic AMP. It is reasonable to assign various functional changes to these cyclic AMP compartments. As shown in Fig. 9 the inotropic effect of milrinone correlates with the changes in cyclic AMP observed at 60–70 s after the addition of the drug. This early increase in cyclic AMP could possibly occur in a compartment which phosphorylates calcium channels in the sarcolemma. This could explain the sensitivity of the milrinone-induced inotropic effect to the extracellular Ca^{++} concentration, and probably the rate and $[K^+]$ dependent changes in inotropy, as well as the cross-tachyphylaxis with Bay K8644.

In support of this would be the recent report of Weishaar et al. [72] discussed previously, which shows that a particulate phosphodiesterase to be especially sensitive to a

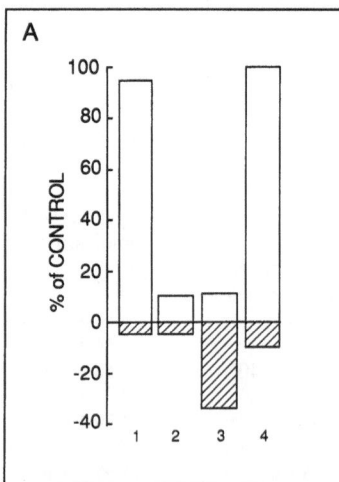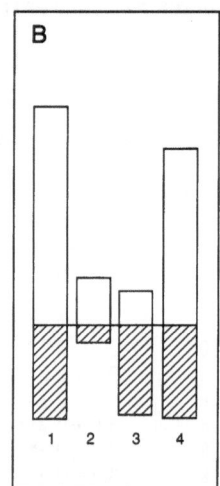

Fig. 11 A, B. Tachyphylactic response of guinea-pig papillary muscles to Bay K8644 and milrinone. **A** 1. Single dose of Bay K8644 0.5 μM; 2. second dose of Bay K8644 after washout; 3. milrinone 25 μM added after washout of Bay K8644; 4. Change of Ca^{++} from 1.8 to 4.5 mM, after Bay K8644. **B** 1. Initial dose of Milrinone 25 μM, 2. Bay K8644 0.5 μM after washout of milrinone dose; 3. second dose of milrinone (25 μM) after washout; 4. norepinephrine (0.3 μM) after washout of milrinone. Upper tracings are the percent-change in contractile force, while the lower tracings indicate the percent-change in the rate of relaxation of the contracting muscle. Each column represents the average of five to seven guinea pig papillary muscles

Table 3. Effect PF $[Ca^{++}]_0$ on contractile force and relaxation time in dog trabeculae

$[Ca^{++}]_0$	Force gms	90% relaxation time ms
0.45	0.37 ± 0.10	240 ± 21
0.9	0.81 ± 0.17	221 ± 16
1.2	1.08 ± 0.20	209 ± 14
1.8	1.68 ± 0.36	186 ± 14
2.7	2.44 ± 0.29	182 ± 20
3.6	3.49 ± 0.34	192 ± 16

number of inotropic phosphodiesterase inhibitors. If these findings can be confirmed, it would be of interest to see whether the particulate phosphodiesterase activity obtained from tachyphylactic tissues is non-responsive to inhibition by agents such as milrinone.

The observation was made that milrinone-induced tachyphylaxis affects mainly the contractile response, while relaxation time is not affected by a previous administration of milrinone (Fig. 11). In a similar manner Schmied and Korth [59] recently reported that carbachol inhibited the contractile effects of isoproterenol and IBMX but had no effect on the reduction in the duration of the contractile response. A similar separation of contractile and relaxation responses was produced by the addition of a calcium channel blocker to a milrinone-induced increase in contractile force [34]. This procedure reduced the contractile response but had no effect on the reduction of relaxation time produced by milrinone. All these instances show that the contraction and relaxation responses are pharmacologically different and are probably controlled by partially independent mechanisms. Cyclic AMP could be involved in both if different compartments for cyclic AMP are postulated, However, other possible mechanisms related to increase in Ca^{++} entry could also be operative [24]. What is clear is that relaxation effects of drugs are not simply a reversal of the factors that cause the increase in contractile force and should be studied as separate entities.

It is generally believed that addition of Ca^{++} to the bath does not affect relaxation time, although it increases the rate and force production. Recently [19] we showed that reducing $[Ca^{++}]_0$ below 1.8 mM produced a prolongation of the relaxation time, which was proportional to the reduction in $[Ca^{++}]_0$ (Table 3). This suggests that Ca^{++}-uptake by the endoplasmic reticulum is not the only process involved in relaxation. Extracellular Ca^{++} may actually be essential for relaxation of cardiac muscle, possibly via a calcium sensitive phosphorylation of phospholamban, independent of cyclic AMP.

References

1. Alousi AA, Canter JM, Montenero MJ, Fort DH, Ferrari RA (1983) Cardiotonic activity of milrinone a new and potent cardiac bipyridine, on the normal and failing heart of experimental animals. J Cardiovasc Pharmacol 5:792–803
2. Alousi AA, Johnson DC (1986) The pharmacology of the bipyridines; amrinone and milrinone. Circulation 73 II:10–24
3. Alousi AA, Stankus GP, Stuart JC, Walton LH (1983) Characterization of the cardiotonic effects of milrinone, a new and potent cardiac bipyridine, on isolated tissues from several species. J Cardiovasc Pharmacol 5:804–811

4. Aranson PS (1985) Kinetic properties of the plasma membrane Na^+-Ca^{2+} exchanger. Ann Rev Physiol 47:545–558
5. Baim DS, McDowell AV, Cherniles J, Monrad ES, Parker JA, Edelson J, Braunwald E, Grossman W (1983) Evaluation of a new bipyridine inotropic agent – Milrinone – in patients with severe congestive heart failure. NEJ Med 309:748–756
6. Beavo JA, Mumby MC (1982) Cyclic AMP dependent protein phosphorylation. In: Nathanson JA, Kabebian JW (eds) Cyclic nucleotides. Handbook of experimental pharmacology, vol 58/I. Springer, Berlin Heidelberg New York, pp 363–392
7. Benotti JR, Grossman W, Braunwald E, Carabello (1980) Effects of amrinone on myocardial energy metabolism and hemodynamics in patients with severe congestive heart failure due to coronary artery disease. Circulation 62:28–34
8. Binah O, Danilo P, Rosen MR (1982) Developmental changes in the effects of amrinone on cardiac contraction. Am J Cardiol 49:993–997
9. Canniff PC, Farah AE, Sperelakis N, Wahler GM (1985) The effects of milrinone (Win 47203) on the in vitro electrophysiological properties of mammalian cardiac tissue. J Cardiovasc Pharmacol 7:813–821
10. Cody SH, Chatterjeie K, Kubo SH, Simonton C, Rutman H (1984) Ascending dose range study of oral milrinone in chronic congestive heart failure. In: Braunwald E, Sonnenblick E, Sonnenblick EH, Chakrin LW, Schwarz RP (eds) Milrinone: investigation of new inotropic therapy for congestive heart failure. Raven Press, New York, pp 109–118
11. Corbin JD, Sugden PH, Lincoln TM, Keely SL (1977) Compartmentalization of adenosine 3′;5′ monophosphate and adenosine 3′;5′ monophosphate dependent protein kinase in heart tissue. J Biol Chem 252:3854–3861
12. De Clerk NM, Brutsaert DL (1984) Effect of amrinone and milrinone on single cardiac cells. Eur Heart J 5 (suppl I):60
13. Di Bianco R, Shabatai R, Kostukne W, Moran J, Schlant R, Wright R, and the milrinone multicenter trial group (1987) Oral milrinone and digoxin in heart failure. Results of a placebo controlled, prospective trial of each agent and the combination. Circulation, 76(Suppl II):1020
14. DiGenero M, Vasalle M (1984) Role of calcium on the action of caffeine in ventricular muscle fibers. J Cardiovasc Pharmacol 6:739–747
15. Earp HS, Steiner AL (1978) Compartmentalization of cyclic nucleotide mediated hormone action. Ann Rev Pharmacol Toxicol 18:431–459
16. Endoh M, Yamashita S, Taira N (1982) Positive inotropic effect of amrinone in relation to cyclic nucleotide metabolism in the canine ventricular muscle. J Pharmacol Exp Ther 221:775–783
17. Endoh M, Yanasigawa T, Taira N, Blinks JR (1986) Effects of new inotropic agents on cyclic nucleotide metabolism and calcium transients in canine ventricular muscle. Circulation 76:III 117–133
18. Fabiato A, Fabiato F (1978) Calcium induced release of calcium from the sarcoplasmic reticulum of skinned cells from adult human, dog, cat, rabbit, rat and frog hearts and from fetal newborn rat ventricles. Ann NY Acad Sci 307:491–552
19. Farah AE, Canniff PC, Bentley R, Frangakis C (1988) The effect of extracellular Ca^{++} and related ions on the cardiac action of milrinone. J Cardiovasc Pharmacol 11:591–601
20. Farah AE, Canniff PC, Bentley R, Kaiser LD (1987) Effects of milrinone on contractility of isolated dog ventricular muscle. J Cardiovasc Pharmacol 10:607–615
21. Fozzard HA, Wasserstrom JA (1985) Voltage dependent of intracellular sodium and control of contraction. In: Zipes DP, Jalife H (eds) Cardiac electrophysiology and arrhythmias. Grune and Stratton, New York, pp 51–60
22. Frangakis CJ, Lasher KP, Alousi AA (1985) Relationship of slow calcium channels to the effects of milrinone in isolated adult myocytes. Fed Proc 44:717
23. Frank JS, Langer GA, Nudel LM, Seraderian K (1972) The myocardial cell surface its histochemistry and the effects of sialic acid and calcium removal on its structure and cellular exchange. Circ Res 45:702–714
24. Hayes JS, Bowling N, Boder B, Kauffman (1984) Molecular basis for the cardiovascular activities of amrinone and AR-L57. J Pharmacol Exp Ther 237:124–132

25. Hayes JS, Brunton LL, Brown JH, Reese JB, Mayer SE (1979) Hormonally specific expression of cardiac protein kinase activity. Proc Natl Acad Sci USA 76:1570–1574
26. Higgings JP, England PJ (1983) Sarcolemmal Phospholamban is phosphorylated in isolated rat hearts perfused with isoprenaline. FEBS 163:297–302
27. Honerjaeger P, Schaefer-Korting, Reiter M (1981) Involvement of cyclic AMP in the direct inotropic action of amrinone. Biochemical and functional evidence. N Sch Arch Pharmakol 318:112–120
28. Iwasa Y, Hosey M (1984) Phosphorylation of cardiac sarcolemma proteins by calcium activated phospholipid dependent protein kinase. J Biol Chem 259:534–540
29. Yaari J, Hamon B, Lux HD (1987) Development of two types of calcium channels in cultured mammalian hippocampal neurons. Science 235:680–682
30. Jaski BE, Fifer MA, Wright RF, Braunwald E, Colucci WS (1985) Positive inotropic and vasodilator actions of milrinone in patients with severe congestive heart failure. J Clin Invest 75:643–649
31. Kabela E, Barcenas L, Farah A (1986) The effect of milrinone (Win 47203) on coronary blood flow and oxygen consumption of the dog heart-lung preparation. Am Heart J 111:702–709
32. Kariya T, Wille LJ, Dage RC (1982) Biochemical studies on the mechanism of cardiotonic activity of MDL 17043. J Cardiovasc Pharmacol 4:509–514
33. Katz AM, Tada M, Kirchberger MA (1975) Control of calcium transport in the myocardium by the cyclic AMP-protein kinase system. Adv Cyclic Nucl Res 5:453–472
34. Kenakin T (personal communication)
35. Kittelson MD, Pipers FS, Knauer KW, Keister DM, Knowlen GG, Miner WS (1985) Efficacy of milrinone in naturally occurring myocardial failure in dogs. In: Braunwald E, Sonnenblick EH, Chakrin LW, Schwarz RP (eds) Milrinone: investigation of new inotropic therapy for congestive heart failure. Raven Press, New York, pp 77–89
36. Kobylarz DC, Staffen RP, Weishaar RE, Evans DB (1985) Relation between inhibition of ventricular peak III phosphodiesterase (PDE) and the in vivo cardiotonic effect. The Pharmacologist 27:683
37. Kubo SH, Cody RJ, Chatterjee K, Simonton C, Rutman H, Leonard D (1985) Acute dose range study on milrinone in congestive heart failure. Am J Cardiol 55:726–730
38. Larsson R, Liedholm H, Anderson KE (1985) Pharmacokinetics and effects on blood pressure of single oral dose of milrinone in healthy subjects and in patients with chronic renal failure. Cardiovasc Pharmacother Int Symp 5(part 2):508
39. Luft LH (1971) Ruthenium red and violet. II. Five structural localizations in animal tissue. Anatomical Record 171:369–392
40. Lullman H, Peters T (1973) Plasmolemmal calcium in cardiac excitation-contraction coupling. Clin Exp Pharmacol Physiol 4:49–57
41. Lullman H, Peters T, Freuner J (1983) Role of the plasmalemma for calcium homeostasis and for excitation-contraction coupling in cardiac muscle. In: Drake-Holland AJ, Noble MM (ed) Cardiac metabolism. Wiley, New York, pp 1–18
42. Malecot CO, Bers DM, Katzung BG (1986) Biphasic contractions induced by milrinone at low temperature in ferret ventricular muscle: role of the sarcoplasmic reticulum and transmembrane calcium fluxes. Cir Res 59:151–162
43. Malecot CO, Katzung BG (1984) Modification of intracellular calcium movements by amrinone in ferret papillary muscle. Biophys J 45:52A
44. Maskin CS, Sinoway L, Chadwick B, Sonnenblick EH, LeJemtel TH (1983) Sustained hemodynamic and clinical effects of a new cardiotonic agent, Win 47203 in patients with severe congestive heart failure. Circulation 67:1065–1070
45. McNeil JH (1979) Cyclic AMP and myocardial contraction. In: Kalsner S (ed) Trends in autonomic pharmacology. Urban & Schwarzenberg, München, pp 421–441
46. McNeil JH, Continko FE, Verma SC (1975) Lack of interaction between norepinephrine or histamine and theophylline on cardiac cyclic AMP. Can J Physiol 52:1095–1101
47. Meisheri DD, Hwang O, Van Breeman C (1981) Evidence for two separate Ca^{++} pathways in smooth muscle plasmalemma. J Memb Biol 59:19–25

48. Monrad ES, Baim DS, Smith HS, Lanone A, Braunwald E, Grossman W (1985) Effect of mil-
    rinone on coronary hemodynamics and myocardial energetics in patients with congestive heart-
    failure. Circulation 71:972–979
49. Morgan JP, Blinks JR (1982) Intracellular [Ca^{++}] transients in the cat papillary muscle. Can
    J Physiol Pharmacol 60:524–528
50. Morkin E, La Raia PJ (1974) Biochemical studies on the regulation of myocardial contractility.
    N Engl J Med 290:445–451
51. Mushlin P, Boerth RC, Wells JN (1981) Selective phosphodiesterase inhibition and alterations
    in cardiac function by alkylated xanthines. J Mol Pharmacol 20:179–189
52. Mylotte KM, Cody V, Davis PJ, Davis FB, Blas SP, Schvenl M (1985) Milrinone and thyroid
    hormone stimulate myocardial membrane Ca^{++}-ATPase activity and share structural homolo-
    gies. Proc Natl Acad Sci USA 82:7974–7978
53. Nilius B, Hess P, Lansman JB, Tsien R (1985) A novel type of cardiac calcium channel in ven-
    tricular muscle. Nature 316:443–446
54. Nowycky MC, Fox AP, Tsien R (1985) Three types of neuronal calcium channels with different
    calcium agonists sensitivity. Nature 316:440–443
55. Pastelin G, Mendez R, Kabela E, Farah A (1983) The search for a digitalis substitute II Milri-
    none (Win 47203): its action on the heart-lung preparation of the dog. Life Sci 33:1787–1796
56. Rapundalo ST, Grupp I, Grupp G, Abdul Matlib M, Solaro RJ, Schwarz A (1986) Myocardial
    actions of milrinone: characterization of its mechanism of action. Circulation 73:III 134–144
57. Rosenthal JE, Ferrier GR (1982) Inotropic and electrophysiological effects of amrinone in un-
    treated and digitalized ventricular tissues. J Pharmacol Exp Ther 221:188–196
58. Schellenberg GD, Anderson L, Swanson P (1983) Inhibition of Na$^+$-Ca^{2+} exchange in rat brain
    by amiloride. Molecul Pharmacol 24:251–258
59. Schmied R, Korth M (1988) Muscarinic receptor-mediated antagonist of cyclic AMP dependent
    effects in guinea-pig myocardium. N Sch Archiv Pharmacol 337(Suppl):R62
60. Schramm M, Thomas G, Towart R, Frankoviak G (1983) Novel dihydropyridine with positive
    inotropic action through activation of Ca^{++} channels. Nature 303:535–537
61. Stroshane RM, Benzinger DP, Edelson J (1984) Pharmacokinetics of milrinone in congestive
    heart failure patients. In: Braunwald E, Sonnenblick EH, Chakrin LW, Schwarz RP (eds) Mil-
    rinone: investigation of new inotropic therapy for congestive heart failure. Raven Press, New
    York, pp 119–131
62. Sutko JL, Willerson JT, Templeton GH, Jones LR, Besh HR (1979) Ryanodines: its alterations
    of cat papillary muscle contractile state and responsiveness to inotropic interactions and sug-
    gested mechanism of action. J Pharm Exp Ther 209:37–47
63. Sutko JL, Willerson JT (1980) Ryanodine alteration of the contractile state of rat ventricular
    myocardium. Comparison of dog, cat and rabbit ventricular tissue. Circ Res 46:332–343
64. Sutko JL, Kenyon JL (1983) Ryanodine, modification of cardiac muscle responses to potassium
    free solutions. Evidence for inhibitions of sarcoplasmic reticulum calcium release. J Gen Physiol
    82:385–404
65. Sutko JL, Kenyon JL, Reeves P (1986) Effect of amrinone and milrinone on calcium influx into
    the myocardium. Circulation 73 II No 3:52–58
66. Sys SU, Brutsaert DL (1984) Does chronic pretreatment with amrinone and milrinone induce
    tachyphylaxis in mammalian heart muscle. Arch Int de Pharmacodynamic et de Therapie
    270:333
67. Sys SU, Brutsaert DL (1984) No tachyphylaxis observed following chronic pretreatment with
    amrinone and milrinone in mammalian cardiac muscle. Eur Heart J 5:60
68. Sys SU, Brutsaert DL (1986) Chronic treatment with amrinone or milrinone induces no tachy-
    phylaxis in mammalian cardiac muscle. In: Erdmann E, Graef K (eds) Cardiac glycosides; 1785–
    1985. Biochemistry, pharmacology and clinical relevance. Springer, Berlin Heidelberg New
    York, pp 229–236
69. Tang CM, Pressor F, Morad M (1988) Ameloride selectively blocks the low threshold (t) cal-
    cium channel. Science 240:213–215
70. Tsien RW (1977) Cyclic AMP and contractile activity of the heart. Adv Cyclic Nucl Res 8:363–
    420

71. Weishaar RE, Cain MH, Bristol JA (1985) A new generation of phosphodiesterase inhibitors: Multiple molecular forms of phosphodiesterase and the potential of drug selectivity. J Med Chem 28:537–545
72. Weishaar RE, Kobylarz-Singer DC, Steffen RP, Kaplan HR (1987) Subclasses of cyclic AMP specific phosphodiesterase in left ventricular muscle and their involvement in regulating myocardial contractility. Circ Res 61:539–547

Author's address:

A. E. Farah, 6001 Pelican Bay Blvd. Apt. 1406, Naples, Florida 33963, USA

Ca-Agonists: a new class of inotropic drugs

M. Bechem, R. Gross, S. Hebisch, M. Schramm

Institut für Pharmakologie, Bayer AG, Wuppertal, FRG

Summary

The basic pharmacology of dihydropyridine Ca-agonists published so far (BAY k8644, CGP 28-392, H 160/51, YC 170, and 202-791) is described. The importance of the potency of the enantiomeres for the effect of a racemic compound is underlined.

The Ca agonist prototype BAY k8644 leads to an increase of the maximal rate of rise of left ventricular pressure (LV(dP/dt)) and an increase of left ventricular stroke work in conscious dogs. When the vascular effects of BAY k8644 are counterbalanced by intravenous injection of sodium-nitroprusside, the left ventricular functions curves show markedly increased stroke work against the same mean arterial blood pressure at the same filling pressure.

BAY k8644 stimulates the heart economically: the net efficiency in isolated working guinea-pig hearts is about 20%, identical to a stimulation by calcium or ouabain. Cardiotonic drugs acting via cAMP-dependent mechanisms like isoprenaline, amrinone, or pimobendane however, stimulate the heart about $1/_3$ less economically.

The mechanism of action of Ca-agonists is explained from electrophysiological findings: Ca-agonistic dihydropyridines increase the open probability of the Ca-channels by a shift of the open-probability curve to more negative membrane potentials. As a consequence, the steady-state inactivation curve of the Ca-channel is also shifted in the same direction. While the effect on open-probability is the underlying mechanism for Ca-agonism, the latter effect results in Ca-antagonism. Therefore, depending on drug concentration and on membrane resting potential, a single chemical compound can act either as a Ca-agonist or a Ca-antagonist. A kinetic model of dihydropyridine action on the Ca-channel is described.

Introduction

The pivotal role of Ca-ions in the intracellular signal transduction [7, 22] and the availability of drugs influencing these processes have directed the interest of both basic and clinical research on Ca-channel-pharmacology. Originally, it started with the clinically successful Ca-antagonists verapamil [15] and nifedipine [37] and became effectively accelerated by two recent developments in this field:

1) With the availability of tritiated dihydropyridines (DHPs) not only the existence of specific DHP-receptors could be proven [4, 12], but in consequence, also the identification and purification of the protein could be achieved [8, 13], leading to determination of the primary structure of the DHP-receptor [32]. This protein is believed to be an essential part of the Ca-channel [13, 36].

2) The discovery of Ca-agonistic 1,4-DHPs, i.e., compounds chemically closely related to nifedipine but with pharmacological properties opposite to Ca-antagonists: instead of lowering, they increase the transmembraneous influx of Ca-ions. Therefore, Ca-agonists not only serve as tools in basic research on the physiology of the Ca-channel,

but they introduced a new pharmacological principle of positive inotropic drug action. This contribution summarizes the basic pharmacology of Ca-agonsits and will help to give more insight into the understanding of the molecular action of DHPs on the Ca-channel.

Methods and material

1) Isolated guinea-pig hearts

Hearts of adult guinea pigs weighing 200–250 g were perfused according to Langendorff at 32 °C in a non-recirculating system with Krebs-Henseleit solution. Calcium concentration was 1.2 mmol/l. The hearts were retrogradely perfused via an aortic cannula at a constant perfusion rate (10 ml/min). Coronary perfusion pressure was measured with a pressure transducer. The solution was gassed with a mixture of 95% O_2 and 5% CO_2. Isovolumetric left ventricular pressure was measured with another pressure transducer connected to a latex balloon which was inserted into the left ventricle via the atrium and the mitral valve. Drugs were dissolved in DMSO, diluted and infused into the aortic cannula at 1% of the perfusion rate.

For the working heart, the basic preparation was similar but the hearts were electrically driven at 180/min by right atrial pacing. The pulmonary artery was cannulated and connected to a flow-meter and an oxygen-probe of the Clark-type. The left atrium was connected to a reservoir to adjust a filling pressure of 5 mm Hg. The aortic cannula, a vertically mounted glass tubing of 5 mm i.D. with a side branch at 32 cm height and a defined resistance by a fine-bore cannula, a pressure transducer, and an electromagnetic flowmeter represented the outflow tract. All variables were simultaneously recorded on a potentiometric strip chart recorder. Drugs were infused into the atrial cannula at a rate of 0.25 ml/min. Measurement of lactate concentration in the coronary effluate gave evidence for only minimal ($< 5\%$) contribution of anaerobic glycolysis to the total energy turnover under these conditions.

2) Conscious dogs

Under sterile conditions mongrel dogs were prepared for chronic studies: after a left thoracotomy in the fifth intercostal space a tygon catheter was placed into the aorta and fixed to the adventitia. An electromagnetic flow probe of appropriate size was implanted around the ascending aorta or the circumflex branch of the left coronary artery. A precalibrated miniature pressure transducer was implanted into the left ventricle. Epicardial pacing electrodes were sutured to the base of the right ventricle. All cables and catheters were subcutaneously led to the animals' neck and exteriorized there. After recovery of the dogs, the experiments were carried out under quiet laboratory conditions. The drugs were dissolved in "placebo solution" (60 g glycerol, 100 g water, polyethylene 400 ad 1129 g) and injected into a forearm vein. All variables were simultaneously recorded on an oscillographic recorder and stored using a pulse-code system on magnetic tape for off-line analog computing of stroke work from left ventricular pressure and aortic blood flow (for details see [14]).

3) Electrophysiological studies

In single, isolated and cultured atrial myocytes from adult guinea pigs ("cardioballs") whole cell voltage clamp experiments were performed using single, tight seal patch pipettes [19]. Sodium and potassium-currents were blocked and inactivation of the calcium currents was attenuated by appropriate bathing and pipette solutions (all concentrations in mmol/l: K_3-citrate: 60, CsCl: 30, $MgCl_2$: 1, cAMP: 1, EGTA: 2, HEPES NaOH pH 7.4: 10, TTX: 0.03, temperature:22–24 °C). The currents-signals were filtered with 3 kHz and sampled at a rate of 20 kHz. The calcium currents were corrected for leakage and capacitive currents (for detailed description see [1]).

Results and discussion

Up to now five Ca-agonistic DHPs have been published: YC 170 [20, 31], CGP 28-392 [9], BAY K 8644 [26], H 160/51 [11], and 202-791 [17]. They are chemically characterized by a non-ester moiety in the 3- or 5-position of the DHP-ring opposite to an ester-function (see Fig. 1). Due to this asymmetry two enantiomeres exist. The Ca-agonistic properties of these compounds are restricted to only one enantiomere, while its optical antipode is a Ca-antagonist. This has been shown for BAY k8644 [10], for 202-91 [17], and for H 160/51 [12].

Since the potencies for the Ca-antagonistic enantiomeres are much weaker than for the Ca-agonistic ones, the pharmacodynamic effect of the racemic compound is almost entirely brought about by the Ca-agonistic enantiomere. The only exception represents 202-791: the potencies of both enantiomeres are almost identical [17]; therefore, the pharmacodynamic effect of the racemic compound can be looked at as a combined effect of both enantiomeres.

The relative potency of the inotropic vs the vasoconstrictive effects differ between the various compounds: while YC 170 exhibits almost only vasoconstrictive properties [20, 31], H 160/51, BAY k8644, and CGP 28-392 exhibit almost identical inotropic and smooth muscle constricting potencies as shown by the similar EC50-values in guinea pig papillary muscle and portal vein preparations (see Fig. 2).

In the isolated guinea pig heart perfused according to Langendorff, there is also no major difference between those three compounds in this respect. In Fig. 3 the concentra-

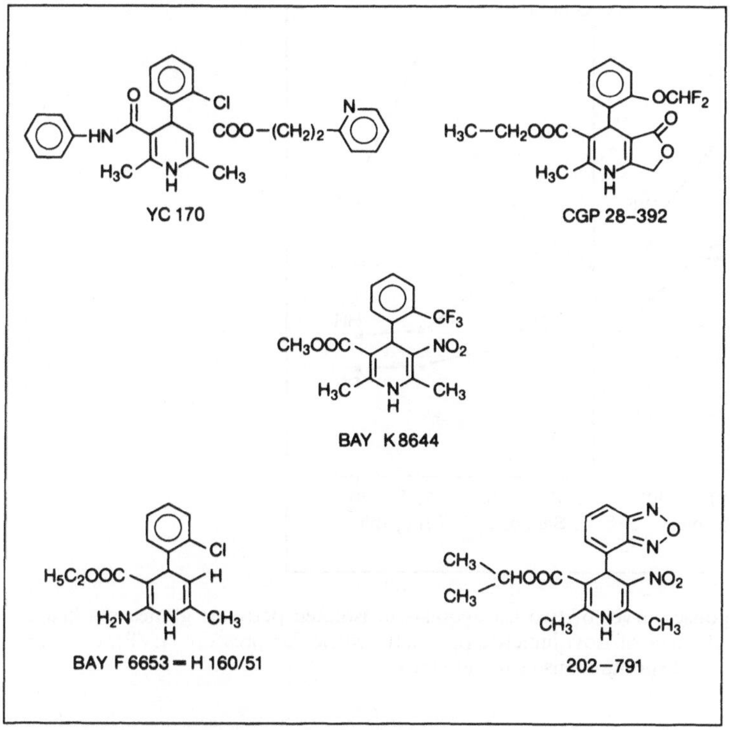

Fig. 1. Chemical structures of the Ca-agonists

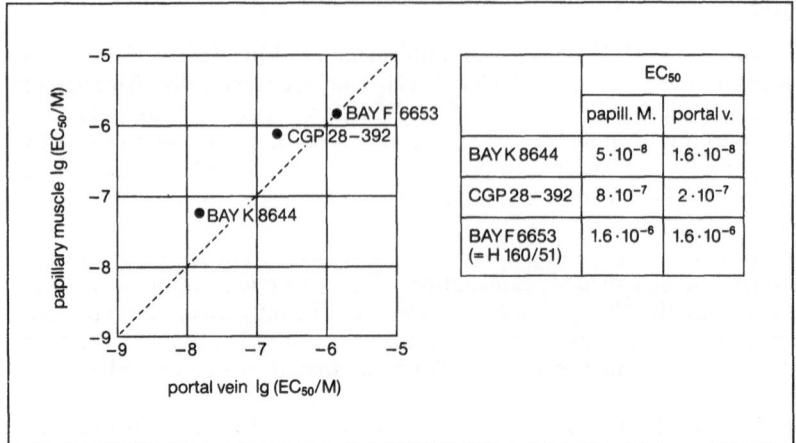

Fig. 2. Half-maximal effective concentrations (EC_{50}) of three Ca-agonists in myocardial and in smooth muscle tissue. Data taken from [5, 6, 11]

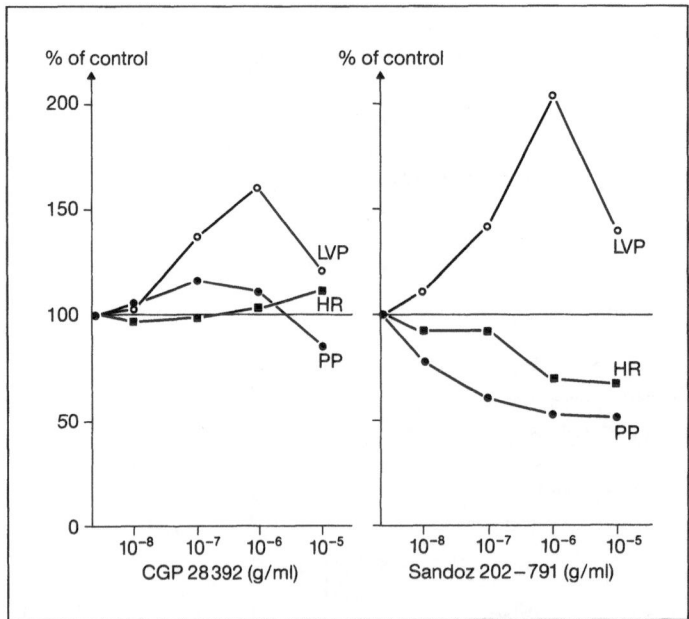

Fig. 3. Concentration-response curves of two Ca-agonists in isolated perfused guinea pig hearts. Depicted are the relative changes of isovolumetric peak left ventricular pressure (LVP), coronary perfusion pressure (PP), and of spontaneous heart rate (HR)

tion-response curves are given for CGP 28-392 and for 202-791: over a concentration range of two decades, there is a pronounced positive inotropic effect as measured from isovolumetric peak left ventricular pressure with both compounds; at higher concentrations the Ca-agonistic effect declines again. Under certain experimental conditions even a reversal to Ca-antagonistic effects can be observed [33].

Comparing changes of coronary perfusion pressure with these two compounds, a striking difference is seen: while CGP 28-392 increases coronary perfusion pressure, this is effectively reduced under 202-791. The latter, at first glance, suggests an ideal ratio between cardiac and smooth muscle effects for use as a cardiotonic drug – but since this pharmacological profile obviously results from the above mentioned combined effect of the two enantiomers (see [17]), therapeutic use is hindered. In addition, the ventricular rate abruptly declines at the concentration with the maximal positive inotropic effect as a consequence of an atrioventricular block.

The studies in vivo confirm this difference as demonstrated in Fig. 4: in unanesthetized, chronically instrumented dogs the relative increase of arterial blood pressure, as compared to the increase of the maximal rate of rise of left ventricular pressure (LV(dP/dt)$_{max}$), gives a crude measure of vascular vs cardiac inotropic effects of the compound. While under the influence of H 160/51 there is a dose-dependent marked increase of arterial blood pressure; under 202-791 the increase of mean arterial blood pressure is almost zero at the lower dose of 0.1 mg/kg i.v., and only little at the higher dose (0.3 mg/kg i.v.) but with a clear increase of pulse pressure and a marked rise of LV(dP/dt)max. The lesser vasoconstriction is probably also the reason for the lack of a pronounced reflex bradycardia as observed under H 160/51 (see Fig. 4).

While there are only minor differences in the ratio of inotropic to vasoconstrictive effects with H 160/51, BAY k8644, and CGP 28-392 in the isolated heart, there seems to exist some graduation in this respect in conscious dogs: CGP 28-392 exerts only little positive inotropic effect while at comparable pressure rises the LV(dP/dt) is markedly elevated under the influence of BAY k8644 [21, 22].

Since measurement of LV(dP(dt)max gives only an estimate of positive inotropic drug action, left ventricular function curves [25] were measured in conscious dogs paced at 150 beat per min. The data presented in Fig. 5 show that under the influence of BAY k8644 given as a priming dose of 16 μg/kg i.v., followed by a continuous infusion of 12 μg/kg/min, the left ventricular work increased. Probably due to the increased total peripheral resistance, mean arterial blood and left ventricular enddiastolic pressure (LVEDP) rose. This resulted in a shift of the ventricular function curve to the upper right.

When under steady-state conditions the vascular effects were counterbalanced by an appropriate intravenous injection of 10 μg/kg nitroprusside it was possible to bring the ventricular function curve back to its starting position. Under these conditions BAY k8644 stimulates the heart to deliver an increased stroke work against an unchanged mean arterial blood pressure at a constant filling pressure (for detailed description see [14]).

In isolated working guinea pig hearts the energy costs for the increased cardiac performance were determined. The efficiency of cardiac contraction was calculated from the ratio of external mechanical work (pressure-volume product) and the oxygen consumption of the heart. Under control conditions the brutto-efficiency amounted to 11%, which is close to the data known from literature [30]. The hearts were then stimulated by increasing concentrations of Ca resulting in about a doubling of cardiac performance. The net-efficiency for the additional work was 20%. This net-efficiency under calcium-stimulation was taken as a reference for all other drugs tested. The drug concentrations were chosen to achieve similar increases of cardiac performance (BAY k8644: 3×10^{-9} to

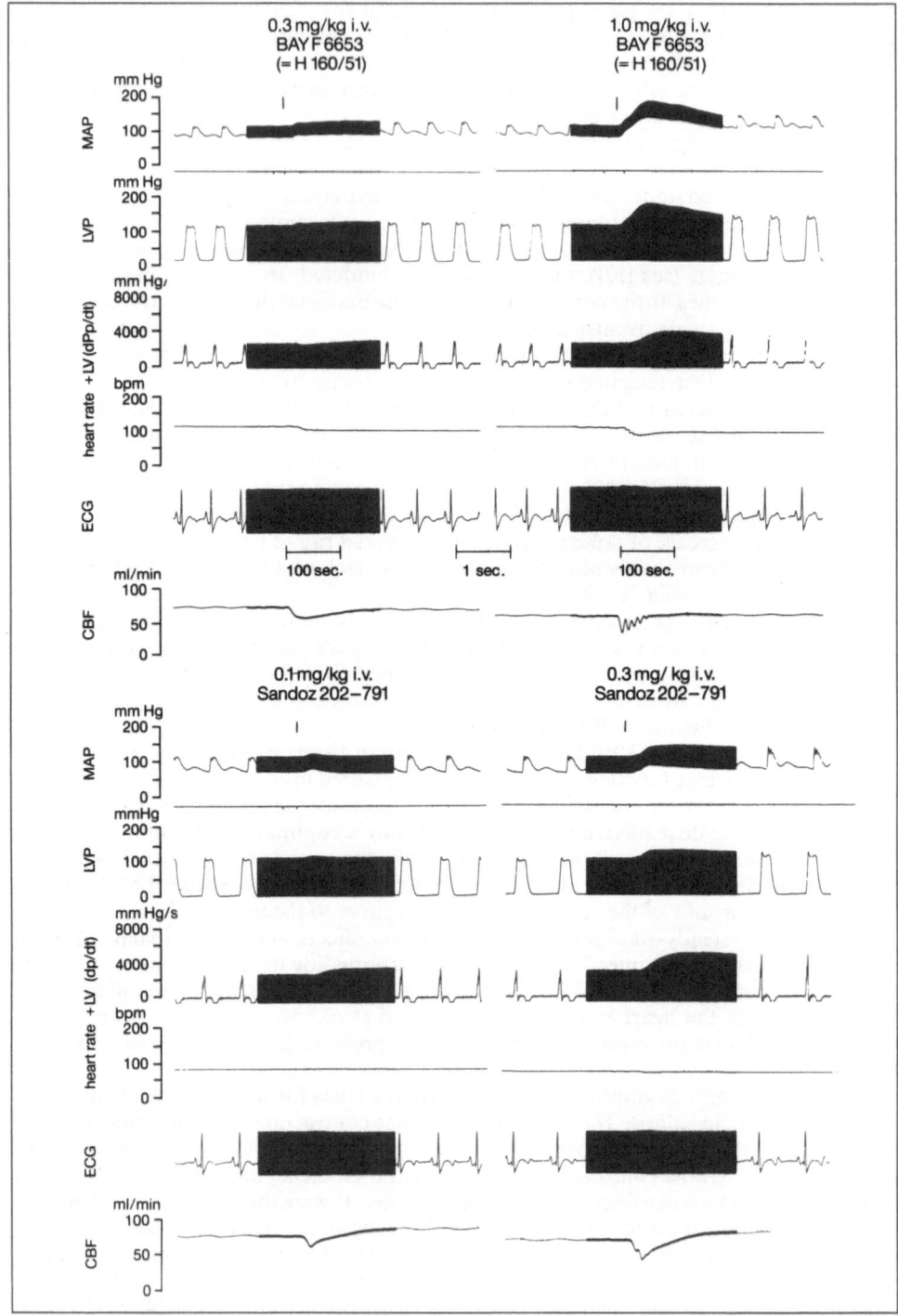

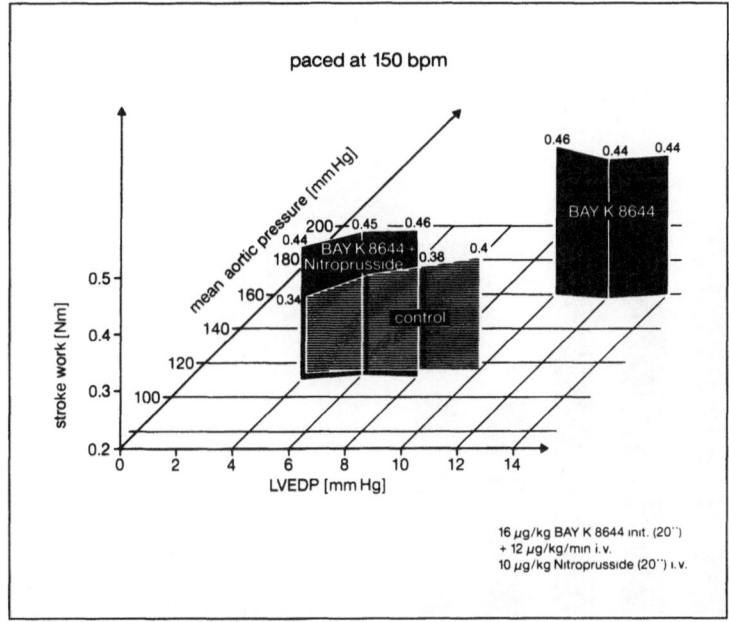

Fig. 5. Ventricular function curves in a conscious dog at rest (control), under the influence of BAY k8644, and under the combined effect of BAY k8644 and sodium nitroprusside. (LVEDP = left ventricular enddiastolic pressure)

3×10^{-7} M, ouabain: 5×10^{-8} to 5×10^{-7} M, isoprenaline: 5×10^{-10} to 5×10^{-8} M, amrinone: 5×10^{-6} to 5×10^{-4} M, and pimobendane: 3×10^{-7} to 3×10^{-6} M). The Ca-agonist and oubain increase cardiac performance at a similar energy consumption as does calcium, while cAMP-dependent drugs like catecholamines (isoprenaline) or PDE-inhibitors (amrinone and in the concentrations used also pimobendane) lead to unproportionate higher energy cost (see Fig. 6). The relative net-efficency declines to only 76% for isoprenaline and 67% for amrinone or pimobendane of the efficiency measured under calcium-stimulation.

It should be mentioned that these hearts were paced at a constant rate of 180/min; under in vivo conditions, the oxygen consumption will be additionally increased by the direct chronotropic effects of those drugs or reduced by the reflex bradycardia under treatment with ouabain or calcium agonists.

Classical pharmacological studies have shown that Ca-agonists exert their positive inotropic and vasoconstrictive effects by an action on the voltage-dependent Ca-channel and not by an interaction with intracellular structures (see [27–29, 35]). Direct insight into

---

Fig. 4. Original recordings of the effect of intravenous injection of 0.3 and of 1.0 mg/kg BAY F 6653 (=H 160/51) (upper panel) and of 0.1 and of 0.3 mg/kg 202-791 (lower panel) on arterial blood pressure (MAP), left ventricular pressure (LVP), maximal rate of rise of left ventricular pressure (negative edge clipped) (+LV(dP/dt), heart rate, electrocardiogram (ECG), and mean coronary blood (CBF) flow in the left circumflex coronary artery of conscious dogs. Paper speed = 25 mm/s and 0.25 mm/s, respectively (see calibration bars)

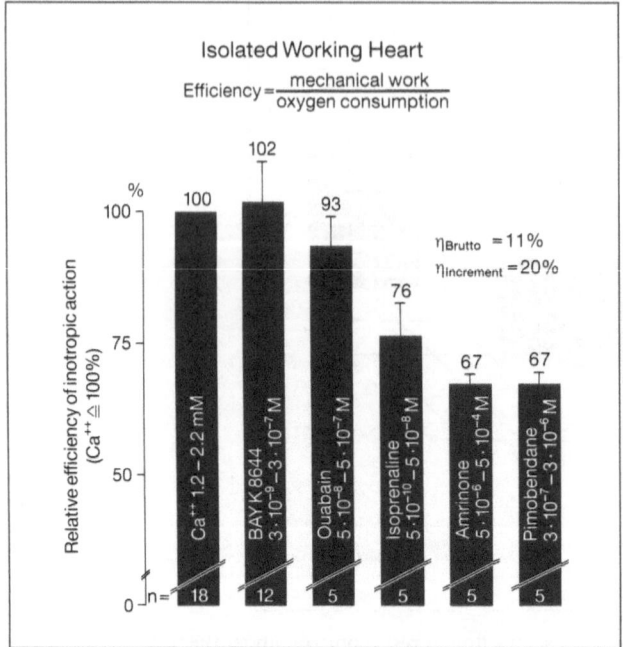

Fig. 6. Net efficiency of the increase of cardiac performance of isolated working guinea pig hearts under inotropic stimulation. The net efficiency under stimulation with calcium is about 20% and taken as reference for the other drugs. n = number of hearts studied in each group

the interaction between the drug molecule and the Ca-channel has been provided by means of electrophysiological studies [34], for review see [2, 3]).

In voltage-clamp studies (whole cell recording patch-clamp technique) BAY k8644 in concentrations which elicit positive inotropic effects (here 30 nmol/l) does not change the current-voltage relationship (IV-curve) at membrane potentials positive to $+10$ mV. However, it increases markedly the Ca-current, mainly between -60 mV and $+10$ mV, with a shift of the peak of the IV-curve to more negative membrane potentials (see Fig. 7). The unchanged linear ("ohmic") part of the IV-curve gives evidence against an influence of BAY k8644 on the number of available Ca-channels and the single channel conductance. Single channel recordings have revealed that Ca-agonists increase the open probability of the single channel, i.e., once the channel opens, it stays open for a longer time [16, 18].

A very intriguing observation is the reversal of the Ca-agonistic effects at very high concentrations. This has been shown early in functional studies [33] and can also be seen in the IV-curve: at 3 μmol/l, BAY k8644 diminishes the Ca-current, while the IV-curve peaks at even more negative membrane potentials. Therefore, the decreased Ca-current is not due to a reversal of the single-channel open probability, but caused by a loss of available channels (see Fig. 7).

In line with that is the observation that under depolarized conditions the Ca-agonistic effect of BAY k8644 is reduced or even reversed to Ca-antagonism as shown in voltage-clamp studies [24]; when IV-curves obtained in a similar way as described above, BAY k8644 (500 nmol/l) increased Ca-current as long as the holding potential from which the

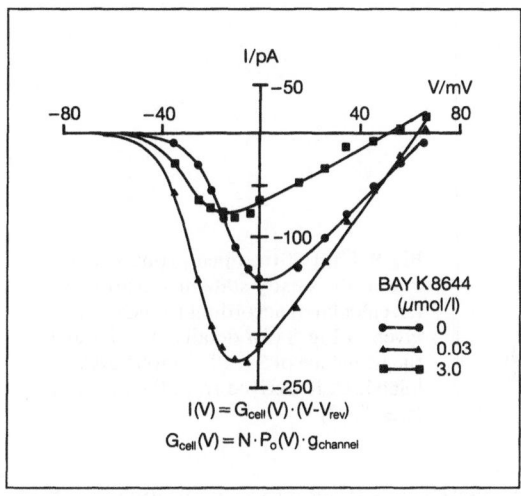

Fig. 7. Influence of BAY k8644 on the peak calcium current-voltage relationship. Holding potential was −70 mV

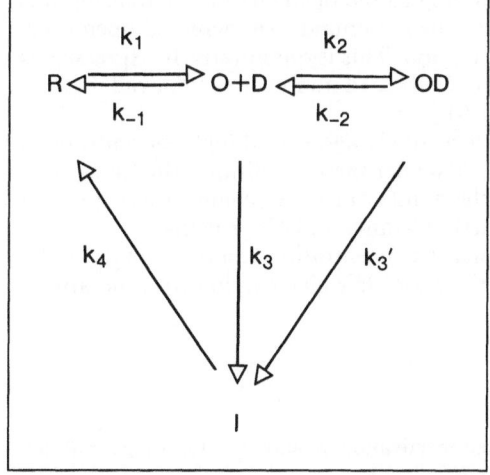

Fig. 8. Reaction model of the DHP action on the Ca-channel. R = rested state, O = open state, OD = open state, drug bound, I = inactivated state

test pulses started was well polarized (−62 mV). If, however, the holding potential was only −45 mV then the same concentration of BAY k8644 markedly depressed the IV-curve below its control.

This finding may have some implications for tissue selectivity (resting membrane potential in the myocardium is more negative than, for example, in vascular smooth muscle) and for the extrapolation of data obtained in partially depolarized tissue or under hypoxic conditions to the in vivo situation.

It could also be interpreted as an in-built safety mechanism in so far as an ischemic or hypoxic cell is not only not further stimulated, but instead, the Ca-agonist, at first glance and paradoxically, protects the cell from Ca-overload.

These electrophysiological findings have led to a kinetic model for the molecular action of DHPs on the Ca-channel (see Fig. 8). The Ca-channel is supposed to exist in three

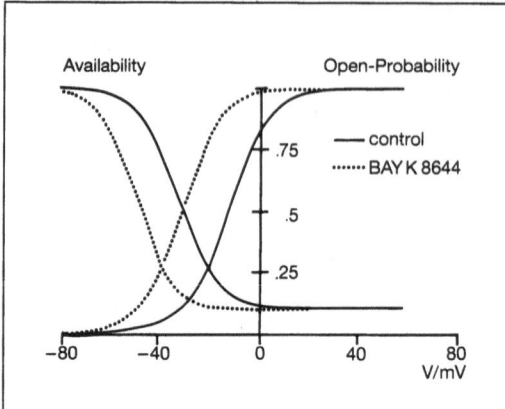

Fig. 9. Shift of the open-probability curve and of the steady-state inactivation curve (as calculated according to the model given in Fig. 8 (for details see [3] under the influence of 3×10^{-6} mol/l BAY k8644. Data derived from the IV-curves (see Fig. 7)

different states: closed (resting), open, or inactivated. Inactivation is assumed to take place preferentially from the open state. The DHPs have a state-dependent action on the channel: they increase open probability by binding to the open channel, stabilizing this state and keeping the channel open as long as they are bound. An increased open time, however, directly results in an enhanced inactivation. This is quantitatively expressed as a shift of the steady-state inactivation curve (percentage of noninactivated, available Ca-channels) to more negative potentials (see Fig. 9).

This easily explains the Ca-antagonistic effects of Ca-agonists at high concentrations or at partially depolarized membrane potentials. Under these conditions, the Ca-agonistic principle is overcome by the reduction of the number of active channels as expressed by the shift of the steady-state inactivation curve (Ca-antagonistic principle).

Such models help us to understand the biphasic concentration-response curves of the DHPs and might be a key for designing new Ca-agonists which can improve therapy of congestive heart failure.

References

1. Bechem M, Pott L (1985) Removal of Ca-current inactivation in dialyzed guinea-pig atrial cardioballs by Ca-chelators. Pflügers Arch 404:10–20
2. Bechem M, Schramm M (1987) Calcium Agonists. J Mol Cell Cardiol 19(Suppl II):63–75
3. Bechem M, Schramm M (1988) The effects of Ca-agonists and Ca-antagonists on the Ca-current. In: Piper HM, Isenberg G (eds) isolated adult cardiomyocytes. CRC Press (in press)
4. Bellemann P, Ferry D, Lübbecke F, Glossmann H (1981) ³H-Nitrendipine, a potent calcium channel antagonist binds with high affinity to cardiac membranes. Arzneim Forsch/Drug Res 31:2064–2067
5. Beyer T, Gansohr N, Gjörstrup P, Ravens U (1986) The effects of the cardiotonic dihydropyridine derivatives BAY k8644 and H160/51 on post-rest adaptation of guinea-pig papillary muscles. Naunyn-Schmiedeberg's Arch Pharmacol 334:488–495
6. Beyer T, Gjörstrup P, Ravens U (1985) Comparison of the cardiac effects of the dihydropyridine-derivative H160/51 with those of the "Ca-agonist" BAY k8644. Naunyn-Schmiedeberg's Arch Pharmacol Suppl 330:R34
7. Campbell AK (1983) Intracellular calcium: its universal role as regulator. Wiley, New York
8. Curtis BM, Catterall WA (1984) Purification of the calcium antagonist receptor of the voltage-sensitive calcium channel from skeletal muscle transverse tubules. Biochemistry 23:2113–2118

9. Erne P, Burgisser E, Bühler FR, Dubach B, Kühnis H, Meier M, Rogg N (1984) Enhancement of calcium influx in human platelets by CGP 28392, a novel dihydropyridine. Biochem Biophys Res Comm 118:842–847
10. Franckowiak G, Bechem M, Schramm M, Thomas G (1985) The optical isomers of the 1,4-dihydropyridine BAY k8644 show opposite effects on Ca-channels. Europ J Pharmacol 114:223–226
11. Gjörstrup P (1985) Effects of H160/51, a new Ca-agonist, and its interaction with felodipine on cardiac and vascular tissue in vitro. Proc Cardiovascular Pharmacotherapy. Intern Symposium Geneva April 22–25, Abstr. 127
12. Gjörstrup P, Harding H, Isaksson R, Westerlund C (1986) The enantiomers of the dihydropyridine derivative H160/51 show opposite effects of stimulation and inhibition. Eur J Pharmacol 122:357–361
13. Glossmann H, Ferry DR, Goll A, Striessnig J, Schober M (1985) Calcium channels: basic properties as revealed by radioligand binding studies. J Cardiovasc Pharmacol 7 Suppl 6:S20–S30
14. Gross R, Kayser M, Schramm M, Taniel R, Thomas G (1985) Cardiovascular effects of the Calcium-agonistic dihydropyridine BAY k8644 in conscious dogs. Arch Internat Pharmacodyn Ther 277:203–216
15. Haas H, Härtfelder G (1962) c-Isopropyl-c-(N-methyl-homoveratryl)-y-aminopropyl-3,4-dimethoxy-phenylacetonitril, eine Substanz mit coronargefäßerweiternden Eigenschaften. Arzneim Forsch 12:549–558
16. Hess P, Lansmann JB, Tsien RW (1984) Different modes of calcium channel gating behaviour favoured by dihydropyridine Ca agonists and antagonists. Nature 311:538–544
17. Hof RP, Rüegg UT, Hof A, Vogel A (1985) Stereoselectivity at the calcium channel: opposite action of the enantiomeres of a 1,4-dihydropyridine. J Cardiovasc Pharmacol 7:689–693
18. Kokobun S, Reuter H (1984) Diyhdropyridines derivatives prolong the open state of Ca channels in cultured cells. Proc Natl Acad Sci USA 81:482–527
19. Marty A, Neher E (1983) Tight-seal whole-cell recording. In: Sakmann B, Neher E (eds) Single channel recording. Plenum, New York, pp 107–122
20. Nakaya H, Hattori Y, Tohse N, Kanno M (1986) Voltage-dependent effects of YC-170, a dihydropyridine calcium channel modulator in cardiovascular tissue. Naunyn-Schmiedeberg's Arch Pharmacol 333:421–430
21. Preuss KC, Brooks HL, Gross GJ, Warltier DC (1985) Positive inotropic actions of the calcium channel stimulator, BAY k8644, in the awake, unsedated dog. Bas Res Cardiol 80:326–332
22. Preuss KC, Chung NL, Brooks HL, Warltier DC (1984) Cardiovascular effects of the nifedipine analog CGP 28392 in the conscious dog. J Cardiovasc Pharmacol 6:949–953
23. Rasmussen H, Barret PQ (1984) Calcium messenger system: an integrated view. Physiol Rev 64:938–984
24. Sanguinetti MC, Kass RS (1984) Regulation of cardiac calcium channel current and contractile activity by the dihydropyridine BAY k8644 is voltage dependent. J Mol Cell Cardiol 16:667–670
25. Sarnoff SJ, Mitchell JH (1961) The control of the function of the heart. In: Handbook of physiology, circulation sect 2, vol 1. Am Physiol Soc, Washington
26. Schramm M, Thomas G, Towart R, Franckowiak G (1983) Novel dihydropyridines with positive inotropication through activation of Ca^{2+} channels. Nature 303:535–537
27. Schramm M, Thomas G, Towart R, Franckowiak G (1983) Activation of calcium channels by novel 1,4-dihydropyridines. Arzneim-Forsch 33:1268–1272
28. Schramm M, Towart R (1985) Modulation of calcium channel function by drugs. Life Sci 37:1843–1860
29. Schramm M, Towart R (1988) Calcium channels as drug receptors. In: Baker PF (ed) Handbook of experimental pharmacology, vol 83. Springer, Berlin Heidelberg New York, pp 89–114
30. Siess M, Stieler K, Seifart HJ (1981) Zur Wirkung von ARL-115 BS auf Funktion und Sauerstoffverbrauch isolierter Meerschweinchenherzvorhöfe im Vergleich zu g-Strophantin und Theophyllin. Arzneim Forsch/Drug Res 31:165–170
31. Takenaka T, Maeno H (1982) A new vasoconstrictor 1,4-dihydropyridine derivative, YC-170. Jap J Pharmacol 132:139P

32. Tanabe T, Takeshima H, Mikami A, Flockerzi V, Takahashi H, Kangawa K, Kojima M, Matsuo H, Hirose T, Numa S (1987) Primary structure of the receptor for calcium channel blockers from skeletal muscle. Nature 328:313–318
33. Thomas G, Groß R, Schramm M (1984) Calcium channel modulation: ability to inhibit or promote calcium influx resides in the same dihydropyridine molecule. J Cardiovasc Pharmacol 6:1170–1176
34. Thomas G, Chung M, Cohen CJ (1985) A dihydropyridine (BAY k8644) that enhances calcium currents in guinea pig and calf myocardial cells. Circ Res 56:87–96
35. Thomas G, Groß R, Pfitzer G, Rüegg JC (1985) The positive inotropic dihydropyridine BAY k8644 does not affect calcium sensitivity or calcium release of skinned cardiac fibres. Naunyn-Schmiedeberg's Arch Pharmacol 328:378–381
36. Triggle DJ, Janis RA (1987) Calcium channel ligands. Rev Pharmacol Toxicol 27:347–369
37. Vater W, Kroneberg G, Hoffmeister F, Kaller H, Meng K, Oberdorf A, Puls W, Schloßmann K, Stoepel K (1972) Zur Pharmakologie von 4-(2'Nitrophenyl)-2,6-dimethyl-1,4-dihydropyridine-3,5-dicarbonsäuredimethylester (Nifedipin, BAY a1040). Arzneim Forsch 22:1–14

Author's address:

Prof. Dr. R. Gross, Institut für Pharmakologie der Bayer AG, Postfach 101709, 5600 Wuppertal 1, FRG

Myofibrillar Ca^{++} activation and heart failure – Ca^{++} sensitization by the cardiotonic agent APP 201-533

J. W. Herzig[1], L. H. Botelho[2], R. J. Solaro[3]

[1] Department Research CVS, Pharmaceutical Division,
Ciba-Geigy Ltd., Basel, Switzerland
[2] Research and Development, Pharmaceutical Division,
Sandoz Inc., East Hanover, New Jersey, USA
[3] Department of Physiology and Biophysics, University of Illinois,
College of Medicine at Chicago, Illinois, USA

Summary

Certain forms of cardiac failure appear to be associated with a decrease in the Ca^{++} sensitivity of the contractile structures, possibly due to troponin I phosphorylation. Interference of cardiotonic drugs with myofibrillar Ca^{++} activation instead of enhancement of Ca^{++} influx may therefore provide a more causal therapeutic concept in the treatment of cardiac insufficiency. APP 201-533 (3-Amino-6-methyl-5-phenyl-2(1H)-pyridinone) (the structure of which is shown below) is a novel cardiotonic agent acting neither via beta adrenoceptor stimulation nor inhibition of Na$^+$/K$^+$ ATPase. In the 100 µM concentration range, it increases the Ca^{++} sensitivity and the Ca^{++} affinity of functionally isolated cardiac contractile structures. This coincides with an inhibitory effect on the cAMP-dependent protein kinase from rat liver. A possible relation with the regulation of troponin I phosphorylation is discussed.

APP 201-533

Introduction

In recent years, increasing evidence has accumulated that the relationship between sarcoplasmic Ca^{++} concentration and myocardial activation is not invariant but is subject to intracellular regulation. The still often used simplification of considering myocardial force as a monitor of the intracellular free Ca^{++} concentration appears therefore increasingly unjustified. Alterations in the Ca^{++} sensitivities of myocardial myofibrillar ATPase [10] or force development of functionally isolated contractile structures from heart muscle [8, 3] have been attributed to corresponding differences in the phosphorylation degree of troponin I (Tn I). This phosphorylation reaction is governed by the opposing actions of a cAMP-dependent protein kinase and a phosphatase, cAMP forming a link with adrenoceptors and the sympathetic nervous system. As cAMP alone, after application to cardiac myofibrils or "skinned", demembranated myocardial cells, induces Tn I phosphorylation and decreases Ca^{++} sensitivity, the cAMP-dependent protein kinase seems to be an intrinsic constituent of the contractile structures [8, 3]. The negative inotropic effect resulting from Tn I phosphorylation is, during adrenergic stimulation of the heart, superimposed by an increase in Ca^{++} influx due to cAMP-dependent phosphorylation of membrane proteins [11].

At the same time, phospholamban phosphorylation induces an increase in the Ca^{++} sequestration rate of the sarcoplasmic reticulum (SR) which is assumed to be responsible for the observed acceleration of relaxation due to catecholamine stimulation [17]. Phosphorylation of both Tn I and phospholamban appears as a kind of defense mechanism limiting energy consumption in the course of adrenergic activation. The relaxant effect of Tn I phosphorylation and concomitant Ca^{++} desensitization is most directly visualized by the ability of catecholamines to relax cardiac K^+ contracture after exclusion of SR pumping activity at constant sarcoplasmic Ca^{++} concentrations [7].

In light of this knowledge it may be hypothesized that at least certain forms of cardiac insufficiency may be the result of a decrease in Ca^{++} sensitivity of the contractile structures rather than of a lack of activator Ca^{++}. Evidence for this hypothesis is derived from the observation that experimental cardiac insufficiency induced by acute administration of isoprenaline in μM concentrations coincides with Tn I phosphorylation [15]. Furthermore, Allen et al. [1] showed in 1982 that heart failure due to hypoxia occurs at unaltered sarcoplasmic Ca^{++} concentration. Consequently, it appears justified to consider mechanisms by which the Ca^{++} sensitivity of the contractile structures is increased (or brought back to normal) as a more causal approach to the treatment of at least certain forms of cardiac failure. APP 201-533 (3-Amino-6-methyl-5-phenyl-2(1H)-pyridinone) is a novel cardiotonic agent which exerts its positive inotropic action [12] neither by beta adrenoceptor stimulation nor via inhibition of the Na^+/K^+ ATPase [4]. We therefore investigated the influence of APP 201-533 upon the responsiveness of functionally isolated contractile structures and myofibrils from heart muscle to Ca^{++} ions at well defined, buffered concentrations.

Materials and Methods

1) Detergent treated myocardial strips

Subendocardial muscle fiber bundles ($\sim$4 mm in length, $\sim$0.2 mm in diameter) were prepared from trabecula septo marginalis from freshly slaughtered pigs. The preparations were then shaken at 4 °C for 24 h in a solution consisting of 50% (v/v) glycerol, 20 mM histidine HCl, 10 mM NaN_3, 0.5% (w/v) Lubrol WX, pH 7.3. The preparations were then stored until use at -18 °C in 50% (v/v) glycerol, 2 mM histidine HCl, 10 mM NaN_3, pH 7.0. For force measurements, the preparations were attached by means of a fast setting glue (cellulose nitrate dissolved in acetone) to an isometric force transducer (AME AE 801 OEM) with a stiffness better than 1 mN/μm. The preparations were relaxed in a solution containing 10 mM $ATPNa_2$, 12 mM $MgCl_2$, 5 mM EGTA, 20 mM imidazole, 5 mM NaN_3, 60 mM KCl, 5 mM phosphoenole pyruvate, 4 U/ml pyruvate kinase from rabbit muscle, pH 6.7 at 22 °C, free Ca^{++} concentration 1.66×10^{-10} M. Contraction was induced by replacing EGTA with Ca-EGTA or by mixing both in distinct ratios. The resulting free Ca^{++} concentrations (up to 4.15×10^{-5} M) were calculated using a computer program which is a modified version of a programme used by D. C. S White, York, England, based on the one published by Perrin and Sayce in 1967 [9].

2) Preparation of cardiac myofibrils

Myofibrils were prepared from dog hearts using Trition X-100 according to Solaro et al. [13].

3) Measurement of myofibrillar ATPase

The Ca^{++} activated, Mg-dependent ATPase of the myofibrils was measured by determination of liberated inorganic phosphate (cf. also the following paragraph).

4) Myofibrillar Ca^{++} binding

Myofibrillar Ca^{++} binding was determined by a centrifugation method using ^{3}H glucose as solute space marker. Solutions for ATPase measurements and Ca^{++} binding studies were identical with the exception that for binding studies 0.3 mCi/l ^{45}Ca^{++} and 0.3 mCi/l ^{3}H glucose were added. Other conditions were: 0.5–1.0 mg myofibrillar protein/ml (determined according to Lowry et al. [6]), 65 mM KCl, 40 mM imidazole, 3 mM MgCl$_2$, 2 mM ATPNa$_2$, 0–1 mM CaCl$_2$, 0–1 mM EGTA, pH 7.0 at 30 °C. Details of the procedures for ATPase measurement and Ca^{++} binding were as described by Solaro and Shiner [14].

5) cAMP-dependent protein kinase assay

The cAMP-dependent protein kinase activity was determined in a 48 000 × g supernatant from homogenized rat liver by a slightly modified version (MES instead of MOPS buffer) of the phosphocellulose absorption method described by Witt and Roskoski in 1975 [18], using histone f$_2$b as substrate.

Results

1) Influence of APP 201-533 on the Ca^{++}-activated force in detergent treated myocardial strips

Fiber bundles from porcine trabecula septo marginalis were extracted and Ca^{++} activated as described in the Methods. According to the free Ca^{++} concentration in the incubation bath which, after a brief equilibration period, equalled the "intracellular" Ca^{++} concentration in the immediate surrounding of the contractile structures, sustained contractions were elicited (Fig. 1 a, inset), resulting in the well known sigmoidal relationship between force and pCa, i.e., the negative logarithm of the molar Ca^{++} concentration (Fig. 1 b). Half maximal activation (EC$_{50}$) was reached at approximately pCa 6.075, full relaxation near pCa 7.5, full activation at pCa 4.8. Maximum Ca^{++}-activated force ranged from 250 to 1000 µN, depending on preparation diameter (100–300 µm). In all experiments, 0.5% (v/v) of 1-Methyl-2-pyrrolidone were present. This organic solvent is necessary to keep APP 201-533 in solution at near neutral pH. Application of APP 201-533 (but not Amrinone, cf. Fig. 1 a, inset) in a concentration of 5×10^{-4} M led to a leftward shift of the Ca^{++} activation curve, displacing the EC$_{50}$ to lower Ca^{++} concentrations by 0.225 pCa units (Fig. 1 b), without significantly affecting relaxation or maximal Ca^{++} activation. This leftward shift was reversible and concentration-dependent in concentrations of APP 201-533 exceeding 10^{-4} M (Fig. 1 a) and did not seem to be saturated at a concentration of 2×10^{-3} M where the upper limit of solubility is reached. Such a Ca^{++} sensitization may be the result of an increase in the Ca^{++} affinity of the regulator protein troponin C. We therefore investigated the influence of APP 201-533 on myofibrillar Ca^{++} binding.

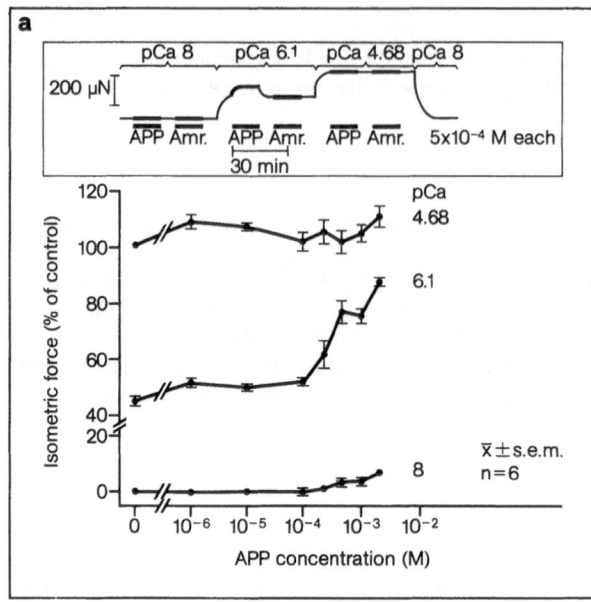

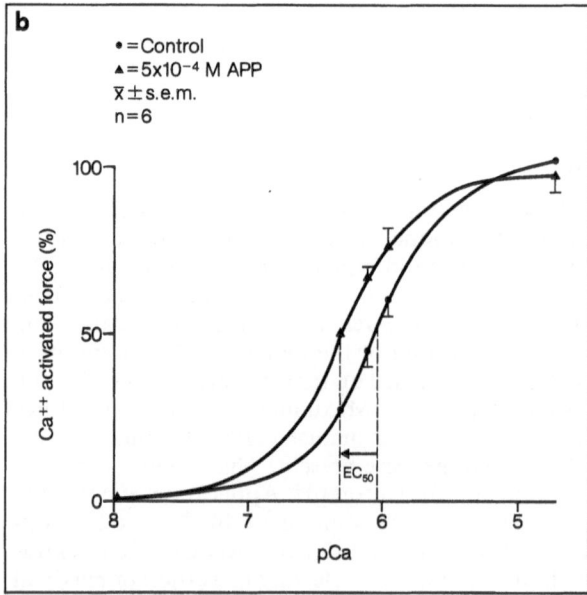

Fig. 1 a, b. Influence of APP 201-533 on the Ca^{++} sensitivity of force in detergent-treated cardiac muscle. **a** Reversible "extra activation" at constant Ca^{++} concentration by APP 201-533. Note: Relaxation (pCa 8) and maximum force (pCa 4.68) are not influenced. Amrinone is ineffective in this model (inset). **b** Left shift of the Ca^{++} activation curve by 5×10^{-4} M APP 201-533. For experimental conditions see Methods.

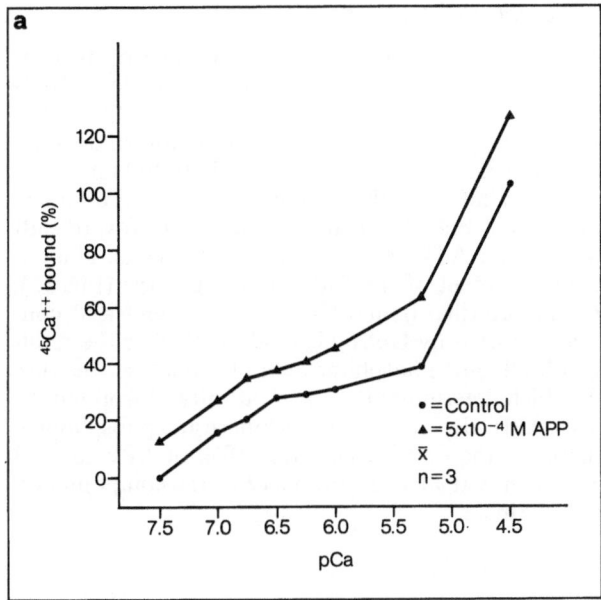

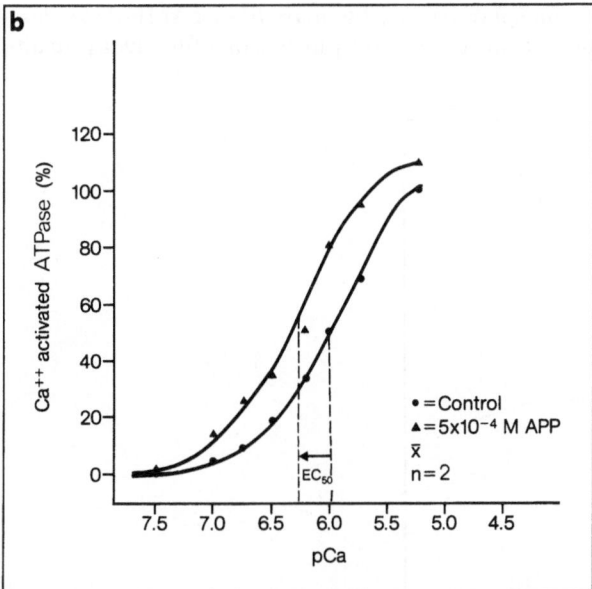

Fig. 2 a, b. Influence of APP 201-533 on the Ca^{++} sensitivities of ^{45}Ca^{++} binding (a) and Ca^{++} activated ATPase (b) in cardiac myofibrils. For experimental conditions see Methods

2) Influence of APP 201-533 on myofibrillar Ca^{++} binding

Canine cardiac myofibrils were isolated as described in the Methods. Using a centrifugation technique, binding of $^{45}Ca^{++}$ to the myofibrils was measured as a function of pCa
(Fig. 2a). In parallel, the Ca^{++}-activated, Mg-dependent ATPase of the myofibrils was
measured in absence and presence of 5×10^{-4} M APP 201-533, as a monitor for drug action. Typical values at pCa 7.5 and pCa 5.25 were 12–15 nmoles and 90–100 nmoles P_i
per mg protein per min at 30 °C. It is shown in Fig. 2b that, similar to the results obtained
in detergent-treated cardiac fiber bundles (cf. Fig. 1b), there was a marked leftward shift
of the Ca^{++} activation curve for myofibrillar ATPase by about 0.25 pCa units. This coincides with a significant leftward displacement of the Ca^{++} binding curve (Fig. 2a),
showing an increase in the amount of myofibrillar bound $^{45}Ca^{++}$ at a given Ca^{++} concentration. A similar observation was reported by Holroyde et al. in 1979 as the result
of treatment of cardiac myofibrils with alkaline phosphoprotein phosphatase. Because
the phosphorylation degree of Tn I, which determines the Ca^{++} affinity of troponin C,
is governed by the opposing actions of cAMP-dependent protein kinase and phosphoprotein phosphatase and because furthermore the Ca^{++} sensitizing effect of APP 201-533
resembles the action of phosphatase, we investigated the influence of the compound on
the activity of cAMP dependent protein kinase.

3) Influence of APP 201-533 on the activity of cAMP-dependent protein kinase from rat liver

Using Histone f_2b as substrate for phosphorylation, the activity of cAMP-dependent
protein kinase from rat liver supernatant was measured by means of a filter assay, in ab-

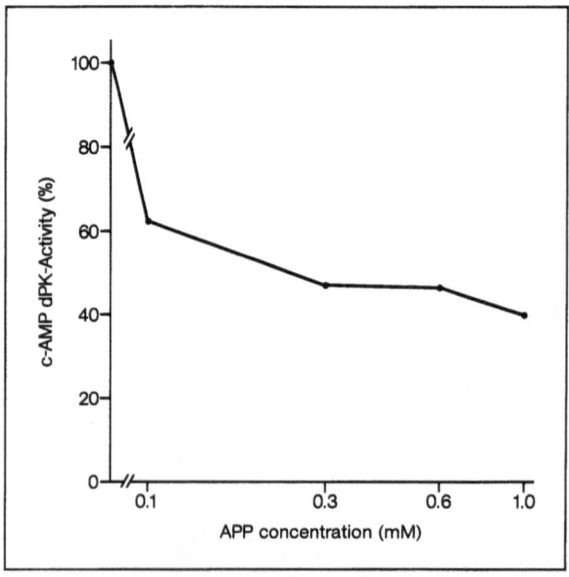

Fig. 3. Inhibitory effect of APP 201-533 on the activity of cAMP-dependent protein kinase (cAMP
dPK) from rat liver supernatant. For experimental details see Methods

sence and presence of various concentrations of APP 201-533. In Fig. 3 it is shown that the compound concentration dependently reduced the activity of the cAMP-dependent protein kinase, although the inhibition was incomplete and amounted to maximally 60% decrease in activity at 10^{-3} M. Half maximal inhibition was reached at a drug concentration smaller than 10^{-4} M.

Discussion

As shown in the present study, APP 201-533 increases the Ca^{++} sensitivities of force and myofibrillar ATPase. This is paralleled by enhanced Ca^{++} binding to the myofibrils. A similar observation has been reported by Solaro and Rüegg [16] for Sulmazole (AR-L 115 BS), another compound with Ca^{++} sensitizing properties [2]. It may be stated that the cardiotonic agents Amrinone and Milrinone which are structurally related to APP 201-533, show no Ca^{++} sensitizing activity in our model (unpublished observations, cf. also Fig. 1 a, inset). Furthermore, APP 201-533 has an inhibitory action on cAMP dependent protein kinase from rat liver. A cAMP-dependent protein kinase is an intrinsic constituent of myocardial contractile structures and is not removed from the contractile system by the detergent treatment used in this study (cf. also [3]).

As cAMP-dependent phosphorylation of Tn I is known to reduce the Ca^{++} affinity of troponin C and vice versa [5], we propose that the Ca^{++} sensitizing action of APP 201-533 may be based on a shift in the relative activities of the enzyme couple of cAMP-dependent protein kinase and phosphoprotein phosphatase, thereby favoring troponin I to be dephosphorylated, i.e., to be in a state where the Ca^{++} affinity of troponin C is enhanced. Assuming that at least certain forms of cardiac insufficiency (e.g., hypoxia or isoprenaline induced cardiac failure) may be due to a decrease in Ca^{++} sensitivity of the contractile structures rather than a lack of Ca^{++} available for activation, the concept of Ca^{++} sensitization of the contractile proteins, possibly via inhibition of Tn I phosphorylation, with compounds of the type of APP 201-533 appears to provide a more causal approach to the treatment of at least certain forms of heart failure.

Recent observations (Herzig and Quast, unpublished) have provided evidence that APP 201-533 does not discriminate between cardiac and skeletal muscle myofibrillar ATPase in terms of Ca^{++} sensitization. As skeletal muscle Tn I lacks the serine residue 20 phosphorylated in cardiac Tn I, our hypothesis of APP's action via inhibition of Tn I phosphorylation is challenged. Further studies will concentrate on clarification of this point.

Acknowledgements. The authors thank Dr. D. C. S. White, Department of Biology, University of York, Heslington, UK, and Dr. R. Hummel, Sandoz Ltd, Basel, Switzerland, for their help in establishing the computer program used in this study to calculate concentrations of free Ca^{++} and various complex species in the experimental solutions.

References

1. Allen DG, Kurihara S, Orchard CH (1982) The effects of hypoxia on intracellular calcium transients in mammalian cardiac muscle. J Physiol (Lond) 328:22–23 P

2. Herzig JW, Feile K, Rüegg JC (1981 a) Activating effects of AR - L 115 BS on the Ca^{++} sensitive force, stiffness and unloaded shortening velocity (V_{max}) in isolated contractile structures from mammalian heart muscle. Arzneim Forsch/Drug Res 31:188–191
3. Herzig JW, Köhler G, Pfitzer G, Rüegg JC, Wölffle G (1981 b) Cyclic AMP inhibits contractility of detergent treated glycerol extracted cardiac muscle. Pflüger's Arch Eur J Physiol 391:208–212
4. Herzig JW, Bornmann G, Botelho L, Erdmann E, Salzmann R, Solaro RJ (1983) Intracellular actions of APP 201-533, a novel cardiotonic agent: increase in Ca^{++} sensitivity and economization of the myocardial contractile process. J Molec Cell Cardiol 15(Suppl 1):244
5. Holroyde MJ, Howe E, Solaro RJ (1979) Modification of the Ca^{++} requirements for activation of cardiac myofibrillar ATPase by cyclic AMP dependent phosphorylation. Biochim Biophys Acta 586:63–69
6. Lowry OH, Rosebrough ND, Farr AC, Randall RJ (1951) Protein measurements with the Folin phenol reagent. J Biol Chem 193:265–271
7. Marban E, Rink TJ, Tsien RW, Tsien RY (1980) Free calcium in heart muscle at rest and during contraction measured with Ca^{++} sensitive microelectrodes. Nature 286:845–850
8. Mc Clellan G, Winegrad S (1978) The regulation of calcium sensitivity of the contractile system in mammalian cardiac muscle. J Gen Physiol 72:737–767
9. Perrin DD, Sayce IG (1967) Computer calculation of equilibrium concentrations in mixtures of metal ions and complexing species. Talanta 14:833–842
10. Ray KP, England PJ (1976) Phosphorylation of the inhibitory subunit of troponin and its effects on the calcium dependence of cardiac myofibrillar ATPase. FEBS Lett 70:11–16
11. Reuter H, Scholz H (1977) The regulation of the calcium conductance of cardiac muscle by adrenaline. J Physiol (Lond) 264:49–62
12. Salzmann R, Bormann G, Herzig JW, Markstein R, Scholtysik G (1985) Pharmacological actions of APP 201-533, a novel cardiotonic agent. J Cardiovasc Pharm 7:588–596
13. Solaro RJ, Pang DC, Briggs FN (1971) The purification of cardiac myofibrils with Triton X-100. Biochim Biophys Acta 245:259–262
14. Solaro RJ, Shiner JS (1976) Modulation of the Ca^{++} control of dog and rabbit cardiac myofibrils by Mg^{++}: comparison with rabbit skeletal myofibrils. Circ Res 39:8–14
15. Solaro RJ, Holroyde MJ, Herzig JW, Peterson JW (1980) Cardiac relaxation and myofibrillar interactions with phosphate and vanadate. Europ Heart J 1(Suppl A):21–27
16. Solaro RJ, Rüegg JC (1982) Stimulation of Ca^{++} binding and ATPase activity of dog cardiac myofibrils by AR-L 115 BS, a novel cardiotonic agent. Circ Res 51, 290–294
17. Tada M, Kirchberger MA, Katz AM (1974) Phosphorylation of a 22 000 dalton component of the cardiac sarcoplasmic reticulum by adenosine 3':5'-monophosphate-dependent protein kinase. J Biol Chem 250:2640–2647
18. Witt JJ, Roskoski R Jr (1975) Rapid protein kinase assay using phosphocellulose paper absorption. Anal Biochem 66:253–258

Author's address:

Prof. Dr. J.W. Herzig, Pharmaceutical Division, Department Research CVS, Bldg. K-125/1058, CH-4002 Basel, Switzerland

Positive inotropic stimulation in the normal and insufficient human myocardium

E. Erdmann, M. Böhm

Medizinische Klinik I, University of Munich, Klinikum Großhadern, München, FRG

Summary

Cardiac α- and β-adrenoceptors and the positive inotropic effects of several adenylate cyclase dependent and independent agents have been measured in papillary muscle strips from patients without, as well as with moderate and severe heart failure. The number of β-adrenoceptors was found to be decreased depending on the degree of heart failure. This does not apply to α-adrenoceptors, which remain unchanged. The antagonist affinity of adrenoceptors for the different ligands did not change in heart failure.

Maximal increases in force of contraction were measured after raising Ca^{++} up to 15 mM in the muscle strips. In healthy human myocardium, isoprenaline, dobutamine, IBMX or cardiac glycosides increase force of contraction to the same maximal values as Ca^{++} does. However, in cardiac tissue from heart failure patients, positive inotropic agents which increase intracellular cAMP or are cAMP-dependent are less effective than Ca^{++}. Furthermore, the results seem to indicate a homologous (agonist specific) downregulation of receptors in moderate heart failure and a heterologous downregulation in severe heart failure. Thus, many well known positive inotropic drugs lose their effectivness just when they are needed most: in severe heart failure.

1. Introduction

Myocardial failure is characterized by an inadequate, low cardiac output and increased filling pressures. Accepted ways of treatment in this situation include drugs that lower preload, stimulate contractility, and decrease afterload. Positive inotropic therapy depends on the possibility to utilize a "contractile reserve" of the diseased cardiac muscle. Obviously, force of contraction of fibrous tissue (for instance after myocardial infarction, aneurysm, etc.) cannot be increased very much by positive inotropic agents. On the other hand, well known inotropic drugs (noradrenaline, dobutamine) fail to be effective in some cardiac diseases in man, although the myocardium at least histologically does not show gross abnormalities (for instance in dilated cardiomyopathy).

The failing heart has been demonstrated to exhibit reduced inotropic response to β-adrenoceptor agonists [1–3, 5–7, 9]. This has been attributed to a decreased number of functional β-adrenoceptors in the cell membranes [2–8]. The pathophysiological explanation for this "downregulation" of receptors seems to be connected to the generally found increased levels of plasma noradrenaline in heart failure [10, 15], which correlate with the degree of myocardial insufficiency. However, in heart failure other positive inotropic drugs, not acting through β-adrenoceptors, have also been demonstrated to be less effec-

Supported by the DFG (Er 65/4-5) and Friedrich-Baur-Stiftung Munich)

tive [2, 3, 9, 12, 13, 18]. The exact mechanism of this loss of activity confined to myocardium from heart failure patients is not clear at present. As there does not exist an accepted animal model of heart failure [19, 21], investigations to elucidate the biochemical mechanisms of this loss of inotropic effectiveness, well known in clinical medicine [1, 11, 22, 23], have to be done on human cardiac muscle from patients with this disease.

Hemodynamic measurements in patients (for instance during cardiac catheterization) usually will not help much, as the prerequisite for evaluation of contractility is to keep constant preload, afterload, and frequency. Obviously, this cannot be achieved in patients during applications of different positive inotropic drugs with all the other known hemodynamic side effects. Therefore, we have performed experiments in isolated papillary muscle strips from patients undergoing mitral valve replacement surgery (termed moderate heart failure) and cardiac transplantation. The explanted diseased hearts without exception were in terminal heart failure (NYHA IV). Some "healthy" donor hearts because of technical reasons could not be implanted. Their papillary muscle strips, etc., served as controls ("normal heart"). The detailed methods have been published elsewere [2–4].

2. β-Adrenoceptors

In order to characterize the functional state of the myocardium, the number of β-adrenoceptors and their affinity for β-adrenoceptor agonists were determined. Figure 1 shows that the total number of β-adrenoceptor in the normal myocardium was measured as 41.6 ± 1.2 fmol/mg protein (n = 3). In moderate heart failure the density of cardiac β-adrenoceptors was 18.0 ± 1.1 fmol/mg protein (n = 16) and in terminal heart failure 9.5 ± 1.6 fmol/mg protein (n = 7) were found. Thus, previous investigations of other groups [1, 5–8, 13, 14] can be supported. Furthermore, these biochemically distinct mea-

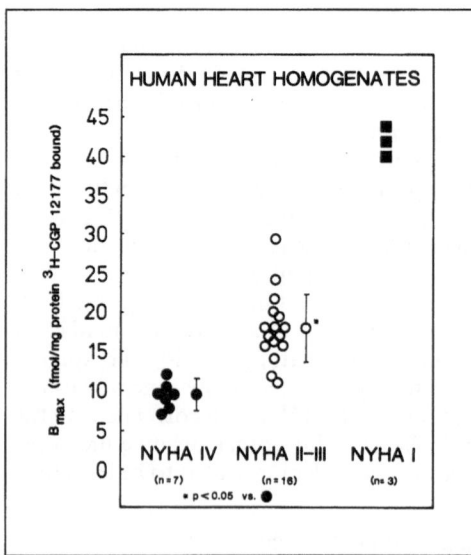

Fig. 1. Density of myocardial β-adrenoceptors in cardiac membranes from patients with severe (NYHA IV) heart failure, moderate (NYHA II–III) heart failure and without (NYHA I) heart failure. Ordinate: maximal number of binding sites B_{max}, fmol (^{3}H)-CGP 12177 bound per mg protein

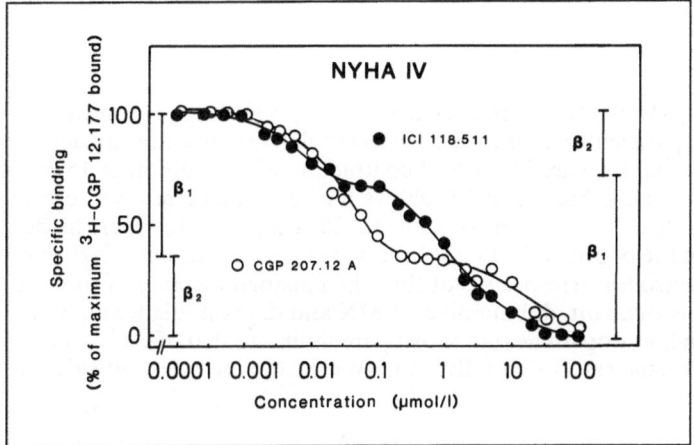

Fig. 2. Inhibition of ^{3}H-CGP 12 177 binding to cardiac β_1 and β_2-adrenoceptors by the selective β_1-adrenoceptor antagonist CGP 207.12 A, and the selective β_2-adrenoceptor antagonist ICI 118.511. Note that the competition curve of the antagonist is biphasic revealing a β_1- to β_2 ratio of 60:40. The same ratio is obtained with both antagonists

surements are in line with the separation of the cardiac muscles as "normal" (control), moderate, and severe heart failure.

There were no changes in affinity of the β-adrenoceptors for ^{3}H-CGP 12.177, thus providing evidence against alterations of the receptor conformation.

2.1 β_2-Adrenoceptors

Since noradrenaline primarily acts on the β_1-adrenoceptor subtype, one would expect that noradrenaline-induced β-adrenoceptor downregulation reduces the number of β_1-adrenoceptors more pronouncedly than the number of β_2-adrenoceptors. Since β_2-adrenoceptors mediate positive inotropic effects in the heart [8], it is mandatory to study the β-adrenoceptor subtypes. This can be done with competition experiments with selective antagonists such as CGP 207.12 A for β_1- and ICI 118.54 for β_2-adrenoceptors. One of these experiments is shown in Fig. 2. In the failing heart, there was a β_1 to β_2 ratio of 60:40 with both, CGP 207.12 A and ICI 118.551. In the nonfailing myocardium the ratio is 80:20 (not shown). These findings show that in the failing heart noradrenaline-induced downregulation affects mainly the β_1-adrenoceptor subtype.

3. α-Adrenoceptors

The number of cardiac α-adrenoceptors, as determined with ^{3}H-prazosin bound, was similar in normal heart 8.1 fmol/mg protein (n = 1), moderate heart failure 6.7 ± 0.8 fmol/mg protein (n = 12), and severe heart failure 7.4 ± 0.9 fmol/mg protein (n = 9). Unfortunately, the number of α-adrenoceptors in human cardiac cell membranes is extremely low, just close to the limit of detection. Nevertheless, our measurements did not show any differences, neither in the number nor in the affinity of the receptors for ^{3}H-prazosin.

4. Force of contraction

4.1 Normal heart

In the papillary muscle strips from three normal hearts the increase in force of contraction by different positive inotropic agents was measured in a cumulative concentration dependent manner (Fig. 3). Maximal increase in force of contraction after addition of 15 mM Ca^{++} was taken as "maximum": This probably shows the "contractile reserve" of the myocardium. Maybe, the "true" maximum is some 10–20% higher. However, higher concentrations of Ca^{++} in the organ bath cannot be tested as Ca^{++} falls out of solution then at the stimulating electrodes. Irrespective of this, the maximal increase of force of contraction of isoprenaline, ouabain, dobutamine, IBMX and digoxin, digitoxin, milrinone, amrinone, noradrenaline, dopamine (not shown) are similar to that of Ca^{++}. Thus, they can use up the "contractile reserve" totally. However, histamine is less effective in the normal heart.

4.2 Diseased myocardium

Figure 4 shows that Ca^{++} increases force of contraction to a similar extent in the failing heart as compared to the normal heart. So do cardiac glycosides, forskolin, and db-cAMP. However, isoprenaline, dobutamine, IBMX, histamine and (not shown) milrinone, amrinone, noradrenaline, dopamine, Bay K 8644, and glucagon cannot make use of the contractile reserve of the muscle. Thus, there is a fundamental difference in healthy and severely diseased human heart in respect to the inotropic response.

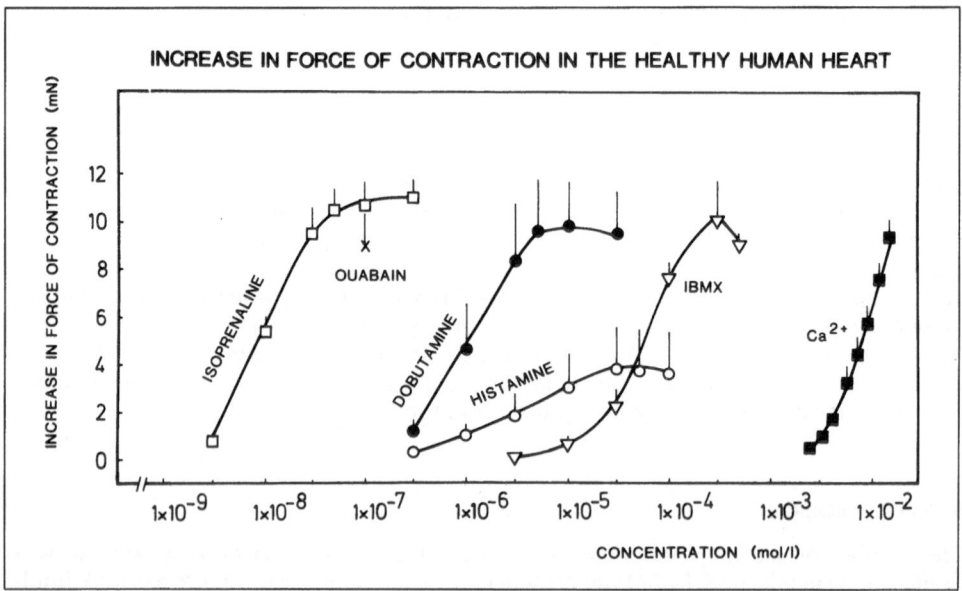

Fig. 3. Concentration-response curves of isoprenaline, dobutamine, histamine, isobutylmethylxanthine (IBMX), ouabain, and Ca^{++} in isolated, electrically driven papillary muscle strips from human nonfailing heart. Abscissa: increase in force of contraction in mN. Ordinate: drug concentration in mol/l. The predrug values were 2.7 ± 0.6 (n=4; isoprenaline), 2.4 ± 0.7 (n=4; dobutamine), 4.1 (n=2; histamine), 3.5 (n=2; IBMX), 1.0 ± 0.2 (n=4; ouabain), and 1.8 ± 0.8 (n=6; Ca^{2+})

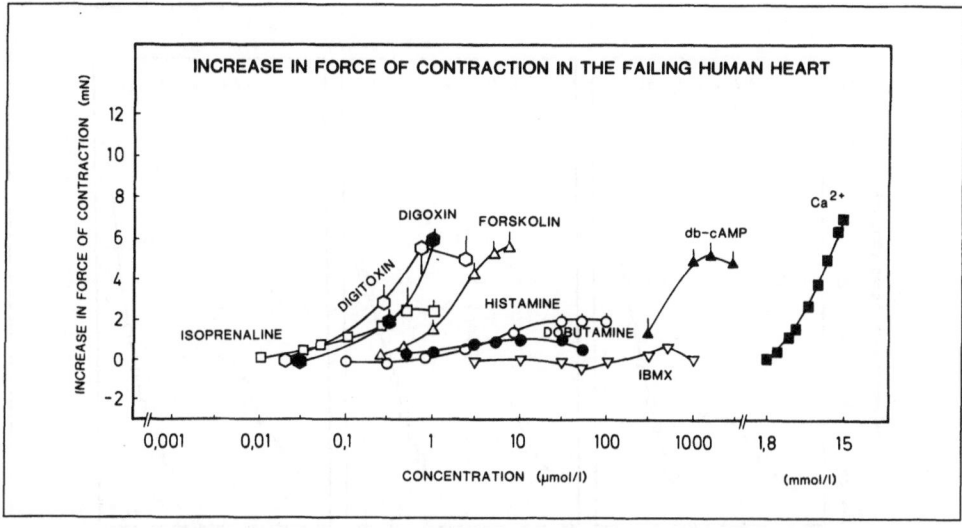

Fig. 4. Concentration-response curves of isoprenaline, dobutamine, histamine, forskolin, db-cAMP, isobutylmethylxanthine (IBMX), ouabain, and Ca^{2+} in isolated, electrically driven papillary muscle strips from human failing hearts. Abscissa: increase in force of contraction in mN. Ordinate: drug concentration in µmol/l or mmol/l. The predrug values were 2.4 ± 0.3 (n=114) mN

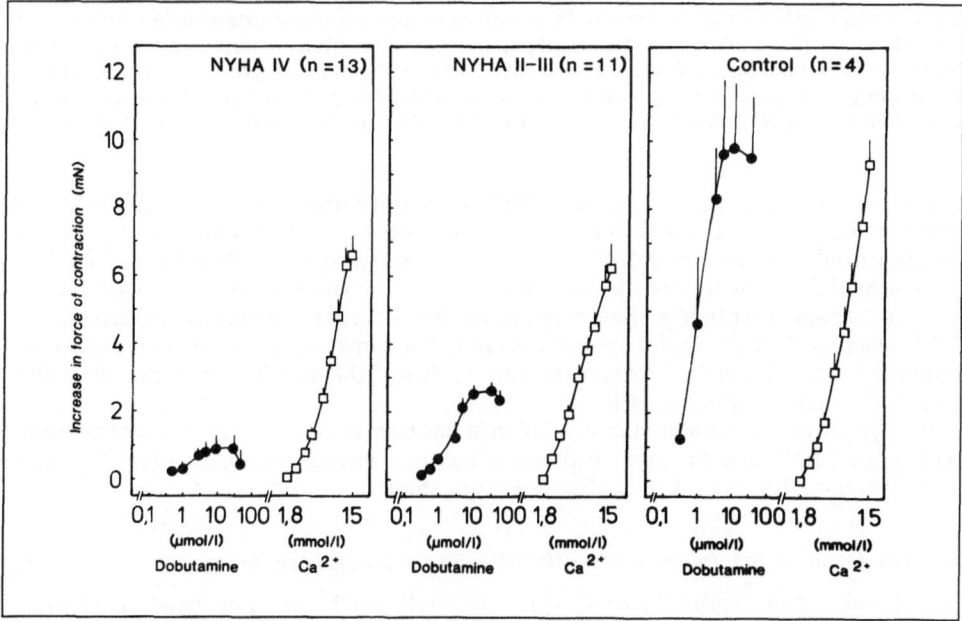

Fig. 5. Concentration-response curves for the positive inotropic effect of dobutamine (0.3–30.0 µ mol/l) and calcium (1.8–15.0 mmol/l) in 13 papillary muscle strips from five patients with severe (NYHA IV) heart failure, in 11 strips from four patients with moderate (NYHA II–III) heart failure, and in four strips from two nonfailing donor hearts (control). Basal force of contraction was 2.9 ± 0.3 mN (n=13; NYHA IV), 3.3 ± 0.1 mN (n=11; NYHA II–III) and 2.4 ± 0.7 mN (n=4; Control)

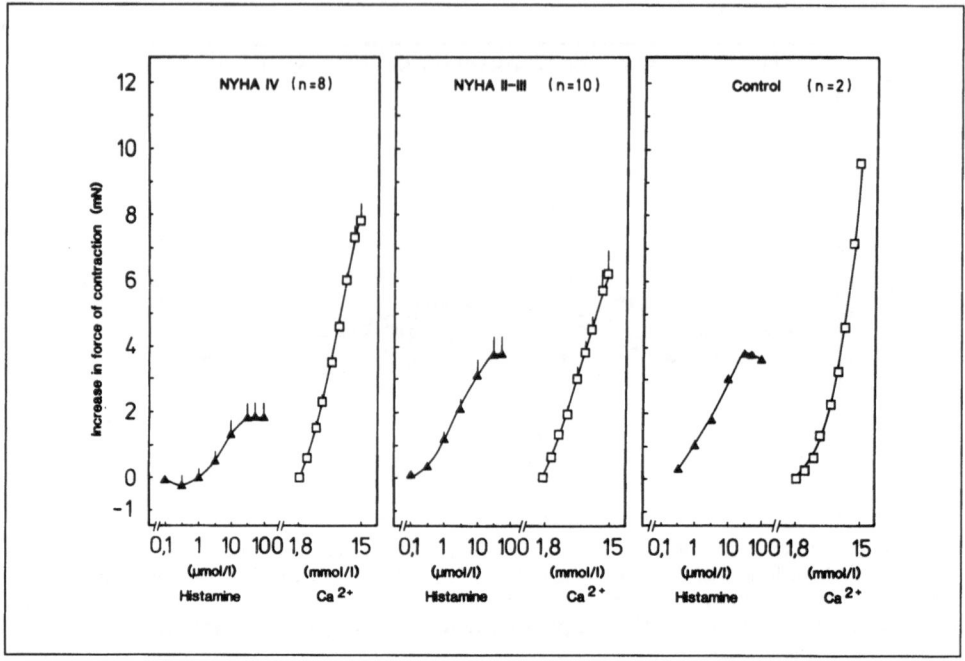

Fig. 6. Concentration-response curves for the positive inotropic effects of histamine (0.1–100.0 μmol l^{-1}) and calcium (1.8–15.0 mmol l^{-1}) in eight papillary muscle strips from three patients with severe (NYHA IV) heart failure, in 10 strips from eight patients with moderate (NYHA II–III) heart failure, and in two strips from one non-failing donor heart (control). Basal force of contraction was 3.5 ± 0.3 mN (n = 8; NYHA IV), 1.4 ± 0.2 mN (n = 10; NYHA II–III) and 4.1 mN (n = 2; control)

Figure 5 shows that the decrease in effectiveness of dobutamine corresponds to the degree of heart failure. In the normal heart, dobutamine is as effective as Ca^{++}, in moderate heart failure it has about 50% of the effectiveness as compared with Ca^{++}, and in severe heart failure, there is hardly any effect. The maximal effect of Ca^{++} is decreased by about 20% in heart failure. Nevertheless, cardiac glycosides are always as effective as Ca^{++}, whereas β-adrenoceptor agonists are not. The same applies to phosphodiesterase inhibitors [2, 3, 13] and Ca^{++}-agonists (Bay K 8644) [18], which cannot use all of the "contractile reserve" of the muscle.

Phenylephrine as an α-adrenoceptor agonist has been tested in the presence of propranolol (1 μmol/l). It gave the same response in the three groups. However, phenylephrine has only about 20% of the Ca^{++}-effectivity (not shown).

4.3 Homologous vs heterologous desensitization of the β-adrenoceptor system

The reduced responses after β-adrenoceptor stimulation in heart failure can be explained by a downregulation of β-adrenoceptors. The reduced effectiveness of phosphodiesterase inhibitors in this situation may be explained by decreased basal cAMP levels in the cardiac cell in heart failure. This view is supported by the finding of full effectiveness of PDE-inhibitors after prestimulation of the cardiac muscle by β-adrenoceptor agonists [13]. The situation is less clear for histamine (Fig. 6). Although the effectiveness of histamine is

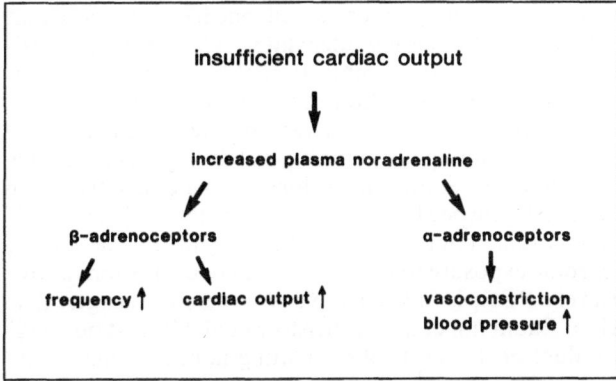

Fig. 7. In acute cardiac failure increased plasma noradrenaline levels maintain organ perfusion. In chronic heart failure, due to β-adrenoceptor downregulation, the positive inotropic effects are lost

much less than that of Ca^{++} or isoprenaline in the normal heart, its effectiveness stays about the same in moderate heart failure. Only in severe heart failure it is significantly decreased.

If the effectiveness of histamine is decreased also in heart failure, a heterologous (agonist nonspecific) desensitization of receptors must be supposed. Our results (Fig. 6) seem to indicate that in moderate heart failure (histamine effectiveness preserved) a homologous desensitization (agonist specific) occurs, whereas in severe heart failure (histamine effectiveness decreased) a heterologous (agonist nonspecific) desensitization takes place. Recently, Reithmann and Werdan [20] clearly demonstrated this concentration-dependent homologous and heterologous desensitization in isolated cardiac cell cultures.

5. Discussion

In heart failure plasma noradrenaline levels rise according to the severity of the disease [10]. Increased hormone levels lead to a downregulation of the respective hormone receptors [16]. Thus, the physiological regulation of cardiac output by noradrenaline (Fig. 7), operating well in acute heart failure, is disturbed in chronic heart failure. Therefore, several positive inotropic drugs loose their efficacy in the chronic situation. α-Adrenoceptors appears to stay unchanged, though. However, the low effectiveness of phenylephrine does not suggest a pharmacological role for these drugs in chronic heart failure. This, of course, is not a promising treatment because of their unwanted vasoconstrictive effects on the coronary arteries. If the α-adrenoceptors also stay unchanged in the vascular tissue, this would explain at least some of the peripheral vasoconstriction in chronic heart failure.

There is no question in the literature that β-adrenoceptor agonists are less effective in chronic heart failure. This applies to experimental [1, 2, 5–7, 12–14, 20] and clinical observations (11, 22, 23). However, there are contradictory reports on the efficacy of histamine [1, 5, 12]. If the downregulation of β-adrenoceptors only involves these receptors, other substances like glucagon, histamine and PGE_1 should be still fully effective. If also these hormones loose their effectiveness, other components of the adenylate cyclase system seem also to be affected in chronic heart failure. Since it is difficult to understand that histamine and glucagon receptors are downregulated as well, one has to suppose that the common pathway of several receptors stimulating adenylate cyclase, e.g., G_s, is altered in this situation [17]. Certainly, an increased concentration of G_i mediating inhibitory effects on adenylate cyclase and force of contraction might be responsible for this, as well. Our finding of unchanged histamine effects in moderate and decreased effectiveness in severe heart failure suggests that homologous (agonist specific) desensitization is induced at first. At higher noradrenaline concentrations or longer time of exposure (severe heart failure) heterologous (agonist nonspecific) desensitization occurs. These findings are very well supported in the cardiac contracting cell-model [20]. In this model, PGE_1 shows less effectiveness at chronic exposure to noradrenaline concentrations above 10^{-8} M. At lower noradrenaline levels, PGE_1 is fully effective. We therefore suggest that the discrepancy between the work of Baumann et al. [1], Bristow et al. [5], and ours [12] depends on the severity of heart failure or the levels of circulating noradrenaline in the patient before surgery.

The experiments furthermore show clearly that positive inotropic drugs may loose their effects in heart failure. This is not due to a completely fibrosed myocardium – as Ca^{++} or other drugs still increase force of contraction – but is rather due to specific biochemical changes connected with the disease. Cardiac muscle from healthy animals or humans do not show this abnormality. Thus, results from animal experiments cannot simply be extrapolated to the situation in human disease – a fact that seems to apply generally.

References

1. Baumann G, Riess G, Erhard WD, Felix SB, Ludwig L, Blümel G, Blömer H (1981) Impaired beta-adrenergic stimulation in the uninvolved ventricle post-acute myocardial infarction: Reversible defect due to excessive circulating catecholamine-induced decline in number and affinity of beta-receptors. Am Heart J 101:569–581
2. Böhm M, Beuckelmann D, Brown L, Feiler G, Lorenz B, Näbauer M, Kemkes B, Erdmann E (1988) Reduction of beta-adrenoceptor density and evaluation of positive inotropic responses in isolated, diseased human myocardium. Eur Heart J 9:844–852
3. Böhm M, Diet F, Feiler G, Kemkes B, Kreuzer E, Weinhold C, Erdmann E (1988) Subsensitivity of the failing human heart to isoprenaline and milrinone is related to β-adrenoceptor downregulation. J Cardiovasc Pharmacol 12:726–732
4. Böhm M, Diet F, Feiler G, Kemkes B, Erdmann E (1988) α-Adrenoceptors and α-adrenoceptor-mediated positive inotropic effects in failing human myocardium. J Cardiovasc Pharmacol 12:357–364
5. Bristow MR, Ginsburg R, Minobe W, Cubicciotti RS, Sageman WS, Lurie K, Billingham ME, Harrison DC, Stinson EB (1982) Decreased catecholamine sensitivity and β-adrenergic-receptor density in failing human heart. N Engl J Med 307:205–211
6. Bristow MR (1984) Myocardial β-adrenergic receptor downregulation in heart failure. Int J Cardiol 5:648–652
7. Bristow MR, Kantrowitz NE, Ginsburg R, Fowler MB (1985) β-Adrenergic function in heart muscle disease and heart failure. J Mol Cell Cardiol 17 (Supplement):41–52
8. Brodde OE (1987) Cardiac beta-adrenergic receptors. ISI Atlas of Science: Pharmacol 1:107–111
9. Brown L, Lorenz B, Erdmann E (1986) Reduced positive inotropic effects in diseased human ventricular myocardium. Cardiovasc Res 20:516–520
10. Cohn JN, Levine TB, Olivari MT, Garberg V, Lura D, Francis GS, Siman AB, Rector T (1984) Plasma norepinephrine as a guide to prognosis in patient with chronic congestive heart failure. N Engl J Med 311:819–823

11. Colucci WS, Denniss AR, Leatherman GF, Quigg RJ, Ludmer PL, Marsh JD, Gauthier DF (1988) Intracoronary infusion of dobutamine to patients with and without severe congestive heart failure. J Clin Invest 81:1103–1110
12. Erdmann E (1988) The effectiveness of inotropic agents in isolated cardiac preparations from the human heart. Klin Wochenschr 66:1–6
13. Feldman MD, Copelas L, Gwathmey JK, Phillips P, Warren SE, Schoen FJ, Grossman W, Morgan JP (1987) Deficient production of cyclic AMP: pharmacologic evidence of an important cause of contractile dysfunction in patients with end-stage heart failure. Circulation 75:331–339
14. Fowler MB, Laser JA, Hopkins GL, Minobe W, Bristow MR (1986) Assessment of the β-adrenergic receptor pathway in the intact failing human heart: progressive receptor down-regulation and subsensitivity to agonist response. Circulation 74:1290–1302
15. Francis G (1985) Plasma catecholamine levels in patients with congestive heart failure. Cardiovasc Rev Rep 6:444–454
16. Lefkowitz RJ, Caron MG, Stiles G (1984) Mechanism of membrane-receptor regulation. N Engl J Med 310:1570–1579
17. Murakami T, Katada T, Yasuda H (1987) Reduction in the activity of the stimulatory guanine nucleotide-binding protein in the myocardium of spontaneously hypertensive rats. J Mol Cell Cardiol 19:199–208
18. Näbauer M, Brown L, Erdmann E (1988) Positive inotropic effects of the calcium channel activator Bay K 8644 on guinea-pig and human isolated myocardium. Naunyn-Schmiedeberg's Arch Phramacol 337:85–92
19. Newman WH (1983) Biochemical, structure and mechanical defects of the failing myocardium. Pharmac Ther 22:215–247
20. Reithmann C, Werdan K (1989) Homologous vs heterologous desensitization of the adenylate cyclase system in heart cells. Eur J Pharmacol 154:99–104
21. Smith HJ, Nutall A (1985) Experimental models of heart failure. Cardiovasc Res 19:181–186
22. Tan LB (1986) Cardiac pumping capability and prognosis in heart failure. Lancet Dec 13:1360–1363
23. Unverferth DV, Blanford M, Kates RE, Leier CV (1980) Tolerance to dobutamine after a 72 hour continuous infusion. Am J Med 69:226–266

Author's address:

Prof. Dr. E. Erdmann, Medizinische Klinik I, University of Munich, Klinikum Großhadern, Marchioninistraße 15, 8000 München 70, FRG

Do human cardiac beta-2 adrenoceptors play a (patho)physiological role in regulation of heart rate and/or contractility?

O.-E. Brodde, H.-R. Zerkowski[1]

Biochem. Research Lab, Med. Klinik & Poliklinik, Div Renal & Hypertensive Diseases and [1] Div Thoracic & Cardiovascular Surgery, University of Essen, FRG

Summary

There can be no doubt that in human heart in addition to β_1-adrenoceptors, functional β_2-adrenoceptors exist. Their (patho)physiological role in regulating heart rate and/or contractility, however, is still an open question. Under normal physiological conditions cardiac β_2-adrenoceptors may not be of functional importance, since heart rate and contractility seem to be under the control of noradrenaline that in the human heart acts nearly exclusively at β_1-adrenoceptors. However, in situations of stress when large amounts of adrenaline are released from the adrenal medulla additional stimulation of cardiac β_2-adrenoceptors may contribute to increases in heart rate and/or contractility. Moreover, in endstage congestive cardiomyopathy where cardiac β_1-adrenoceptors are selectively down-regulated, cardiac β_2-adrenoceptors may substitute for the loss of β_1-adrenoceptors to maintain (at least partially) contractility; under these conditions β_2-adrenoceptor agonists, like dopexamine, may be of beneficial therapeutic effect. While a decrease in cardiac β-adrenoceptor function appears to be a general phenomenon in all kinds of heart failure, it is not always due to a selective reduction in cardiac β_1-adrenoceptors: in mitral valve disease both cardiac β_1- and β_2-adrenoceptors decline concomitantly in relation to the degree of heart failure. It is, therefore, doubtful whether under these conditions β_2-adrenoceptor agonists may also be useful to support the failing heart.

Introduction

During the past few years a vast body of evidence has accumulated that in human heart both β_1- and β_2-adrenoceptors coexist. Mainly by the use of radioligand binding studies numerous authors have shown that in human right and left atria about 30–35% of the total β-adrenoceptor population is of the β_2-subtype, while in the right and left ventricles the amount of β_2-adrenoceptors is about 20–25% (for references see Table 1). Both β_1- and β_2-adrenoceptors are coupled to the adenylate cyclase [5, 22, 26, 25] and contribute to the positive inotropic effects of β-adrenergic agonists on the isolated electrically driven right atria as well as right and left papillary muscles (for references see [6]). Among the catecholamines isoprenaline and adrenaline mediate their positive inotropic effects via β_1- *and* β_2-adrenoceptor stimulation, whereas noradrenaline – the main transmitter of the sympathetic nervous system – produces its positive inotropic effect predominantly through β_1-adrenoceptor stimulation [22, 26, 32, 7, 3].

Table 1. Distribution of β_1- and β_2-adrenoceptors in human cardiac tissues

Tissues	β_1:β_2-Ratio	References
Right atrium	82:18	Brodde et al. (1983) Circ Res 53:752–758
	75:25	Golf et al. (1985) Cardiovasc Res 19:636–641
	74:26	Stiles et al. (1983) Life Sci 33:467–473
	70:30	Michel et al. (1986) J Hypertension 4 (Suppl 6):215–218
	63:37[a]	Brodde et al. (1986) J Cardiovasc Pharmacol 8:1235–1242
	59:41	Buxton et al. (1987) Br J Pharmacol 92:299–310
	50:50	Robberecht et al. (1983) Mol Pharmacol 24:169–173
	43:57	Hedberg et al. (1985) J Pharmacol Exp Ther 234:561–568
Left atrium	66:34[a]	Brodde et al. (1986) J Cardiovasc Pharmacol 8:1235–1242
	65:35	Heitz et al. (1983) Br J Pharmacol 80:711–717
Right ventricle	87:13	Brodde et al. (1982) Naunyn-Schmied Arch Pharmacol 321:R 242
	77:23	Bristow et al. (1986) Circ Res 59:297–309
	75:25	Golf et al. (1985) Cardiovasc Res 19:636–641
	74:26[a]	Brodde et al. (1986) J Cardiovasc Pharmacol 8:1235–1242
	62:38[b]	Bristow et al. (1986) Circ Res 59:297–309
Left ventricle	86:14	Stiles et al. (1983) Life Sci 33:467–473
	77:23	Bristow et al. (1986) Circ Res 59:297–309
	75:25	Golf et al. (1985) Cardiovasc Res 19:636–641
	73:27	Kaumann and Lemoine (1987) Naunyn-Schmiedeberg's Arch Pharmacol 335:403–411
	74:26[a]	Brodde et al. (1986) J Cardiovasc Pharmacol 8:1235–1242
	69:31	Vago et al. (1984) Biochem Biophys Res Commun 121:346–354
	65:35	Heitz et al. (1983) Br J Pharmacol 80:711–717
	64:36	Buxton et al. (1987) Br J Pharmacol 92:299–310
	63:37	Lemoine et al. (1988) Br J Pharmacol 95:55–66
	62:38[b]	Bristow et al. (1986) Circ Res 59:297–309
	41:59	Hedberg et al. (1985) J Pharmacol Exp Ther 234:561–568

[a] Determined in explanted hearts obtained from heart transplant recipients who suffered from end-stage idiopathic dilated cardiomyopathy
[b] Determined in explanted hearts obtained from heart transplant recipients who suffered from severe end-stage biventricular failure due to idiopathic dilated cardiomyopathy or from isolated right ventricular failure due to primary pulmonary hypertension

Role of beta-2 adrenoceptors in human archaic tissues

On isolated electrically driven human right atria the stimulation of β_1-, as well as of β_2-adrenoceptors, produces about the same maximum positive inotropic effect [32, 28]. The same seems to hold true for isolated human left atrium (Fig. 1): on this preparation the highly selective β_1-adrenoceptor antagonist CGP-20712 A (3×10^{-7} M) [18] and the highly selective β_2-adrenoceptor antagonist ICI-118,551 (3×10^{-8} M) [2] – both used in a concentration that occupies more than 90% of β_1- and β_2-adrenoceptors, respectively – shifted the concentration-response curve for the positive inotropic effect of isoprenaline to the right to about the same degree. Both antagonists, however, produced a shift that was less than could be predicted from interaction with only one β-adrenoceptor subtype, indicating that both β_1- and β_2-adrenoceptors are involved in the positive inotropic action of isoprenaline to about the same degree. This assumption is supported by the fact

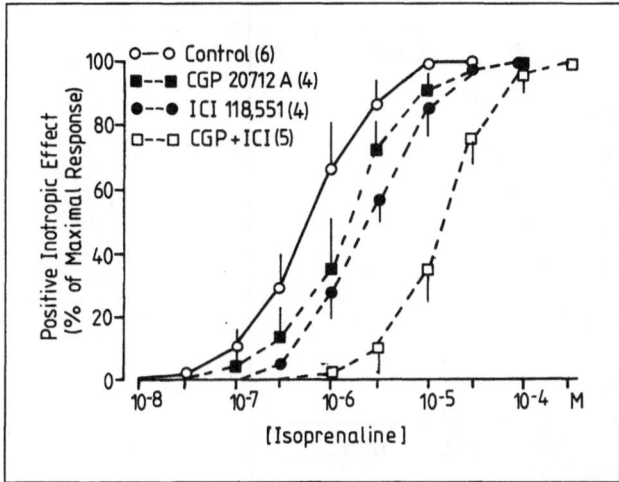

Fig. 1. Effects of 3×10^{-7} M CGP 20712 A and 3×10^{-8} M ICI 118,551 on the positive inotropic effect of isoprenaline on the isolated electrically driven (1.0 Hz, 37 °C) left atrium derived from patients suffering from mitral valve disease (NYHA class III–IV). Ordinate: positive inotropic effect in percent of maximal response. Abscissa: molar concentrations of isoprenaline. Given are means ± SEM; number of experiments in parentheses (unpublished observations)

that combination of both ICI-118,551 and CGP-20712 A resulted in a further marked shift to the right of the concentration-response curve of isoprenaline.

In contrast to right and left atria in isolated electrically driven human left papillary muscle, ICI-118,551 shifted predominantly the lower part of the concentration-response curve for isoprenaline to the right, while CGP-20712 A affected mainly the upper part (Fig. 2). Only the combination of both antagonists produced a parallel shift of the concentration-response curve to the right. These results favor the idea that in the left papillary muscle β_2-adrenoceptor stimulation causes only submaximal positive inotropic effects of catecholamines. A similar submaximal positive inotropic effect mediated by β_2-adrenoceptor stimulation was recently observed in human right papillary muscle where the selective β_2-adrenoceptor agonist zinterol produced an increase in contractile force which amounted to about 40% of that of isoprenaline [3].

The functional role of β_2-adrenoceptors present in human heart is still not completely understood. Besides mediating positive inotropic effects, right atrial β_2-adrenoceptors may be also capable of mediating positive chronotropic effects. Several investigators have demonstrated that selective β_1-adrenoceptor antagonists inhibit isoprenaline-induced tachycardia less effectively than non-selective β-adrenoceptor antagonists when both are given in doses that are equipotent in inhibiting exercise-induced tachycardia [12, 1, 8]. Moreover, in healthy volunteers the selective β_2-adrenoceptor antagonist ICI-118,551 antagonized isoprenaline-induced tachycardia more potently than the selective β_1-adrenoceptor antagonists atenolol [1] or bisoprolol [8], but was equipotent to propranolol [1]. These results and the fact that noradrenaline acts in the human heart predominantly at β_1-adrenoceptors (see above) are compatible with the view that under normal conditions only β_1-adrenoceptors are involved in the control of heart rate and contractility. However, in situations of stress, when large amounts of adrenaline that non-selectively stim-

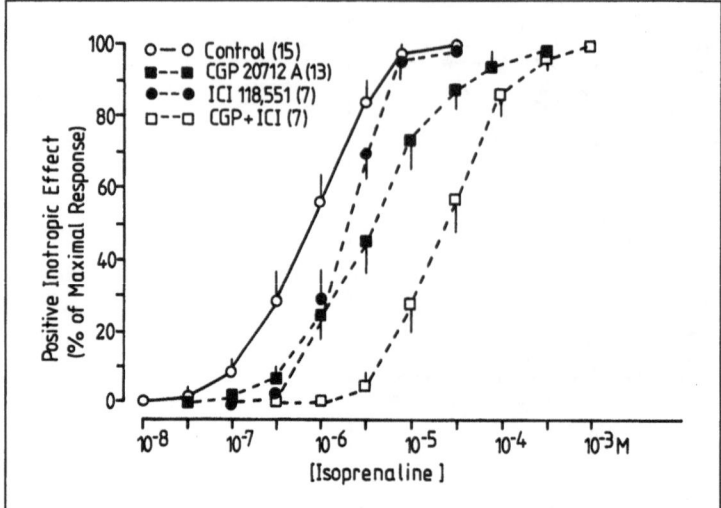

Fig. 2. Effects of 3×10^{-7} M CGP 20712 A and 3×10^{-8} M ICI 118,551 on the positive inotropic effect of isoprenaline on the isolated electrically driven (1.0 Hz, 37 °C) left papillary muscle derived from patients suffering from mitral valve disease (NYHA class III–IV). Ordinate: positive inotropic effect in percent of maximal response. Abscissa: molar concentrations of isoprenaline. Given are means ± SEM; number of experiments in parentheses (unpublished observations)

ulate both cardiac β_1- and β_2-adrenoceptors (see above) are released from the adrenal medulla, additional stimulation of β_2-adrenoceptors may contribute to increases in heart rate or force of contraction in human heart.

Subtype selective decrease in beta-1 adrenoceptors

In addition, recent evidence suggests that cardiac β_2-adrenoreceptors may play an important role in severe heart failure. It is generally believed that in patients suffering from heart failure an increase in the activity of the sympathetic nervous system seems to be a compensatory mechanism to support the failing heart [19, 20]. In patients with congestive cardiomyopathy plasma noradrenaline levels are elevated [15, 31, 21, 16] and cardiac noradrenaline spillover rate to plasma is exaggerated [24]. Thus, the heart is chronically exposed to elevated (endogenous) noradrenaline concentrations. This can lead to a desensitization of cardiac β-adrenoceptors, as was first demonstrated by Bristow et al. [4]. Noradrenaline, however, is a rather selective β_1-adrenoceptor agonist that has an about 10–30 times higher affinity to β_1- than β_2-adrenoceptors [27]. Taking into account that β_1- and β_2-adrenoceptors undergo a subtype selective regulation by β-adrenoceptor agonists and -antagonists [30, 9, 10, 29] it might be possible that the decrease in cardiac β-adrenoceptors in end-stage congestive cardiomyopathy is due to a selective reduction in cardiac β_1-adrenoceptors. This seems to be indeed the case: Bristow et al. [3] and we ([11]; see Fig. 3) recently demonstrated that in patients with end-stage congestive cardiomyopathy (NYHA class IV) in right atria as well as in right and left ventricles there is a selective decrease in β_1-adrenoceptor density, while β_2-adrenoceptors are only marginally af-

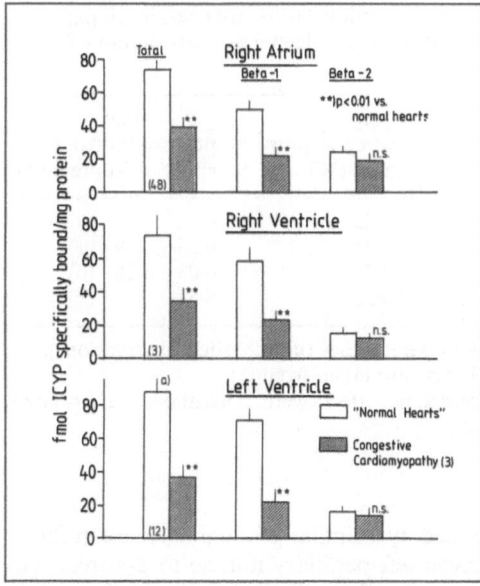

Fig. 3. Human myocardial β-adrenoceptors: comparison of total β-, β_1-, and β_2-adrenoceptor densities in membranes of right atria, as well as of right and left ventricles from normally functioning hearts with those derived from three patients with end-stage congestive cardiomyopathy. Ordinate: total β-, β_1-, and β_2-adrenoceptor density in fmol ICYP specifically bound/mg protein. Given are means $\pm$ SEM; number of experiments in parentheses. (From [11] with slight modifications; data for normal left ventricles are taken from [3])

fected. This subtype selective decrease in β_1-adrenoceptors was accompanied by a similar reduction in the contractile response of isolated right papillary muscles to β_1-adrenoceptor stimulation: the positive inotropic effect of isoprenaline (mediated by β_1- and β_2-adrenoceptors) was markedly reduced and that of the selective β_1-adrenoceptor (partial) agonist denopamine nearly abolished. On the other hand, the increase in contractile force evoked by the selective β_2-adrenoceptor agonist zinterol was not significantly different from that in non-failing normal hearts [3]. Under these conditions the positive inotropic effect of zinterol amounted to 60% of the maximum isoprenaline-effect, while in non-failing hearts it was only 40%. It might be possible, therefore, that in the severely failing human heart cardiac β_2-adrenoceptors may substitute for the loss of β_1-adrenoceptors in order to maintain (at least a part of) the contractility.

Beta-1 and beta-2 adrenoceptors in mitral valve disease

In order to find out whether, in general, in advanced heart failure β-adrenoceptor density is decreased and if this is due to a selective reduction in cardiac β_1-adrenoceptors, we determined left ventricular β-adrenoceptor number, subtype distribution, and function (contractile response to isoprenaline) in patients undergoing mitral valve replacement due to mitral stenosis or mitral regurgitation (NYHA class III to IV). β-adrenoceptor density was determined by $(-)$-$[^{125}I]$iodocyanopindolol (ICYP) binding, β-adrenoceptor subtype distribution by computer analysis of ICI-118,551 competition curves with ICYP binding using the iterative curve fitting program LIGAND (for details see [29]). In these patients left ventricular β-adrenoceptor density gradually declined when the degree of heart failure increased from NYHA functional class III to IV (Table 2). In patients with NYHA class III, β-adrenoceptor density was reduced by more than 50% when compared with recently reported data obtained in non-failing hearts [3]. This gradual de-

Table 2. β-Adrenoceptor density, subtype-distribution and functional responsiveness in left papillary muscles from patients with mitral valve disease in relation to the degree of heart failure (NYHA functional class)

	β-Adrenoceptor density (fmol ICYP bound/mg protein)	$\beta_1:\beta_2$-Adrenoceptor ratio (%)	pD$_2$-Value[b] for the positive inotropic effect of isoprenaline
Control[a]	88.0±7.3 (12)	77:23	–
NYHA class III	26.9±2.9 (15)	76:24	6.36±0.09 (11)
NYHA class III–IV	19.9±1.8 (4)	77:23	6.03±0.18 (5)
NYHA class IV	14.0±3.8 (5)	78:22	5.53 (1)

Given are means ± S.E.M. Number of experiments in parentheses (unpublished observations).
[a] Data for normal left ventricles from non-failing hearts are taken from [3].
[b] pD$_2$-Values for isoprenaline-induced positive inotropic effect were determined on isolated electrically driven (1.0 Hz, 37 °C) papillary muscles.

crease in β-adrenoceptor density was accompanied by a similar gradual decrease in functional response of the isolated, electrically driven left papillary muscle to β-adrenergic stimulation: the pD$_2$-value for the positive inotropic effect of isoprenaline declined with the increase in NYHA functional class (Table 2). However, the left ventricular $\beta_1:\beta_2$-adrenoceptor ratio was relatively constant, irrespective of the degree of heart failure and it was not different from that obtained in non-failing hearts (Table 2). Therefore, in contrast to congestive cardiomyopathy (see above) in mitral valve disease the decrease in left ventricular β-adrenoceptor function seems to be due to a concomitant loss of β_1- *and β_2*-adrenoceptors.

Beta-1 and beta-2 adrenoceptor stimulation by dopexamine

β-Adrenoceptor agonists have been widely studied as positive inotropic agents in the treatment of patients suffering from congestive cardiomyopathy. While the therapeutic use of the catecholamines isoprenaline, adrenaline, and noradrenaline is of very limited value, dopamine is well established in the management of acute and chronic heart failure [17]. Dopamine causes its positive inotropic effects through stimulation of cardiac β_1- adrenoceptors; in addition, at low doses it stimulates renal vascular DA$_1$-receptors, thereby increasing renal blood flow and promoting diuresis [23]. However, at higher doses dopamine stimulates α_1-adrenoceptors resulting in increased peripheral vasoconstriction. This vasoconstrictor effect limits the usefulness of dopamine as a positive inotropic agent in patients with severe heart failure. Recently, dopexamine, a structural analogue of dopamine, was introduced. In vitro and in vivo experiments have shown that this drug activates postjunctional DA$_1$-receptors, has substantial β_2-adrenoceptor-agonist-activity, has no β_1-adrenoceptor-agonist-activity, and unlike dopamine is inactive at α-adrenoceptors [13, 14]. It was, therefore, of interest to study the properties of dopexamine at human cardiac β_1- and β_2-adrenoceptors.

To determine the affinity of dopexamine to human β_1- and β_2-adrenoceptors we assessed inhibition of ICYP binding to human right atrial membranes in the presence of 50 nM ICI-118,551 (= homogeneous population of β_1-adrenoceptors) or in the presence of 300 nM GCP-20712 A (= homogeneous population of β_2-adrenoceptors) and to hu-

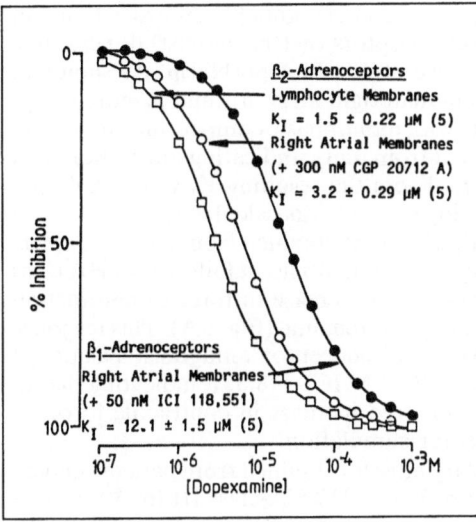

Fig. 4. Inhibition of ICYP binding to human right atrial membranes (derived from patients undergoing coronary artery bypass grafting with mild heart failure, NYHA class I–II) and human lymphocyte membranes by do-pexamine. Membranes were incubated with ICYP (50 pM) and 15–18 concentrations of dopexamine in the presence of 10 μM Gpp(NH)p and specific binding was deter-mined. 100% Inhibition refers to inhibition of specific binding by 1 μM (±)-GCP 12177. Given are means of five experiments with an SEM <10% (unpublished observations)

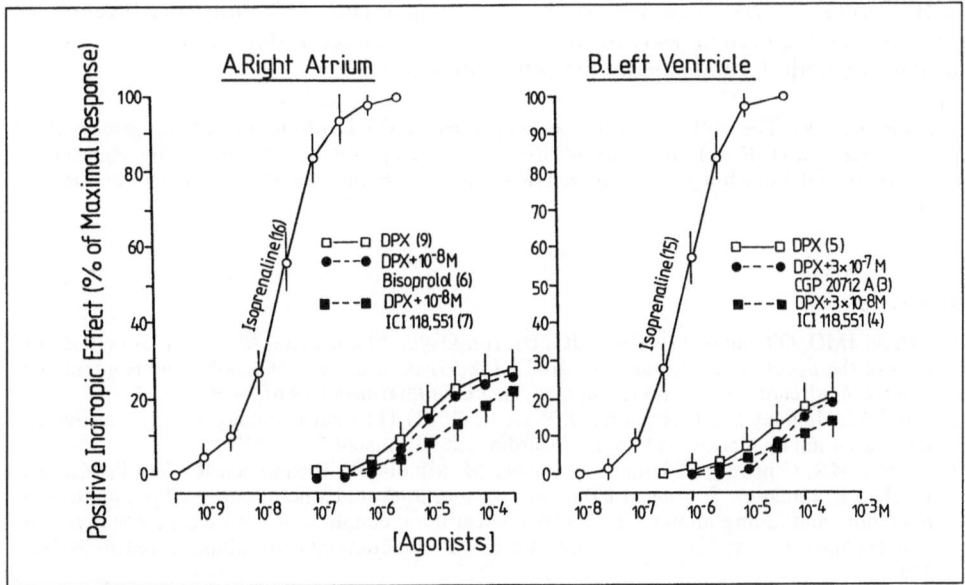

Fig. 5. Positive inotropic effect of dopexamine (DPX) in relation to isoprenaline on isolated electri-cally driven (A) right atria derived from patients undergoing coronary artery bypass grafting with very mild heart failure (NYHA class I–II) and (B) left papillary muscles derived from patients with mitral valve disease (NYHA class III). Ordinate: positive inotropic effect in percent of maximal re-sponse induced by saturating concentrations of isoprenaline. Given are means ± SEM; number of experiments in parentheses (unpublished observations)

man lymphocyte membranes that contain only β_2-adrenoceptors [9]. In order to avoid induction of the high affinity state of the β-adrenoceptors by (the agonist) dopexamine these experiments were performed in the presence of 10 µM Gpp(NH)p. As shown in Fig. 4, dopexamine inhibited ICYP binding to human right atrial β_1- and β_2-adrenoceptors as well as to lymphocytic β_2-adrenoceptors with monophasic competition curves and pseudo-Hill coefficients not significantly different from unity, indicating interaction with a homogeneous population of β-adrenoceptors. From the resulting K_I-values a $\beta_1 : \beta_2$-adrenoceptor ratio of approximately 5–10 for dopexamine was calculated.

On the isolated electrically driven human right atrium obtained from patients undergoing coronary artery bypass grafting with very mild (if at all) heart failure (NYHA class I to II) dopexamine produced a concentration-related increase in force of contraction reaching 25–30% of the peak response produced by isoprenaline (Fig. 5A). This response was only marginally affected by the selective β_1-adrenoceptor antagonist bisoprolol (10^{-8} M) [7] but was significantly antagonized by 10^{-8} M ICI-118,551, indicating that in the (nearly) non-failing heart the dopexamine-induced increase in contractile force on right atria is mediated mainly by β_2-adrenoceptor stimulation.

On the isolated electrically driven left papillary muscle (obtained from patients undergoing mitral valve replacement with severe heart failure, NYHA-class III to IV) the dopexamine-induced increase in contractile force was less than in atria and was related to the degree of heart failure: in patients with NYHA class III it amounted to about 20% of the maximum isoprenaline effect (Fig. 5B), while in patients with NYHA class III–IV it was only about 10–15%. With this preparation 3×10^{-8} M ICI-118,551 shifted mainly the lower part of the concentration-response curve of dopexamine to the right, while 3×10^{-7} M GCP-20712 A shifted mainly the upper part (Fig. 5B). Thus, on left ventricles of severely failing human heart the positive inotropic effect of dopexamine seems to be mediated by both β_1- and β_2-adrenoceptor stimulation.

Acknowledgements. The authors' work was supported by the Deutsche Forschungsgemeinschaft (DFG Ze 218/1-3; H.-R. Z.), the Sandoz-Stiftung für Therapeutische Forschung, the Minister für Wissenschaft und Forschung des Landes NRW, and the Fonds der Chemischen Industrie (O.-E. B.).

References

1. Arnold JMO, O'Connor PC, Ridell JG, Harron DWG, Shanks RG, McDevitt DG (1985) Effects of the β_2-adrenoceptor antagonist ICI 118,551 on exercise tachycardia and isoprenaline-induced β-adrenoceptor responses in man. Br J Clin Pharmacol 19:619–630
2. Bilski AJ, Halliday SE, Fitzgerald JD, Wale JL (1983) The pharmacology of a β_2-selective adrenoceptor antagonist (ICI-118,551). J Cardiovasc Pharmacol 5:430–437
3. Bristow MR, Ginsburg R, Umans V, Fowler M, Minobe W, Rasmussen R, Zera P, Menlove R, Shah P, Jamieson S, Stinson EB (1986) β_1- and β_2-Adrenergic-receptor subpopulations in nonfailing and failing human ventricular myocardium: coupling of both receptor subtypes to muscle contraction and selective β_1-receptor down-regulation in heart failure. Circ Res 59:297–309
4. Bristow MR, Ginsburg R, Minobe W, Cubicciotti RS, Sageman S, Lurie K, Billingham ME, Harrison DC, Stinson EB (1982) Decreased catecholamine sensitivity and β-adrenergic-receptor density in failing human hearts. N Engl J Med 307:205–211
5. Brodde O-E, O'Hara N, Zerkowski H-R, Rohm N (1984) Human cardiac β-adrenoceptors: both β_1- and β_2-adrenoceptors are functionally coupled to the adenylate cyclase in right atrium. J Cardiovasc Pharmacol 6:1184–1191
6. Brodde O-E (1987) Cardiac beta-adrenergic receptors. ISI Atlas Sci: Pharmacology 1:107–112

7. Brodde O-E (1986) Bisoprolol (EMD 33512), a highly selective β_1-adrenoceptor antagonist: in vitro and in vivo studies. J Cardiovasc Pharmacol 8 (Suppl 11):S29–S35

8. Brodde O-E, Daul A, Wellstein A, Palm D, Michel MC, Beckeringh JJ (1988) Differentiation of β_1- and β_2-adrenoceptor-mediated effects in humans. Am J Physiol 254:H199–H206

9. Brodde O-E, Beckeringh JJ, Michel MC (1987) Human heart β-adrenoceptors: a fair comparison with lymphocyte β-adrenoceptors? Trends Pharmacol Sci 8:403–407

10. Brodde O-E (1988) The functional importance of beta$_1$ and beta$_2$ adrenoceptors in the human heart. Am J Cardiol 62:24C–29C

11. Brodde O-E, Schüler S, Kretsch R, Brinkmann M, Borst HG, Hetzer R, Reidemeister JChr, Warnecke H, Zerkowski H-R (1986) Regional distribution of β-adrenoceptors in the human heart: coexistence of functinoal β_1- and β_2-adrenoceptors in both atria and ventricles in severe congestive cardiomyopathy. J Cardiovasc Pharmacol 8:1235–1242

12. Brown JE, McLeod AA, Shand DG (1986) In support of cardiac chronotropic beta$_2$ adrenoceptors. Am J Cardiol 57:11F–16F

13. Brown RA, Dixon J, Farmer JB, Hall JC, Humphries Rg,'Ince F, O'Connor SE, Simpson WT, Smith GW (1985) Dopexamine: a novel agonist at peripheral dopamine receptors and β_2-adrenoceptors. Br J Pharmacol 85:599–608

14. Brown RA, Farmer JB, Hall JC, Humphries RG, O'Connor SE, Smith GW (1985) The effects of dopexamine on the cardiovascular system of the dog. Br J Pharmacol 85:609–619

15. Chidsey CA, Harrison DC, Braunwald E (1962) Augmentation of the plasma norepinephrine response to exercise in patients with congestive heart failure. N Engl J Med 267:650–654

16. Cohn JN, Levine TB, Olivari MT, Garberg V, Lura D, Francis GS, Simon AB, Rector T (1984) Plasma norepinephrine as a guide of prognosis in patients with chronic congestive heart failure. N Engl J Med 311:819–823

17. Colucci WS, Wright RF, Braunwald E (1986) New positive inotropic agents in the treatment of congestive heart failure. N Engl J Med 314:290–299

18. Dooley DJ, Bittiger H, Reymann NC (1986) CGP 20712 A: an useful tool for quantitating β_1- and β_2-adrenoceptors. Eur J Pharmacol 130:137–139

19. Francis GS (1985) Neurohumoral mechanisms involved in congestive heart failure. Am J Cardiol 55:15A–21A

20. Francis GS, Cohn JN (1986) The autonomic nervous system in congestive heart failure. Annu Rev Med 37:235–247

21. Francis GS, Goldsmith SR, Ziesche SM, Cohn JN (1982) Response of plasma norepinephrine and epinephrine to dynamic exercise in patients with congestive heart failure. Am J Cardiol 49:1152–1156

22. Gille E, Lemoine H, Ehle B, Kaumann AJ (1985) The affinity of (−)-propranolol for β_1- and β_2-adrenoceptors in human heart. Differential antagonism of the positive inotropic effects and adenylate cyclase stimulation by (−)-noradrenaline and (−)-adrenaline. Naunyn-Schmiedeberg's Arch Pharmacol 331:60–70

23. Goldberg LI (1974) Dopamine – clinical uses of an endogenous catecholamine. N Engl J Med 291:707–710

24. Hasking GJ, Esler MD, Jennings GL, Burton D, Johns JA, Korner PI (1986) Norepinephrine spillover to plasma in patients with congestive heart failure: evidence of increased overall and cardiorenal sympathetic nervous activity. Circulation 73:615–621

25. Ikezono K, Michel MC, Zerkowski H-R, Beckeringh JJ, Brodde O-E (1987) The role of cyclic AMP in the positive inotropic effect mediated by β_1- and β_2-adrenoceptors in isolated human right atrium. Naunyn-Schmiedeberg's Arch Pharmacol 335:561–566

26. Kaumann AJ, Lemoine H (1987) β_2-Adrenoceptor-mediated positive inotropic effect of adrenaline in human ventricular myocardium. Quantitative discrepancies with binding and adenylate cyclase stimulation. Naunyn-Schmiedeberg's Arch Pharmacol 335:403–411

27. Lands AM, Arnold A, McAuliff JP, Luduena FP, Brown TG (1967) Differentiation of receptor systems activated by sympathomimetic amines. Nature 214:597–598

28. Lemoine H, Schönell H, Kaumann AJ (1988) Contribution of β_1- and β_2-adrenoceptors of human atrium and ventricle to the effects of noradrenaline and adrenaline as assessed with (−)-atenolol. Br J Pharmacol 95:55–66

29. Michel MC, Pingsmann A, Beckeringh JJ, Zerkowski H-R, Doetsch N, Brodde O-E (1987) Selective regulation of β_1- and β_2-adrenoceptors in the human heart by chronic β-adrenoceptor antagonist treatment. Br J Pharmacol 94:685–692
30. O'Donnell SR, Wanstall JC (1987) Functional evidence for differential regulation of β-adrenoceptor subtypes. Trends Pharmacol Sci 8:265–268
31. Thomas JA, Marks BH (1978) Plasma norepinephrine in congestive heart failure. Am J Cardiol 41:233–243
32. Zerkowski H-R, Ikezono K, Rohm N, Reidemeister JChr, Brodde O-E (1986) Human myocardial β-adrenoceptors: demonstration of both β_1- and β_2-adrenoceptors mediating contractile responses to β-agonists on the isolated right atrium. Naunyn-Schmiedeberg's Arch Pharmacol 332:142–147

Author's address:

Dr. Otto-Erich Brodde, Biochemisches Forschungslabor, Medizinische Klinik & Poliklinik, Abtlg. für Nieren- & Hochdruckkranke, Universitätsklinikum, Hufelandstrasse 55, D-4300 Essen 1, FRG

III. Clinical experience
with positive inotropic substances

Influence of isoproterenol on myocardial energetics. Experimental and clinical investigations

G. Hasenfuss[1], C. Holubarsch[1], E. M. Blanchard[2], L. A. Mulieri[2], N. R. Alpert[2], Hj. Just[1]

[1] Medizinische Universitätsklinik, Innere Medizin III, Kardiologie, Universität Freiburg, FRG
[2] Department of Physiology and Biophysics, University of Vermont, Burlington, Vermont, USA

Summary

The influence of isoproterenol on myocardial performance and energetics was investigated in normal guinea pig myocardium and in patients with normal left ventricular function.

The in vitro experiments were performed by simultaneous isometric force and heat measurements using sensitive antimony-bismuth thermopiles. Following the application of isoproterenol (10^{-8} M) isometric peak twitch tension and tension-time integral increased significantly by 185% and 142%, respectively. Tension-independent heat which reflects high energy phosphate hydrolysis of excitation-contraction coupling increased by 183%. Tension-dependent heat reflecting the high energy phosphate hydrolysis of the crossbridges increased by 417%. The ratio of tension-dependent heat to tension-time integral increased by 131%. The recovery/initial heat ratio, reflecting the efficiency of the recovery metabolism, and the resting metabolism did not significantly change.

In the patients the effect of isoproterenol on myocardial energetics was evaluated in terms of myocardial oxygen consumption per left ventricular systolic stress-time integral and external myocardial efficiency. Following isoproterenol administration, left ventricular systolic stress-time integral decreased by 49% due to reductions in end-diastolic pressure, end-diastolic volume and duration of systole. Pressure-volume work remained unchanged. Myocardial oxygen consumption per minute increased in proportion to heart rate. The ratio of myocardial oxygen consumption per beat to left ventricular systolic stress-time integral increased significantly by 95%. External myocardial efficiency was unaltered.

Thus, isoproterenol increases the energy turnover of excitation-contraction coupling and increases the energy consumption of the crossbridges disproportionately to developed tension-time integral in the guinea pig heart. Likewise, in the working human heart, the increase in oxygen consumption per left ventricular systolic stress-time integral is considered to represent the isoproterenol induced changes in excitation contraction coupling and crossbridge energetics.

Introduction

Stimulation of myocardial beta-receptors increases intracellular concentration of cyclic AMP effecting excitation contraction coupling and contraction. Cyclic AMP as a second messenger, increases calcium influx through sarcolemmal calcium channels, increases calcium release and the rate of reuptake by the sarcoplasmic reticulum and reduces the sensitivity of troponin C to calcium ions [1, 5, 13, 22, 24]. As a consequence, these changes in excitation-contraction coupling may increase the total number of crossbridges recruited, as well as the rate of crossbridge activation and deactivation. In addition, cyclic

AMP may exert direct effects on the contractile proteins, resulting in increased cycling frequency of the cross-bridges [12] and an altered crossbridge duty-cyle [8, 9].

The effects of beta-receptor stimulation on excitation-contraction coupling and the contractile proteins are expected to increase myocardial energy consumption and to decrease myocardial efficiency and economy. In addition, catecholamines may influence recovery and/or resting metabolism [14].

Although catecholamines have been extensively investigated in experimental and clinical studies, their influence on myocardial energetics is still a matter of controversy [6–9, 14, 15, 17, 23, 26, 27, 29]. Some previous studies evaluating the effects of isoproterenol on work and energy consumption of the myocardium revealed that efficiency is not significantly changed [6, 7, 15, 27]. On the other hand, myocardial oxygen consumption was shown to increase disproportionately to pressure-time index or tension-time integral, suggesting that isoproterenol decreases the economy of force production [8, 15, 23].

The purpose of the present investigations was: 1) to evaluate the influence of isoproterenol on the energy consumption of isometric force development, on energetics of excitation-contraction coupling, and on recovery and resting metabolism under in vitro conditions; and 2) to evaluate the influence if isoproterenol on the ratio of myocardial oxygen consumption to systolic stress-time integral and on external myocardial efficiency in the intact human heart. The animal experiments were carried out in right ventricular guinea pig papillary muscles by means of isometric heat and force measurements. This method facilitates detailed analysis of energetics of subcellular systems. Guinea pig myocardium was chosen because it shows a pronounced inotropic response to isoproterenol. This is in contrast to rat myocardium in which the effects of isoproterenol have been investigated previously [8]. The clinical data were obtained in patients with normal left ventricular function undergoing cardiac catheterization.

Methods

Animal experiments

The myothermal experiments were performed in right ventricular guinea pig papillary muscles ($n = 5$) using sensitive antimony-bismuth thermopiles [19]. The papillary muscles were removed and attached to the apparatus consisting of the thermopile and an isometric force transducer; the muscles were fixed by means of silk loops as described previously [19]. The muscle was in contact with the active region of the thermopile. The muscle and thermopile were then submerged in Krebs-Ringer solution at 21 °C and the muscle was stimulated end-to-end at 0.2 HZ, by means of a square wave pulse (3 ms) 20% above threshold. After a 90-min-equilibrium period the muscle was stretched gradually until maximum twitch force was reached. Peak twitch tension to resting tension ratio was 3.98 ± 2.19. Thereafter, heat and mechanical measurements were performed as described previously [2]. Heat and mechanical measurements were repeated 30 min after the application of isoproterenol 10^{-8} M. The heat liberated from the isometric contracting papillary muscles was partitioned into initial heat, tension-dependent heat, tension-independent heat, recovery heat, and resting heat. Initial heat is liberated during the isometric contraction phase and is composed of tension-independent heat and tension-dependent heat. The former reflects high energy phosphate hydrolysis of excitation contraction coupling, the latter of crossbridge interaction. The partitioning of initial heat was performed by the shortening method [7]. Recovery heat, for the most part, reflects the resynthesis of high energy phosphates. Resting heat reflects the energy turnover for maintaining the cellular integrity of the resting cell (ionic and chemical gradients, protein synthesis as well as futile crossbridge cycling) [3].

Clinical investigations

Patients and study protocol

The influence of isoproterenol on left ventricular hemodynamics and myocardial oxygen consumption was investigated in 10 patients with atypical chest pain undergoing routine diagnostic cardiac catheterization. Only patients in whom coronary artery disease could be excluded by coronary angiography were included in the study. The study group consisted of two women and eight men (mean age 48 ± 8 years). All patients were in sinus rhythm. Previous medication – nifedipine in 5 patients and sotalol in two patients – was withheld 48 h before the study. Written informed consent was obtained from the patients.

After coronary angiography had been performed, myocardial blood flow was measured by the argon method and blood samples were taken from the aorta and the coronary sinus for oxygen saturation measurements. Thereafter left ventricular angiography with simultaneous pressure measurement was done using Millar microtip catheters. Upon completion of basal measurements isoproterenol was infused intravenously beginning with a dose of 1 µg/min. The dose was increased by 1 µg/min at intervals of 3 min until heart rate had increased by 50% (2.7 ± 0.7 µg/min isoproterenol). Hemodynamics and myocardial blood flow measurements were repeated during steady state conditions.

Hemodynamic measurements

Left ventricular angiography at 50 frames per second was performed by power injection of 40 ml nonionic contrast solution. The projection was a 10° caudally angulated 45° right anterior oblique view. Left ventricular pressure was measured simultaneously. Maximum rate of left ventricular pressure rise was determined from continuous differentiation of left ventricular pressure tracings. During one cardiac cycle, left ventricular volumes were calculated from each cine frame at intervals of 20 ms using the Sandler and Dodge method [25]. Left ventricular wall thickness and muscle mass were obtained according to a modification of the method of Rackley et al. [20]. Instantaneous circumferential wall-stress values were calculated using the ellipsoid model of Mirsky [18]. The systolic stress-time integral was calculated by integrating instantaneous stress values from the time of end-diastole to the time of the last cine frame before left ventricular volume increased. Pressure-volume work was calculated as the area of the pressure-volume loop obtained by relating instantaneous pressure to volume every 20 ms and normalized for 100 g wall mass.

Myocardial energetics

Myocardial blood flow was measured by the argon method [21]. Myocardial oxygen consumption was determined as the product of myocardial blood flow times arterial-coronary-sinus oxygen content difference. Myocardial oxygen consumption per beat was calculated from myocardial oxygen consumption per minute and heart rate. The ratio of myocardial oxygen consumption to systolic stress-time integral was calculated by dividing myocardial oxygen consumption per beat by left ventricular systolic stress-time integral. External myocardial efficiency was obtained as the ratio of pressure-volume work (in cal/g) to myocardial oxygen consumption (in cal/g) assuming that 1 mmHg·ml = 31.79 µcal and 1 l O_2 = 5 kcal [7].

Values are expressed as mean ± standard deviation. Comparisons between measurements before and after application of isoproterenol were made using the paired t-test. Probability values of less than 0.05 were accepted as significant.

Results

Animal experiments

Following the application of isoproterenol, peak twitch tension and tension-time integral increased significantly from 1.36 ± 0.52 to 3.88 ± 0.33 g/mm^2 and 0.95 ± 0.24 to 2.30 ± 0.16 g·s/mm^2, respectively. Initial heat increased from 0.43 ± 0.12 to 1.87 ± 0.20 mcal/g. Partitioning of initial heat into its two components tension-independent heat and tension-dependent heat, revealed an increase of both components from 0.24 ± 0.05 to 0.68 ± 0.13 mcal/g and from 0.23 ± 0.07 to 1.19 ± 0.23 mcal/g, respectively. The ratio of tension-dependent heat to tension-time integral increased from 2.24 ± 0.60 to 5.18 ± 0.89 µcal/g·cm·s) (Fig. 1). The recovery/initial heat ratio (1.24 ± 0.23 before and 1.39 ± 0.11 with isoproterenol) and resting heat rate (2.12 ± 0.61 mW/g before and 1.86 ± 0.83 mW/g with isoproterenol) did not significantly change.

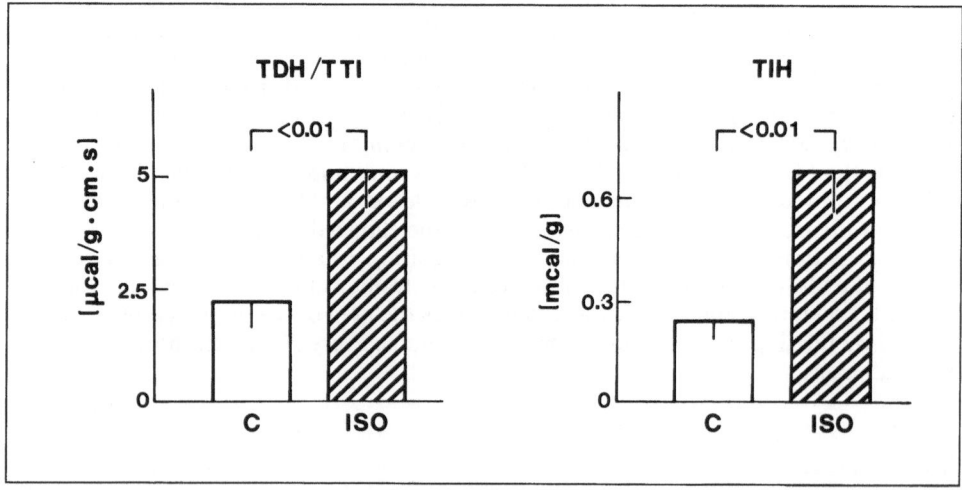

Fig. 1. Influence of isoproterenol on tension-independent heat (TIH) and on tension-dependent heat normalized for tension-time integral (TDH/TTI). C=control; ISO=isoproterenol (10^{-8}) M; n=5

Clinical investigations

During control conditions left ventricular end-diastolic pressure, end-diastolic volume, and ejection fraction were in the normal range. The application of isoproterenol resulted in a significant increase in heart rate and maximum rate of left ventricular pressure rise and in a decrease of left ventricular end-diastolic pressure and volume. Left ventricular systolic stress-time integral decreased by 49%. Left ventricular peak systolic wall stress and left ventricular pressure-volume work did not significantly change (Table 1).

Myocardial blood flow and myocardial oxygen consumption per minute increased significantly from 85 ± 22 to 161 ± 45 and from 10.8 ± 2.9 to 16.2 ± 3.9 ml/min/100 g, respectively, following the application of isoproterenol. Myocardial oxygen consumption per beat did not significantly change (145 ± 21 before and 137 ± 34 with isoproterenol) (Fig. 2). Isoproterenol increased the ratio of myocardial oxygen consumption per beat to

Table 1. Influence of isoproterenol on left ventricular hemodynamics in patients with normal left ventricular function

Parameter	Control	Isoproterenol	p-Value
HR (beats/min)	74 ± 14	119 ± 16	<0.001
dp/dt_{max} (mmHg/s)	1488 ± 216	3129 ± 793	<0.001
LVEDP (mmHg)	10 ± 5	7 ± 3	<0.05
LVEDV (ml)	174 ± 51	138 ± 49	<0.001
EF (%)	76 ± 7	80 ± 9	>0.05
LVS_{peak} (10^3 dyn/cm^2)	236 ± 24	218 ± 71	>0.05
LVSTI (10^3 dyn $\cdot$ s/cm^2)	73 ± 7	37 ± 12	<0.001
LVPV-work (mHg $\cdot$ l/100 g)	6.4 ± 1.4	5.3 ± 1.6	>0.05

dp/dt_{max} = maximum rate of left ventricular pressure rise; EF = ejection fraction; HR = heart rate; LVEDP = left ventricular enddiastolic pressure; LVEDV = left ventricular end-diastolic volume; LVPV-work = left ventricular pressure-volume work; LVS_{peak} = left ventricular peak systolic wall stress; LVSTI = left ventricular systolic stress-time integral.

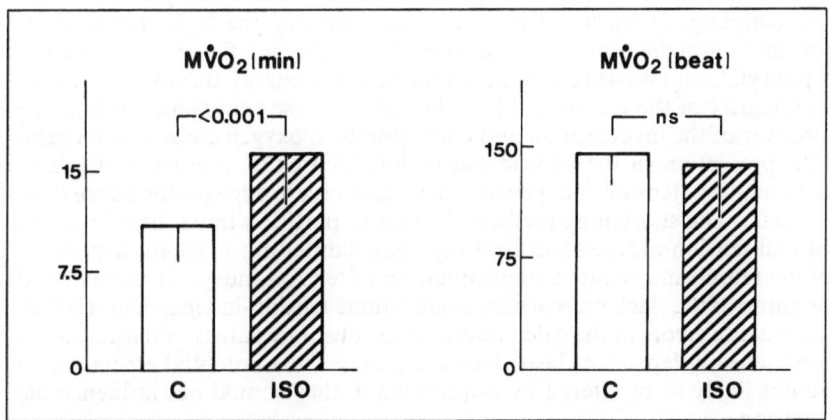

Fig. 2. Influence of isoproterenol on myocardial oxygen consumption per minute (MV̇O$_2$[min]) and per beat (MV̇O$_2$[beat]. C = control; ISO = isoproterenol)

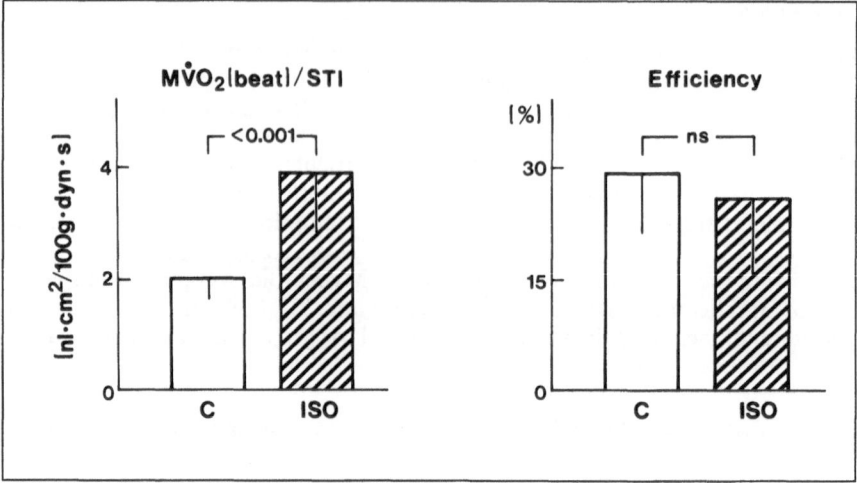

Fig. 3. Influence of isoproterenol on the ratio of myocardial oxygen consumption per beat to left ventricular systolic stress-time integral (left panel) and on external myocardial efficiency (right panel). C = control; ISO = isoproterenol

systolic stress-time integral from 2.0 ± 0.35 to 3.9 ± 1.1 nl·cm^2/100 g·dyn·s. External myocardial efficiency did not significantly change (29 ± 8 before and $26 \pm 10\%$ with isoproterenol) (Fig. 3).

Discussion

Heat production of isometrically contracting guinea pig papillary muscles was measured by means of antimony-bismuth thermopiles simultaneously with developed force. Since no external work is performed in isometric contractions all the energy is liberated as heat representing the following subcellular energy consuming and energy releasing reactions: 1) tension-independent heat reflecting the high energy phosphate hydrolysis for excitation-contraction coupling: 2) tension-dependent heat reflecting the high energy phosphate hydrolysis of contractile proteins: 3) recovery heat reflecting the energy release of oxidative phosphorylation; and 4) resting heat reflecting the energy turnover for maintaining cellular integrity of the resting cell [3]. The sum of these heat values reflects the total energy turnover of the myocardium and corresponds to oxygen consumption measurements. In the present paper the interest was mainly focused on energetics of excitation-contraction coupling (tension-independent heat) and crossbridge performance (tension-dependent heat). The shortening method [7] used to partition initial heat into tension-dependent and tension-independent heat has some limitation: 1) length dependent deactivation of the muscle may cause underestimation of tension-independent heat; and 2) internal shortening of the slack muscle may overestimate tension-independent heat [4]. The artefacts may cause errors in the calculations of the absolute values of tension-independent heat and tension-dependent heat. However, since these potential errors are so diverse and are not likely to be altered by isoproterenol, they should not influence the present interpretation.

Isoproterenol increased tension-independent heat by a factor of 3 and tension-dependent heat by a factor of 5. The increase in tension-independent heat indicates that isopro-

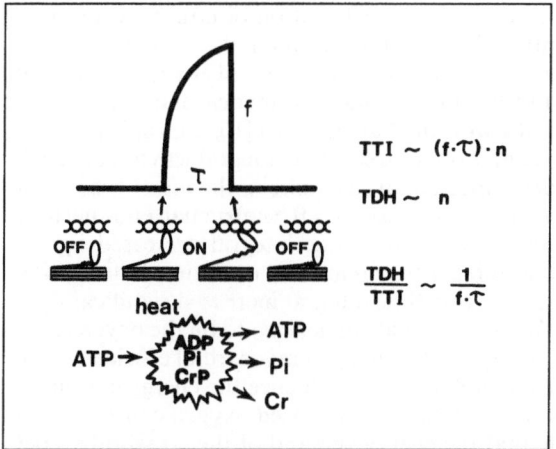

$$TTI \sim (f \cdot \tau) \cdot n$$

$$TDH \sim n$$

$$\frac{TDH}{TTI} \sim \frac{1}{f \cdot \tau}$$

Fig. 4. Schematic of one crossbridge cycle consisting of the "on-time" (τ) during which the crossbridge produces force (f) and the "off-time" during which no force is produced. One molecule of high energy phosphates is hydrolyzed during one crossbridge cycle. Twitch tension-time integral (TTI) results from the summation of the force-time integrals of all individual crossbridge interactions during one twitch. Tension-dependent heat (TDH) represents the number of high energy phosphate molecules hydrolyzed during one twitch, and thus, the number of all crossbridge cycles during one twitch (n). The ratio of TDH/TTI is inversely proportional to the force-time integral of the single crossbridge

terenol tripled total calcium turnover and this can be attributed to increased calcium influx, increased calcium release and reuptake by the sarcoplasmic reticulum and decreased calcium-sensitivity of the contractile proteins [1, 5, 13, 22, 24]. The increase in tension-dependent heat with isoproterenol indicates an increase in high energy phosphate consumption of the contractile proteins by a factor of 5.

Tension-dependent heat is considered to reflect the high energy phosphate hydrolysis of all crossbridge interactions during one twitch and isometric twitch tension-time is the summation of the force-time integrals of all individual cross-bridge interactions (Fig. 4). Assuming that the high energy phosphate consumption per crossbridge cycle and the enthalpy of high-energy phosphate hydrolysis are constant, the ratio of tension-dependent heat to tension-time integral is inversely related to the force-time integral of the single crossbridge interaction. Therefore, the increase in tension-dependent heat per tension-time integral indicates a decrease in the force-time integral of the single cross-bridge interaction. A reduced crossbridge force-time integral may result from reduced crossbridge force or reduced duration of force production. The latter may result in an abbreviation of the crossbridge cycle which would be in accord with the recent findings of Hoh et al. who observed an increased rate of cycling of crossbridges following catecholamine application [12].

In patients with normal left ventricular function the influence of isoproterenol on myocardial energetics was investigated by relating myocardial oxygen consumption per beat to pressure-volume work and to left ventricular systolic stress-time integral. The ratio of pressure-volume work to myocardial oxygen consumption describes the efficiency of the left ventricle as a pump (external myocardial efficiency). Compared to control, with isoproterenol a similar amount of pressure-volume work was performed during a shorter ejection period while myocardial oxygen consumption per beat remained unchanged. Accordingly, since ejection-time is not considered in the calculation of efficiency, external myocardial efficiency did not significantly change with isoproterenol.

Stress-time integral reflects the force per unit crossection area of myocardium over the duration of systole and was shown to be closely related to myocardial oxygen consumption in experimental studies and in the working human heart [10, 15, 16, 28]. Therefore, the ratio of myocardial oxygen consumption to stress-time integral was used to evaluate the energy consumption per unit developed auxotonic force in the intact heart. Isoproterenol reduced the duration of systole and decreased left ventricular end-diastolic pressure and end-diastolic volume. Thus, systolic stress-time integral decreased significantly. Since myocardial oxygen consumption per beat remained unchanged the ratio of myocardial oxygen consumption to systolic stress-time integral increased significantly.

The different effects of isoproterenol on myocardial efficiency and on the oxygen consumption-stress-time integral ratio are intriguing. Unchanged myocardial efficiency with isoproterenol would indicate that isoproterenol does not influence the energy consumption per unit external crossbridge work. In contrast, the increased oxygen consumption-stress-time integral ratio would indicate that the energy demand of the crossbridges per unit developed force is increased. However, we hesitate to draw the conclusion that isoproterenol has different effects on force and external work production by crossbridges. In accord with the myothermal data obtained in guinea pig myocardium we assume that the increased oxygen-stress-time integral ratio represents the isoproterenol-induced changes in excitation contraction coupling and crossbridge energetics. Efficiency as calculated by the ratio of pressure-volume work to myocardial oxygen consumption describes the efficiency of the left ventricle as a pump. Since pressure-volume work is not closely related to myocardial oxygen consumption [10] and since efficiency depends on preload and afterload, both of which are altered by isoproterenol, the ratio of pressure-volume work to myocardial oxygen consumption may be insensitive to isoproterenol induced changes in the efficiency of the crossbridges.

Myothermal measurements with beat to beat evaluation of tension-dependent heat, work, and force are required to compare the effects of isoproterenol on crossbridge economy and efficiency.

Acknowledgements. The in vitro experiments of this paper were supported by PHS Grant No. 28001-07.

References

1. Allen DG, Kurihara S (1980) Calcium transients in mammalian ventricular muscle. Eur Heart J 1:5–15
2. Alpert NR, Mulieri LA (1982) Increased myothermal economy of isometric force generation in compensated cardiac hypertrophy induced by pulmonary artery constriction in the rabbit. Circ Res 50:491–500
3. Alpert NR, Mulieri LA (1984) Hypertrophic adaptation of the heart to stress: A myothermal analysis. In: Zak R (ed) Growth of the heart in health and disease. Raven Press, New York, pp 363–379
4. Alpert NR, Blanchard EM, Mulieri LA (1989) Tension-independent heat in rabbit papillary muscle. J Physiol (in press)
5. England PJ (1986) The phosphorylation of cardiac contractile proteins. In: Rupp H (ed) The regulation of heart function. Basic concepts and clinical applications. Thieme, Stuttgart New York, pp 223–233
6. Gibbs CL, Gibson WR (1972) Isoprenaline, propranolol, and the energy output of rabbit cardiac muscle. Cardiovasc Res 6:508–515
7. Gibbs CL (1978) Cardiac energetics. Physiol Rev 58:174–254

8. Hasenfuss G, Holubarsch Ch, Just H, Blanchard E, Mulieri LA, Alpert NR (1987) Energetic aspects of inotropic interventions in rat myocardium. Basic Res Cardiol 82 [Suppl 2]:252–259
9. Hasenfuss G, Holubarsch Ch, Blanchard EM, Mulieri LA, Alpert NR (1989) Influence of inotropic interventions on crossbridge economy during isometric force development in guinea pig papillary muscles. Biophys J 55:44310
10. Hasenfuss G, Holubarsch Ch, Heiss HW, Meinertz Th, Bonzel T, Wais U, Lehmann M, Just H (1989) Myocardial energetics in patients with dilative cardiomyopathy. Influence of nitroprusside and enoximone. Circulation (in press)
11. Holubarsch Ch, Hasenfuss G, Heiss HW, Bonzel T (1987) The relation between myocardial oxygen consumption and systolic stress-development in patients with normal hearts and with idiopathic dilative cardiomyopathy. Circulation 76 [Suppl IV]:161
12. Hoh JFY, Rossmanith GH, Kwan LJ, Hamilton AM (1988) Adrenaline increases the rate of cycling of crossbridges in rat cardiac muscle as measured by pseudo-random binary noise-modulated pertubation analysis. Circ Res 62:452–461
13. Katz A (1983) Cyclic adenosine monophosphate effects on the myocardium: a man who blows hot and cold with one breath. J Am Col Cardiol 2:143–149
14. Klocke FJ, Kaiser GA, Ross JR, Braunwald E (1965) Mechanism of increase of myocardial oxygen uptake produced by catecholamines. Am J Physiol 209:913–918
15. Krasnow N, Rolet EL, Yurchak PM, Hood WB, Gorlin R (1964) Isoproterenol and cardiovascular performance. Am J Med 37:514–525
16. Laskey WK, Reichek N, Sutton MS, Untereker WJ, Hirshfeld JW (1983) Myocardial oxygen consumption in left ventricular hypertrophy and its relation to left ventricular mechanics. Am J Cardiol 52:852–858
17. Lekven J, Brunsting LA, Jessen ME, Abd-Elfatth AS, Doherty N, Wechsler AS (1988) Myocardial oxygen use during epinephrine administration to ischemically injured canine hearts. Circulation 78 [Suppl III]:125–136
18. Mirsky I (1979) Elastic properties of the myocardium: a quantitative approach with physiological and clinical applications. In: Berne RM (ed) Handbook of physiology. The cardiovascular system. American Physiological Society, Washington DC, pp 497–531
19. Mulieri LA, Luhr G, Trefry J, Alpert NR (1977) Metal-film thermopiles for use with rabbit right ventricular papillary muscles. Am J Physiol 2:C146–156
20. Rackley CE, Dodge HT, Coble YD, Haya RE (1964) A method for determining left ventricular mass in man. Circulation 29:666–671
21. Rau G (1969) Messung der Koronardurchblutung mit der Argon-Fremdgasmethode. Arch Kreislauf-Forsch 58:322–398
22. Reuter H, Scholz H (1977) The regulation of calcium conductance of cardiac muscle by adrenaline. J Physiol 264:49–62
23. Rooke GA, Feigl EO (1982) Work as a correlate of canine left ventricular oxygen consumption, and the problem of catecholamine oxygen wasting. Circ Res 50:273–286
24. Rüegg JC (1986) The vertebrate heart: Modulation of calcium control. In Ruegg JC (ed) Calcium in muscle activation. Springer, Berlin Heidelberg New York London Paris Tokyo, pp 165–201
25. Sandler H, Dodge HT (1968) The use of single plane angiocardiograms for the calculation of left ventricular volume in man. Am Heart J 75:325–334
26. Sonnenblick EH, Ross J, Covell JW, Kaiser GH, Braunwald E (1965) Velocity of contraction as a determinant of myocardial oxygen consumption. Am J Physiol 209:919–927
27. Suga H, Hisano R, Goto Y, Yamada O, Igarashi Y (1983) Effect of positive inotropic agents on the relation between oxygen consumption and systolic pressure volume area in canine left ventricle. Circ Res 53:306–318
28. Weber KT, Janicki (1977) Myocardial oxygen consumption: The role of wall force and shortening. Am J Physiol 233:421–430
29. Winegrad S (1984) Regulation of cardiac contractile proteins. Correlation between physiology and biochemistry. Circ Res:565–514

Author's address:

Gerd Hasenfuss, Medizinische Universitätsklinik, Innere Medizin III, Kardiologie, Universität Freiburg, Hugstetter Strasse 55, D-7800 Freiburg, FRG

Effects of long-term xamoterol therapy on the left ventricular mechanical efficiency in patients with ischemic heart disease

H. Pouleur, C. van Eyll, J. Etienne, H. van Mechelen, A. Vuylsteke,
M. F. Rousseau

Departments of Physiology and Cardiology, University of Louvain,
Brussels, Belgium

Summary

Myocardial oxygen uptake and an index of mechanical left ventricular efficiency were determined in basal conditions or during prolonged therapy with the new β_1-adrenoceptor partial agonist xamoterol in 16 patients with mild to moderate ischemic heart failure. During xamoterol therapy, left ventricular end-diastolic pressure decreased from 24.4 ± 6.5 to 17.8 ± 8.6 mm Hg (P<0.01) and the isovolumic index of inotropic state (dP/dt)/DP40 increased by 14% (P<0.01). The heart rate increased slightly and the mean systolic and peak systolic wall stress also tended to increase (+ 7%; NS) but myocardial oxygen uptake (14.1 vs 14.7 ml/min; NS) and the index of efficiency (8.77 ± 3.44 to 8.82 ± 4.27; NS) were not significantly modified. In conclusion, prolonged therapy with xamoterol was not accompanied by a deterioration in the mechanical efficiency of the ventricle, even in patients with ischemic heart disease.

Introduction

Mechanical efficiency is an important variable of ventricular performance. An index of the efficiency of the left ventricle can be calculated by measuring the external work performed and relating it to the oxygen consumption [8]. This approach focuses on the efficiency of the heart as a pump. On the other hand, efficiency can also be evaluated by relating the total force generated by the ventricular walls to the total oxygen consumption [1, 3, 5]. This last approach more closely describes the efficiency of the muscular contraction [5, 22].

The concept of mechanical efficiency becomes particularly important when trying to select optimal therapies for the failing heart. Indeed, more than two decades ago, it was shown that the oxygen cost of augmentation of contractility was substantial [5]. In some animal models of heart failure, inotropic stimulation caused a depletion in high energy phosphate stores [10] which might be related to excessive oxygen cost [6]. Thus, in the presence of a limited oxygen supply, a positive inotropic stimulation may have deleterious effects on the mechanical efficiency of the myocardium which, in the long-term may deteriorate, instead of improving, the cardiac function [2, 17].

Xamoterol (ICI 118, 587) is a new β_1-adrenoceptor partial agonist [9] which, in basal state, acts as a positive inotrope [16, 19]. As xamoterol is now being evaluated for the treatment of heart failure [11, 14, 15, 18], the aim of our study was, therefore, to examine its effects on the mechanical efficiency of the left ventricle during prolonged therapy.

Patients and Methods

In previous trials of xamoterol in patients with mild to moderate ischemic heart failure, a total of 36 patients with coronary artery disease and a previous myocardial infarction were enrolled [14, 15, 18]. In all these patients left ventricular (LV) angiography was performed at baseline using a Millar micromanometer to record LV pressure; 24 of these patients were selected for the present study because they had simultaneous determination of coronary sinus blood flow and myocardial oxygen uptake by the thermodilution technique. Sixteen of these patients were randomized to xamoterol (200 mg twice daily) and eight were randomized to placebo. The same measurements as in the baseline study were then repeated 3 months later under therapy. A full description of the methods and patient characteristics was published previously [11, 12, 15]. Briefly, in all patients, cardioactive drugs were stopped for at least 3 days before the control study, which was performed between 4 and 8 weeks after the acute myocardial infarction. Catheterization of the left side of the heart was performed by the femoral approach with the patient in the fasting state and without premedication. Aortic pressure was measured through a No. 8F pigtail catheter connected to a Statham P23ID stain gauge. Left ventriculography was performed by single-plane 35 mm cineangiography at 50 frames/s in the right anterior oblique projection (Philips Polydiagnost C, Philips Electronic Instruments), during which high-fidelity LV pressure was recorded with a No. 5F micromanometer-tipped catheter (Millar Instruments). Opacification of the cavity was accomplished by the injection of 0.7 to 0.8 ml/kg of sodium meglumine diatrizoate at a rate of 18 ml/s. The hemodynamic variables and a cinemagraphic frame marker were recorded on magnetic tape (Honeywell 110) and on paper (Honeywell 1858). LV pressure during ventriculography and at the time of the peak of the R wave were also sampled synchronously with frame exposure and displayed in digital form on the corresponding cinemagraphic frame (Cine Data, Philips Electronic Instruments, Inc.).

In addition, a No. 7F Swan-Ganz thermodilution catheter (Webster Laboratories) was placed in the midcoronary sinus through an antecubital vein. The position of this catheter was confirmed angiographically and by oxygen saturation. Coronary sinus blood flow was determined by the thermodilution method [19] and arterial and coronary venous blood samples were obtained for determination of oxygen content. These metabolic measurements were obtained before diagnostic left ventriculography and coronary angiography.

Data analysis

All data were analyzed in a blind fashion. Analog hemodynamic data filtered at 100 Hz were digitized every 2 ms and processed off-line by means of a Hewlett-Packard A900 computer as described previously [10, 16]. For the analysis of the angiographic data, both premature and postpremature beats were excluded. Ventricular silhouettes were outlined frame by frame with a light pen on a video screen. The digitized contours were preprocessed by a computer system (LVV Philips 100) that derived the correction factor for radiographic magnification and calculated volumes applying Simpson's rule. The preprocessed data were then directed to an HP 21 MX computer for smoothing by a cubic spline method [21] and for the computation of the various indexes of LV function. As indices of inotropic state, we used (dP/dt) measured and normalized at a developed pressure of 40 mm Hg (dP/dt/DP40) [15]; the time-constant T_1 of the early isovolumic pressure fall was used as an index of relaxation [20].

Midwall circumferential stress was calculated by the formula of Sandler and Dodge. End-systole was defined as the frame with the maximal pressure-volume ratio. Myocardial oxygen consumption was calculated from the equation: (arterial O_2 content-coronary sinus O_2 content) x coronary sinus flow. The index of mechanical efficiency (in arbitrary units) was calculated as follows:

(Integral of systolic wall stress) $\times$ (Heart rate)/(Oxygen uptake per min). Data presented are means $\pm$ SD. Comparisons before and during therapy were made using a paired t-test. Comparison between groups was made using a Mann-Whitney test.

Results

Tables 1 and 2 summarize the relevant hemodynamic variables before and during therapy in the two groups of patients. At baseline, LV function tended to be more depressed in the xamoterol group than in the placebo group and the index of efficiency was slightly, but not significantly better in the latter group.

During therapy with xamoterol, the most consistent change was a decrease in LV end-diastolic pressure averaging -27%. The isovolumic index of inotropic state (dP/dt)/DP40 which is less sensitive than peak($+$)dP/dt to preload changes, also rose slightly but significantly ($+14\%$) and the ejection time was reduced, confirming the sustained positive inotropic effect of xamoterol. End-systolic volume tended to decrease ($P < 0.12$) despite a slight increase in systolic wall stress and a decrease in end-diastolic volume. Myocardial oxygen uptake and the mechanical efficiency index were unchanged during xamoterol therapy.

Table 1. Left ventricular function indices before and after xamoterol

	Baseline		Xamoterol 3 months
Heart rate (bpm)	70 ± 10	<0.03	77 $\pm$ 10
LVEDP (mmHg)	24.4 $\pm$ 6.5	<0.01	17.8 $\pm$ 8.6
LVSP (mmHg)	127 ± 20	NS	136 $\pm$ 19
Ejection time (ms)	313 ± 24	<0.01	289 $\pm$ 24
T_1 (ms)	59 $\pm$ 9	<0.01	53 $\pm$ 10
(dP/dt)/DP40 (l/s)	22.4 $\pm$ 8.0	<0.01	25.6 $\pm$ 6.3
CSF (ml/min)	111 ± 31	NS	112 $\pm$ 42
MVO_2 (ml/min)	14.1 $\pm$ 4.4	NS	14.7 $\pm$ 4.9
Angiographic data			
EDVI (ml/m^2)	155 ± 29	<0.04	144 $\pm$ 40
ESVI (ml/m^2)	74 ± 31	NS	70 $\pm$ 36
Mean syst stress (kdyne/cm^2)	317 ± 68	NS	333 ± 100
Peak syst stress (kdyne/cm^2)	374 ± 55	NS	400 $\pm$ 71
Efficiency index	8.77$\pm$ 3.44	NS	8.82$\pm$ 4.27

Data are mean $\pm$ SD n = 16.
Abbreviations: LVEDP, LVSP: left ventricular end-diastolic and systolic pressure; CSF: coronary sinus flow; MVO_2: myocardial oxygen uptake; EDVI, ESVI: end-diastolic and end-systolic volume index; T_1, (dP/dt)/DP40: see Methods for definition.

Table 2. Left ventricular function indices before and after placebo

	Baseline			Placebo 3 months		
Heart rate (bpm)	73	±15	NS	76	±	15
LVEDP (mmHg)	17.9	± 6.8	NS	17.2	±	6.0
LVSP (mmHg)	126	±15	NS	125	±	22
Ejection time (ms)	326	±44	NS	339	±	31
T_1 (ms)	59	± 8	<0.01	52	±	7
(dP/dt)/DP40 (l/s)	25.6	± 6.5	NS	25.3	±	5.3
CSF (ml/min)	104	±16	NS	105	±	27
MVO_2 (ml/min)	13.4	± 3.5	NS	13.1	±	4.6
Angiographic data						
EDVI (ml/m²)	138	±32	NS	132	±	27
ESVI (ml/m²)	66	±23	<0.05	56	±	23
Mean syst stress (kdyne/cm²)	327	±83	NS	298	±	104
Peak syst stress (kdyne/cm²)	421	±54	NS	393	±	94
Efficiency index	10.65	± 3.75	NS	10.50	±	4.96

Data are mean ± SD n = 8.
Abbreviations: same as in Table 1.

In the placebo group, the LV end-diastolic pressure, the ejection time and the index of inotropic state all remained unchanged. Further, despite a tendency for systolic wall stress and end-systolic volume to decrease, the efficiency index also remained unaffected (Table 2).

Discussion

The most consistent changes reported during xamoterol therapy are a decrease in cardiac filling pressure together with a downward shift of the diastolic LV pressure-volume relation [14, 15, 19] and with a modest increase in the indices of inotropic state [11, 15, 19]. The present analysis confirmed these findings and indicates, in addition, that these effects of xamoterol were not accompanied by a deterioration in the mechanical efficiency of the ventricle, or by an increase in myocardial oxygen consumption.

It has been shown previously that positive inotropic drugs may, in-vivo, alter the efficiency of myocardial contraction by two different mechanisms: their direct effect on contractile state and intracellular metabolism on one hand and their indirect effects on the determinants of oxygen uptake such as heart rate and afterload [7] on the other hand. Xamoterol has no vasodilatory action [9] and systolic wall stress increased slightly during therapy. Heart rate increased significantly in this small study group. Thus, global myocardial oxygen uptake per min was unchanged despite these two factors which should tend to increase it. This suggests a small decrease in oxygen uptake per beat during xamoterol therapy in our patients.

However, before interpreting the clinical significance of these observations, the limitations of the methods must be examined. First, all patients had a previous myocardial infarction. Wall-stress distribution during systole is not uniform in this setting [12] and we may have underestimated the systolic stress integral. Indeed, regional systolic stress in hypokinetic areas is generally greater than the average value predicted by the Sandler and Dodge formula because the wall is thinner and the radius of curvature greater in the hypokinetic zones [12]. Second, all patients had coronary artery disease and it has been shown that even in the absence of clinical signs of ischemia, anaerobic glycolysis was used in these ventricles [4]. Thus, by measuring a global index of oxygen uptake, we may have underestimated the total energy utilization, thereby overestimating the efficiency. Accordingly, the index used in this study must be interpreted with caution. Despite these limitations, it is nevertheless reassuring to see that this index did not deteriorate during xamoterol therapy, possibly because its effects on diastolic filling are more pronounced clinically than its inotropic effects.

Acknowledgements. The authors thank Mrs. Isabelle Mottard for her careful secretarial assistance.

References

1. Braunwald E (1971) Control of myocardial oxygen consumption. Am J Cardiol 27:416–432
2. Davidson S, Maroko PR, Braunwald E (1974) Effects of isoproterenol on contractile function of the ischemic and anoxic heart. Am J Physiol 227:439–443
3. Ford LE (1980) Effect of afterload reduction on myocardial energetics. Circ Res 46:161–166
4. Gertz EW, Wisneksi JA, Neese R, Bristow JD, Searle GL, Hanlon JT (1981) Myocardial lactate metabolism: evidence of lactate release during net chemical extraction in man. Circulation 63:1273–1279
5. Graham TP, Covell JW, Sonnenblick EH, Ross J JR, Braunwald E (1968) Control of myocardial oxygen consumption: relative influence of contractile state and tension development. J Clin Invest 47:375–385
6. Gunning JF, Coleman HN (1973) Myocardial oxygen consumption during experimental hypertrophy and congestive heart failure. J Molec Cell Cardiol 5:25–38
7. Hasenfuss G, Holubarsch C, Heiss HW, Bonzel T, Meinertz T, Just H (1986) Influence of enoximone and isoproterenol on myocardial energetics in the human heart, Abstract. Circulation 74(II):II-398
8. Nichols AB, Pearson MH, Sciacca RR, Cannon PJ (1986) Left ventricular mechanical efficiency in coronary artery disease. J Am Coll Cardiol 7:270–279
9. Nuttall A, Snow HM (1982) The cardiovascular effects of ICI 118, 587: A β_1-adrenoceptor partial agonist. Br J Pharmac 77:381–388
10. Pouleur H, Marechal G, Balasim H, Van Mechelen H, Ries A, Rousseau MF, Charlier AA (1983) Effects of dobutamine and sulmazol (AR-L115 BS) on myocardial metabolism, coronary, femoral and renal blood flows: a comparative study in normal dogs and in dogs with chronic volume overload. J Cardiovasc Pharmacol 5:861–867
11. Pouleur H, Rousseau MF, Hanet C, Marlow HF, Charlier AA (1987) Left ventricular sensitivity to β-adrenoceptor stimulating drugs in patients with ischemic heart disease and varying degrees of ventricular dysfunction. Circ Res 61(I):I91–I95
12. Pouleur H, Rousseau MF, van Eyll C, Charlier AA (1984) Assessment of regional left ventricular relaxation in patients with coronary artery disease: importance of geometric factors and changes in wall thickness. Circulation 69:696–702
13. Pouleur H, Rousseau MF, van Eyll C, Gurné O, Hanet C, Charlier AA (1986) Impaired regional diastolic distensibility in coronary artery disease: relations with dynamic left ventricular compliance. Am Heart J 112:721–728
14. Pouleur H, van Eyll C, Cheron P, Hanet C, Charlier AA, Rousseau MF (1986) Changes in left ventricular filling dynamics after long-term xamoterol therapy in ischemic left ventricular dysfunction. Heart Failure 2:176–184
15. Pouleur H, van Eyll C, Hanet C, Cheron P, Charlier AA, Rousseau MF (1988) Long-term effects of xamoterol on left ventricular diastolic function and late remodeling: a study in patients with anterior myocardial infarction and single-vessel disease. Circulation 77:1081–1089

16. Pouleur H, Van Mechelen H, Balasim Habib, Rousseau MF, Charlier AA (1984) Comparisons of the inotropic effects of the β_1-adrenoceptor partial agonists SL 75.177.10 and ICI 188,587 with digoxin on the intact canine heart. J Cardiovasc Pharmacol 6:720–726
17. Pozen R, DiBianco R, Katz R, Bortz R, Myerburg R, Fletcher R (1981) Myocardial metabolic and hemodynamic effects of dobutamine in heart failure complicating coronary artery disease. Circulation 63:1279–1285
18. Rousseau MF, Cheron P, Nannan M, Vincent MF, Lavenne F, Pouleur H (1985) Place des agonistes partiels des récepteurs β-adrénergiques dans l'insuffisance ventriculaire gauche d'origine ischémique. Intérêt du Xamotérol (ICI 118,587, Corwin). Ann Med Int 136:247–250
19. Rousseau MF, Pouleur H, Vincent MF (1983) Effects of a cardioselective beta-1-partial agonist (Corwin) on left ventricular function and myocardial metabolism in patients with previous myocardial infarction. Am J Cardiol 51:1267–1274
20. Rousseau MF, Veriter C, Detry JMR, Brasseur LA, Pouleur H (1980) Impaired early left ventricular relaxation in coronary artery disease. Effects of intracoronary Nifedipine. Circulation 62:764–772
21. van Eyll C, Rousseau MF, Pouleur H, Charlier AA, Brasseur LA (1982) Algorithms for wall motion analysis: importance of data smoothing and correct determination of end-systole. In Computers in Cardiology Seattle, IEEE Computer Society, pp 413–416
22. Weber KT, Janicki JS (1979) The metabolic demand and oxygen supply of the heart: physiologic and clinical considerations. Am J Cardiol 44:722–729

Author's address:

H. Pouleur, M.D., University of Louvain, School of Medicine, Avenue Hippocrate 55 HEDY/5560, 1200 Brussels, Belgium

Modulation of the autonomic control of the failing heart

H. M. Snow

Bioscience II, ICI Pharmaceuticals, Macclesfield, Cheshire, U.K.

Summary

The failing heart operates with an abnormal combination of heart rate, stroke volume, and enddiastolic volume. This mismatch becomes more evident during exercise of patients with heart failure, when an increase in cardiac output is achieved with higher heart rate, a lower stroke volume and a higher enddiastolic volume. Using the β_1-adrenoceptor partial agonist xamoterol which lacks β_2-adrenoceptor agonism the response of the heart to sympathetic stimulation can be modulated. At rest and low levels of exercise xamoterol provides an inotropic support of the heart, whereas it reduces inappropriate tachycardia at higher levels. Thereby, xamoterol tends to normalize the balance of the inotropic and chronotropic control of the failing heart, because cardiac output is increased with a more normal combination of heart rate, stroke volume, and filling pressure. The beneficial effects of xamoterol are discussed as being especially important for failing ischemic hearts, because the balance between energy supply and energy demand may be improved by xamoterol.

A characteristic of the failing heart is its inability to increase and in some cases even maintain stroke volume as heart rate increases during exercise (Fig. 1). In the normal

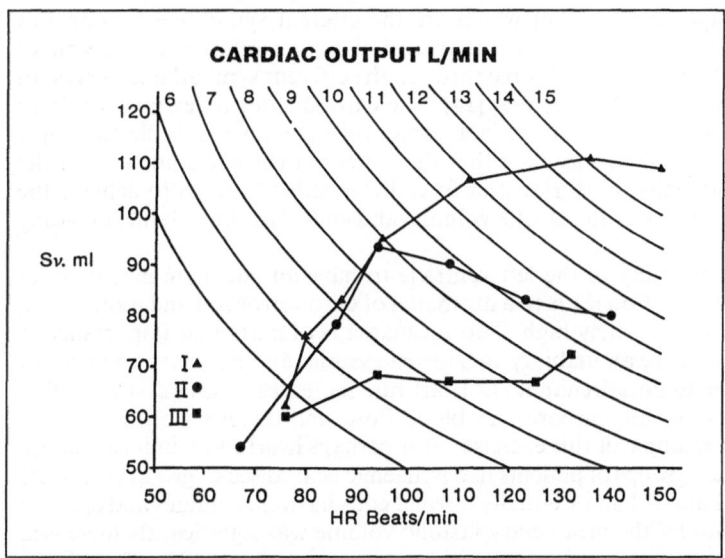

Fig. 1. The relationship between heart rate, stroke volume and cardiac output in three groups of patients with increasing severity of ischaemic heart disease (I < II < III) from [16]

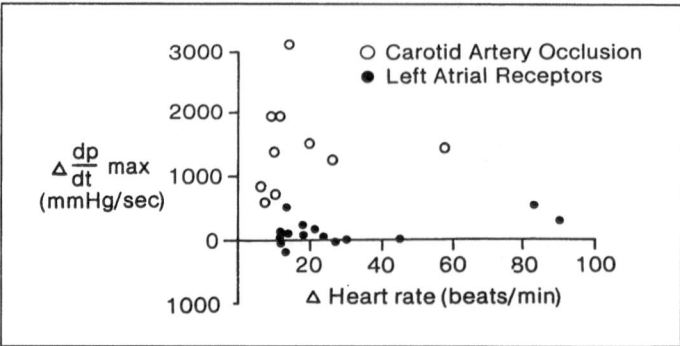

Fig. 2. Relative inotropic and chronotropic effects on the heart during occlusion of the carotid arteries (o) and stimulation of left atrial receptors (●). Stimulation of atrial receptors has only chronotropic effects

healthy subject cardiac output increases from about 5 to 25 l/min on going from rest to exercise, heart rate (HR) from 70 to 190 beats/min and stroke volume (SV) from 70 to 130 ml [1]. Typical values in a failing heart during maximum exercise could be SV 70 ml, HR 140 beats/min and maximum cardiac output reduced to about 10 l/min.

The cardiac determinants of stroke volume are end-diastolic volume (Frank-Starling law) and the inotropic state, in particular that of the left ventricle.

Heart rate and inotropic state are under the influence of the autonomic nervous system and from this point of view the control of the output of the heart is about matching heart rate and inotropic state to the venous return so as to maintain end-diastolic volume within narrow limits. Such ideas were stated explicitly by Starling [32] and Bainbridge [2]. The mechanisms which operate to perform this matching are presummed to be the cardiovascular reflexes, important parts of which are the efferent sympathetic control of chronotropism and inotropism. It is known that cardiovascular reflexes are capable of modifying independently both rate and force through the efferent sympathetic nerves. In Fig. 2 are shown results obtained in the dog [14]. The carotid sinus reflex involves both chronotropic and inotropic response; the atrial receptor reflex involves only chronotropic responses. It seems reasonable to speculate that the system of cardiac control is, at the very least, capable of adjusting both rate and force independently so as to achieve the most efficient matching of heart rate, stroke volume, and end-diastolic volume, on going from rest to exercise.

In heart failure, the inability of the left ventricle to have an adequate inotropic response to sympathetic stimulation leads to a mismatch of chronotropism and inotropism, and the heart rate is inappropriately high. This mismatch is of particular importance in ischaemic heart disease; where an inability to shorten systole (an important component of the inotropic response to noradrenaline) as heart rate increases, causes systole to further encroach upon diastole, impair coronary blood flow, and thereby worsen left ventricular dysfunction. An example of this exercise −, (or perhaps heart rate), induced failure is shown in Fig. 1. All these groups of patients had ischaemic heart disease, group I < II < III [16]. Patients in groups I and II had essentially normal end-diastolic volumes and ejection fractions at rest. In group III the mean end-diastolic volume was significantly increased and the ejection fraction decreased with respect to groups I and II. The ability to increase and maintain stroke volume is related to the severity of the ischaemic heart disease, as is also the ability to achieve a high heart rate. These results are but an example of those previously reported by others [6, 9], who also measured pressure in the pulmonary circulation and obtained indirect evidence that the filling pressure of the left ventricle was increased above the normal values observed [17].

In summary the failing heart operates with an abnormal combination of heart rate, stroke volume, and end-diastolic volume. During exercise a particular cardiac output is achieved with a higher heart rate, a lower stroke volume, and a higher end-diastolic volume. Is it possible to pharmacologically manipulate the cardiac response to noradrenaline in order to achieve a more normal balance between inotropism and chronotropism in the failing heart?

Effects of xamoterol on autonomic control of the heart

The cardiovascular pharmacology of xamoterol in the dog is summarised in Table 1. Xamoterol has no agonist activity at the β_2-adrenoceptor despite having significant effects at the β_1-adrenoceptor, and it was selected from a series of compounds as the compound having the greatest agonist activity at the β_1-adrenoceptor whilst having no agonist activity at the β_2-adrenoceptor [3]. In this respect xamoterol is unique among those β-adrenoceptor drugs which have been used in the chronic treatment of heart failure, for example both prenalterol and pirbuterol have significant β_2-adrenoceptor agonist activity. Experiments carried out in an anaethetised dog preparation, depleted of catecholamines and in which cardiovascular reflexes were prevented from occurring, showed that xamoterol produces about 43% of the maximum increase in heart rate brought about by a full agonist (isoprenaline) with an ED_{50} of 3.2 μg/kg iv (Table 1, Fig. 3) [24]. No effects were observed on hind limb perfusion pressure and it was concluded that even though xamoterol occupied β_2-adrenoceptors on the smooth muscle of arterial resistance vessels, it had no agonist activity.

Agonist effects of xamoterol on the heart may be measured either as chronotropic or inotropic effects. In this areflexic dog preparation, the relations between effects on rate and force brought about by either xamoterol or isoprenaline are the same (Fig. 4a). Further, the ED_{50} for both the chronotropic and inotropic effects of xamoterol are similar.

Table 1. Summary of pharmacological properties of xamoterol in the dog. K_A and α are from agonist dose response curves on heart rate and hind limb perfusion pressure. K_β from displacement of agonist dose response curves by xamoterol. Partial agonist activity (PAA) is calculated as the maximum response to xamoterol as a percentage of the maximum response to a full agonist, e.g. isoprenaline

β_1:	PAA,	43%
	ED50, K_A	2.3 μg/kg
	K_β (against noradrenaline)	3.0 μg/kg
	K_β (against isoprenaline)	8.7 μg/kg
β_2:	PAA,	0%
	K_β (against isoprenaline)	71 μg/kg

Orally active	Bioavailability 5%
Hydrophillic	Log P −0.3 (octanol/water)
Renal excretion	
Terminal blood half life 14 h	

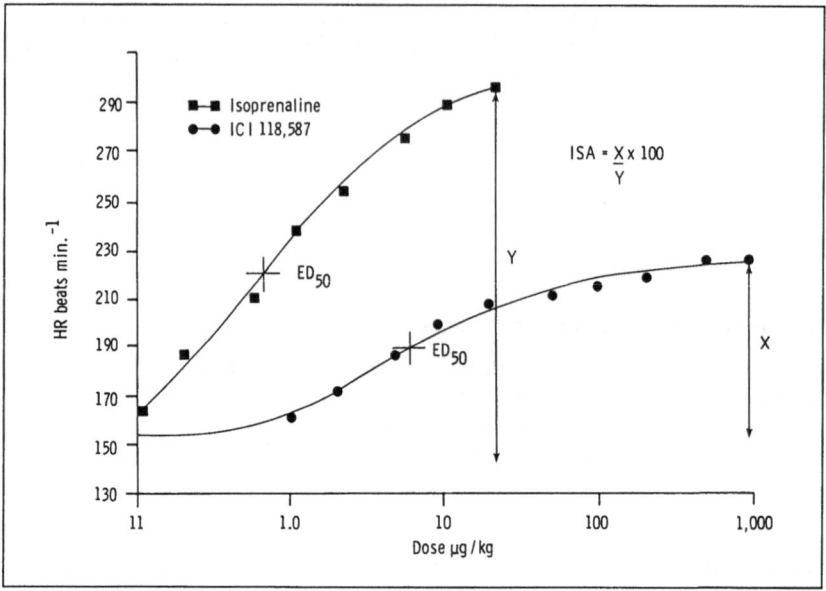

Fig. 3. Dose-response curves in a single dog, relating heart rate to the dose of isoprenaline (■) and xamoterol (ICI 118587) (●). The partial agonist activity or ISA of xamoterol is calculated as the maximum change in heart rate (X) as a percentage of the maximum caused by isoprenaline (Y)

These results support the conclusion of Furnival et al. [13] that the β-adrenoceptors mediating rate and force in the sinu-atrial node and the muscle of the left ventricle are also similar.

Direct evidence relating inotropic effects on the left ventricle to the blood concentration of xamoterol in man was obtained by Pouleur et al. [27] in patients suffering from ischaemic heart disease. The ED_{50} for this inotropic effect (3.7 ± 0.8 µg/kg iv) is similar to that found in dogs. The magnitude of the inotropic effect in terms of LV dp/dt max, about 1000 mm Hg/s, is smaller than the 2000–3000 mm Hg/s observed in healthy beagle dogs.

Chronic dosing with xamoterol in patients at 200 mg bd produces average peak and through blood concentrations of 56 ng/ml and 175 ng/ml. These concentrations produce stimulant effects at rest between 75% and 100% of the partial agonist activity of xamoterol throughout the day. In animals chronic dosing with xamoterol does not appear to induce receptor down regulation. Barnett and Maguire [4] compared the effects of 6 days' dosing with isoprenaline and xamoterol on the binding of iodo-pindolol to ventricular cell membranes in the rat. They concluded that maximal binding capacity was reduced by 42% by isoprenaline and not affected by xamoterol and suggested that isoprenaline but not xamoterol also caused some uncoupling of the production of C-AMP from β_2-adrenoceptors. This lack of an effect may be related to both the level of partial agonist activity at β_1 adrenoceptors and the absence of any β_2-adrenoceptor activity.

The effect of xamoterol on the response of an areflexic dog heart to stimulation of the sympathetic nerves is shown in Fig. 5. This result demonstrates not only the ability of xamoterol to antagonise the effects of noradrenaline but also the failure of the sympathetic nerves, even at high rates of stimulation, to release sufficient noradrenaline to

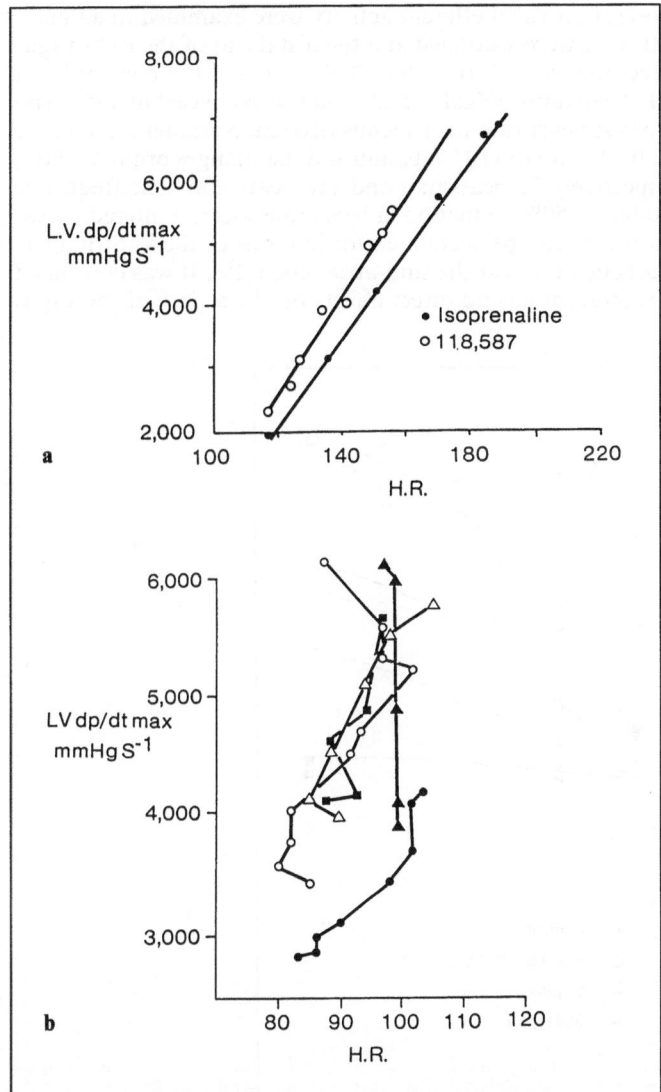

Fig. 4. a The effects of cumulative doses of isoprenaline (●) and xamoterol (ICI 118587) (○) on heart rate and LV dp/dt max in an anaesthetised areflexic dog preparation and **b** of xamoterol in five resting conscious dogs. In the presence of cardiovascular reflexes xamoterol has a predominantly inotropic effect

displace the required amount of xamoterol from the β_1-adrenoceptor and achieve the control maximum response. Therefore, as the dose of xamoterol is increased, the range of the response of the heart to sympathetic nerve stimulation becomes increasingly narrowed (Fig. 5). The overall effect is to stabilize both the inotropic and chronotropic responses of the heart to sympathetic nerve stimulation about a level determined by the partial agonist activity of xamoterol.

The direct effects of xamoterol on vagal efferent activity were examined in an anesthetised dog preparation. Both vagi were sectioned and the distal end of the right vagus electrically stimulated. In three dogs xamoterol (50 µg/kg) increased the control heart rate from 144 beats/min to 207 beats/min. Vagal stimulation was increased in a stepwise manner so as to obtain the slowest heart rate with a sinus rhythm. Xamoterol increased this slowest heart rate from 72 beats/min to 110 beats/min and the changes brought about by vagal stimulation were respectively 72 beats/min and 110 beats/min. The frequency of stimulation required to produce a 50% reduction in heart rate was not altered by xamoterol. These changes in heart rate can be accounted for in terms of the known interaction of the vagal and sympathetic nerves at the sinu-atrial node [19]. It was concluded from these experiments that xamoterol has no direct effects on the ability of the vagus

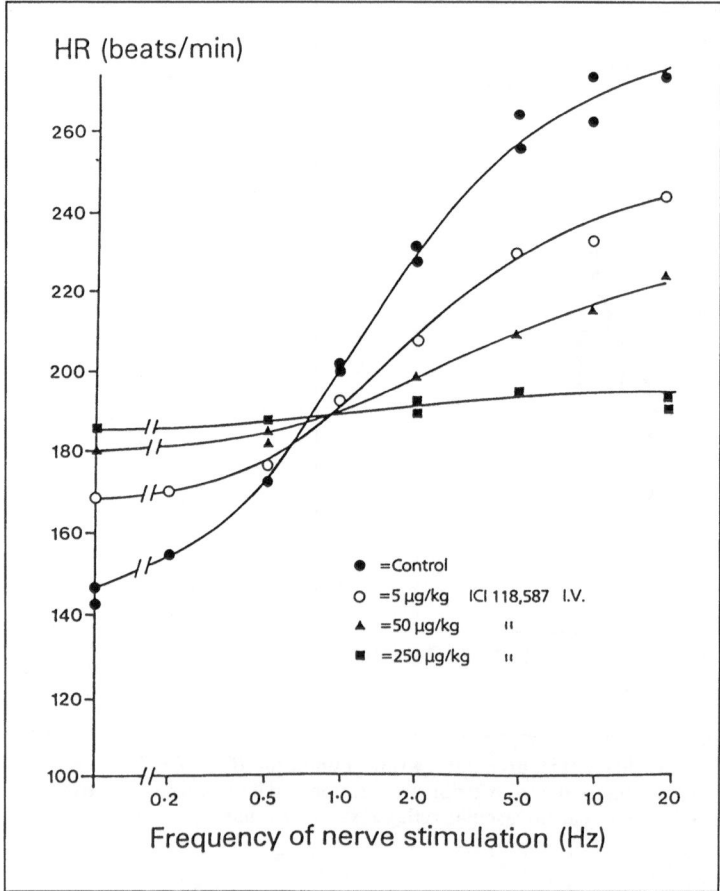

Fig. 5. Frequency response curves relating heart rate to the frequency of stimulation of the right ansa subclavia in an anaesthetised dog, in the absence and presence of 5 (○), 50 (▲) and 250 (■) µg/kg of xamoterol (ICI 118587). The rotation of the curves about a fixed point is characteristic of the action of a partial agonist. The depression of the maximum response results from the inability of the sympathetic nerves to release sufficient noradrenaline. Xamoterol acts to stabilize the response of the heart at about a level equivalent to its own intrinsic activity

to produce changes in heart rate and is unlikely to interfere with either the release or activity of acetylcholine at muscarinic receptor sites in the heart.

In contrast to effects in the areflexic dog, xamoterol has little or no effect upon heart rate in the conscious dog (Fig. 4 b). However, the inotropic effects are similar. The dissociation of an effect on rate from that on force in the intact animal probably results from the lack of a vasodilating (β_2-adrenoceptor) effect with xamoterol. Therefore, systolic blood pressure is increased and the baroreceptor reflexes, operating principally through the vagal nerves to the heart, preferentially slow heart rate. In general the apparent ability of some β-adrenoceptor agonists to have differing effects on rate and force, e.g. noradrenaline and isoprenaline, resides in their opposing actions on the peripheral circulation and the consequent operation of baroreceptor reflexes, and not in the direct action of compounds at the β-adrenoceptors in the sinu-atrial node and muscle of the left ventricle [20].

A lack of a significant chronotropic response to xamoterol has also been observed in normal man at rest, but is present after atropine [18, 30].

The ability of xamoterol to modulate the response of the heart to sympathetic stimulation is central to an understanding of its pharmacological action in the intact animal. Thus, whether xamoterol acts to increase or decrease the response of the heart to sympathetic stimulation depends upon the ongoing level of endogenous sympathetic activity (Fig. 5). Consequently in the intact animal it is not sufficient just to consider the direct action of xamoterol on the heart, but also its interaction with sympathetic activity and secondary effects from cardiac reflexes. The primary interaction between xamoterol and noradrenaline at the β_1-adrenoceptor results in an alteration in the pattern of cardiac control at different levels of exercise. The cardiac output is achieved with a different combination of heart rate, stroke volume and end-diastolic volume. Evidence for this effect on cardiovascular control in normal dog and man and in patients with heart failure will now be considered.

Effects of xamoterol on the determinants of cardiac output at rest and during exercise

Measurements of the effects of xamoterol on the heart and circulation at rest and during exercise, have been made in both man and dog [9, 25, 18, 31]. In man, Jennings et al. [18] measured heart rate, blood pressure and cardiac output at rest and at three levels of exercise, before and at various times up to 6 h after xamoterol 100 µg/kg iv (Fig. 6). At rest, xamoterol had no effect upon heart rate and diastolic blood pressure in either the sitting or the supine position; systolic blood pressure and cardiac index were increased, the latter by between 10% and 20%. During exercise no significant effects were observed at 25% of the maximum sustained work capacity. At work rates of 50% and 75% of maximum the only significant effects of xamoterol were a reduction in both heart rate and systolic blood pressure. All these effects were time-dependent and present 6 h after dosing. Apart from the 10–20% increase at rest, xamoterol had little effect upon cardiac output at each level of exercise, however the manner in which a particular cardiac output was achieved in terms of HR and SV was different (Fig. 7). At rest and at low levels of exercise the resultant effect on cardiac control is an increase in stroke volume and at maximum exercise a reduction in heart rate.

This pharmacological manipulation of the autonomic control of the heart during exercise in the dog is shown in Fig. 8. These results, taken from Ohaygi et al. [25], show that the relation between LVEDP (an index of heart size) and LV dp/dt max (an index of the

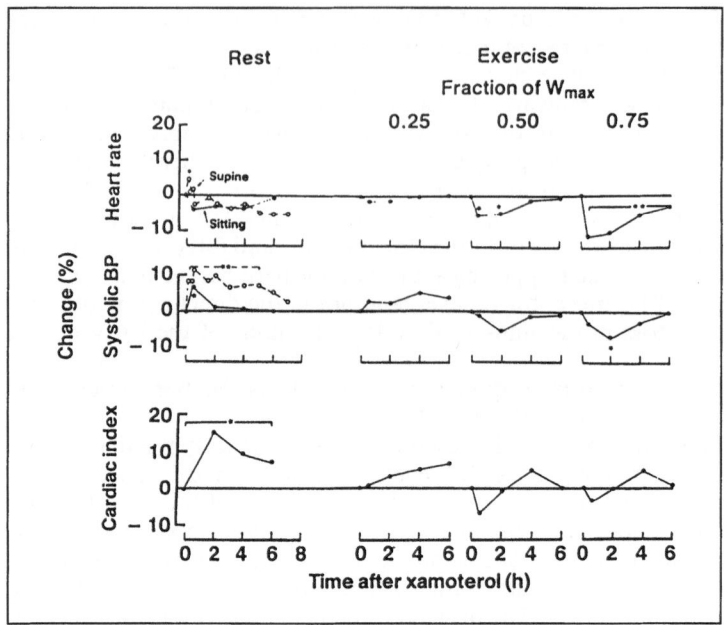

Fig. 6. Effects of xamoterol (100 μg/kg iv) on heart rate, systolic blood pressure and cardiac index. Measurements were made before and at 2, 4, 6, and 8 h after xamoterol, at rest and at increasing work loads of 0.25, 0.5, and 0.75 of a previously measured maximum sustained work load. Supine measurements of heart rate and blood pressure were made at rest and are shown as O–O–O. * $p < 0.05$; ** $p < 0.01$

inotropic response to sympathetic stimulation) is altered by both propranolol and xamoterol. Heart rate was increased at rest and decreased during maximum exercise by xamoterol and decreased at all levels of activity by propranolol. Cardiac output was not measured in these experiments but on the assumption that output was similar in the presence of placebo and each drug, at corresponding levels of exercise, then it may be seen that the manner in which the output was achieved is quite different in the three situations. In the normal dog, in the absence of any drug interference, autonomic control in terms of inotropism (LV dp/dt max) and chronotropism deals with the venous return from the exercising muscles and maintains end-diastolic volume (LVEDP) relatively constant. Near complete removal of sympathetic control (propranolol) leaves the heart with only the Frank-Starling mechanism and small increases in heart rate, because of vagal withdrawal, and as a means of increasing output. In the presence of a β_1-adrenoceptor partial agonist, normal sympathetic control is again altered (Fig. 8). At rest the inotropic state of the heart is increased, resulting in a reduction in LVEDP, and at the highest level of exercise these effects are reversed.

Thus in the normal heart xamoterol leads to a reduction in the range of the response of the heart to sympathetic stimulation, and even though maximum cardiac output may not be reduced it is achieved with a slower heart rate and higher end-diastolic and stroke volumes (Fig. 8).

In patients with left ventricular dysfunction, as discussed earlier, an important element of the reflex control of the heart is impaired, namely, the ability of the left ventricle

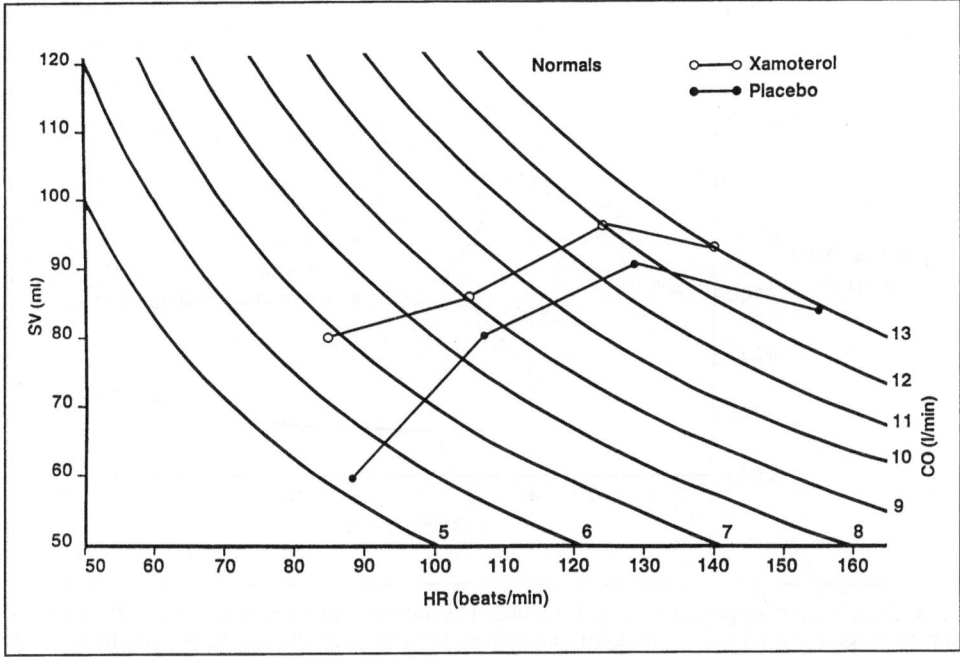

Fig. 7. The effect of xamoterol on the relationship between heart rate, stroke volume and cardiac output at rest and three levels of increasing exercise on a bicycle ergometer in normal subjects. Xamoterol alters the relation such that at rest and low levels of exercise stroke volume is increased with little change in heart rate and during more severe exercise heart rate is decreased

to respond adequately to sympathetic stimulation. Both contraction and relaxation are impaired. Consequently, as the severity of exercise increases, the relation between heart rate, stroke volume and end-diastolic volume become abnormal. In ischaemic heart disease this mismatching of a high heart rate with a poor inotropic response and a consequent inability to shorten systole and preserve diastole further depresses left ventricular dysfunction. In these circumstances the same directional changes as produced by xamoterol in the normal heart rate have a beneficial effect and a more normal pattern of response is observed.

This effect was first demonstrated by Detry et al. [9], who showed that at both rest and exercise pulmonary capillary wedge pressure was consistently reduced by xamoterol, despite the reduction in maximum exercise heart rate. These findings are in contrast to those observed at rest and during submaximal exercise with the β-antagonists propranolol, practolol or atenolol, when decreases in cardiac output are associated with either no change or an increase in the filling pressure of the left ventricle [11, 26, 12, 29, 28, 6]. This ability of xamoterol to cause a reduction in the filling pressure of the diseased left ventricle at rest and throughout all levels of exercise was confirmed by Detry et al. [10] in patients with angina pectoris and by Sato et al. [31] in patients with heart failure (Fig. 9). The results obtained by Sato et al. [31] provide an important link between the effects of xamoterol on the response of the heart to sympathetic stimulation and the mode of action in heart failure patients. They measured both plasma noradrenaline concentrations and haemodynamic changes, at rest and during three levels of exercise and were, therefore,

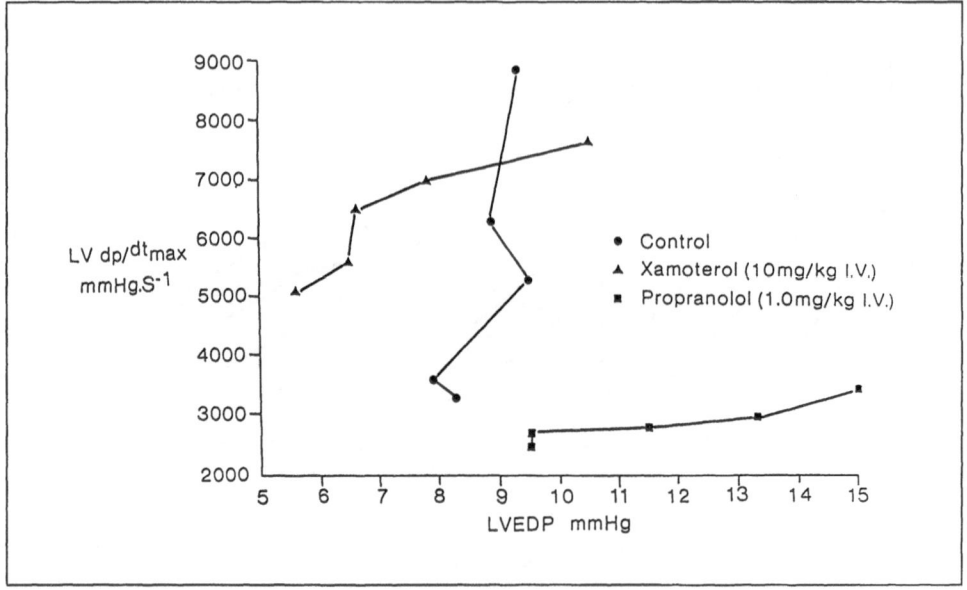

Fig. 8. The effects of propranolol (■) and xamoterol (▲) on the relation between LV dp/dt max (inotropic changes) and LVEDP (Frank-Starling effect), at rest (lying, then standing) and three levels of exercise on a treadmill. Control measurements (●). Results are mean values obtained in 10 dogs. Both xamoterol (β_1 partial agonist) and propranolol (β_1/β_2 antagonist) alter the normal balanced contribution of inotropism and Starling's law to the manner in which the cardiac output at a particular level of exercise is achieved

able to examine the effects of xamoterol on the relation between an index of the stimulus and response. The changes produced by xamoterol are as expected. At low levels of stimulus xamoterol adds to the effect of noradrenaline and the net effect is seen as an increase in heart rate, cardiac output and systolic blood pressure (Fig. 9). During higher levels of stimulus the effects of noradrenaline are attenuated. However, more noradrenaline is produced and the resultant cardiac output, during maximum exercise, is only slightly reduced.

Importantly, the filling pressures at all corresponding levels of either exercise or noradrenaline concentration are reduced. Therefore, when the relation between filling pressure and cardiac index is examined (Fig. 10) it is seen that, both at rest and during exercise, the cardiac output is achieved at a lower filling pressure in the presence of xamoterol. It is important to realise that these effects during maximal exercise are achieved through a slowing of heart rate and a more normal matching of rate and force, and not through a direct inotropic effect of xamoterol upon the left ventricle.

Conclusion

In terms of cardiac control left ventricular dysfunction may be defined as the inability of the left ventricle to respond to sympathetic stimulation. This blunting of the inotropic response of the left ventricle leads to an imbalance in the chronotropic and inotropic control of the heart at rest and in particular during exercise. The consequent inability of the diseased heart to shorten systole results in a failure to maintain filling of the left ventricle

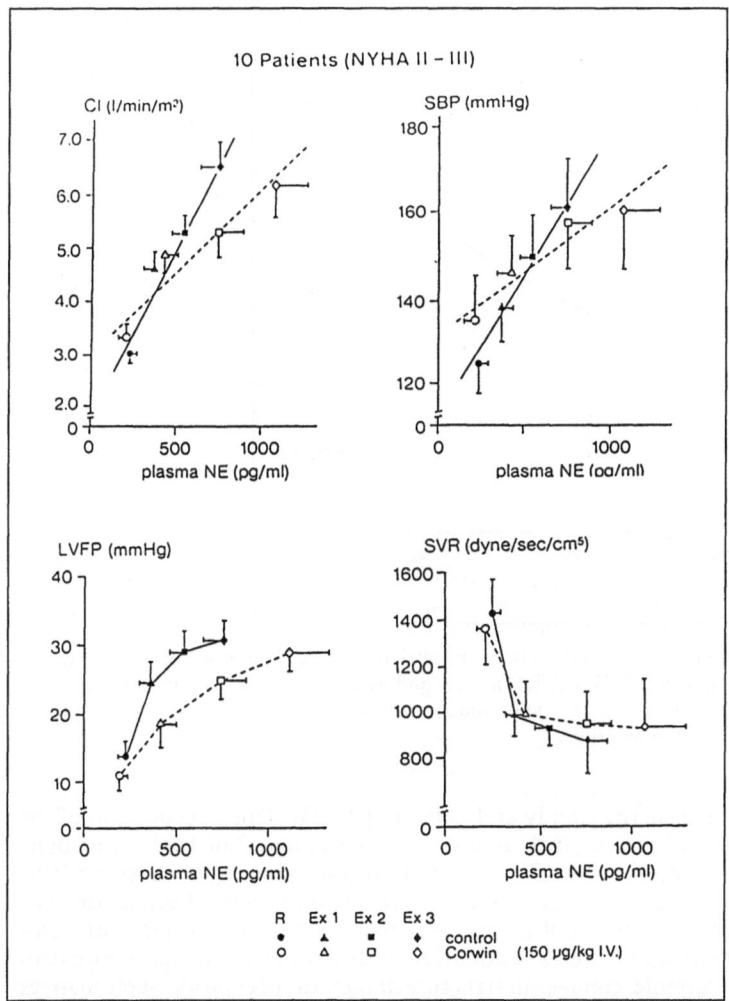

Fig. 9. Effects of xamoterol on the relation between cardiac index (CI), systolic blood pressure (SBP), pulmonary wedge pressure (LVFP), systemic vascular resistance (SVR) and plasma noradrenaline concentration (NE) in 10 patients with heart failure (NYHA II/III). Observations were made in the presence of xamoterol ($\Diamond$) and placebo ($\blacklozenge$) at rest and at three levels of supine exercise using a cycle ergometer

and in those patients with coronary artery disease to increase coronary blood flow adequately.

By modulating the response of the heart to sympathetic stimulation and therefore providing inotropic support during low levels of exercise and reducing the inappropriate tachycardia at higher levels, xamoterol tends to normalise the balance of the inotropic and chronotropic control of the failing heart. The cardiac output is achieved with a more normal combination of heart rate, stroke volume and filling pressure. These changes during exercise are of particular benefit to the failing ischaemic heart, since they tend to im-

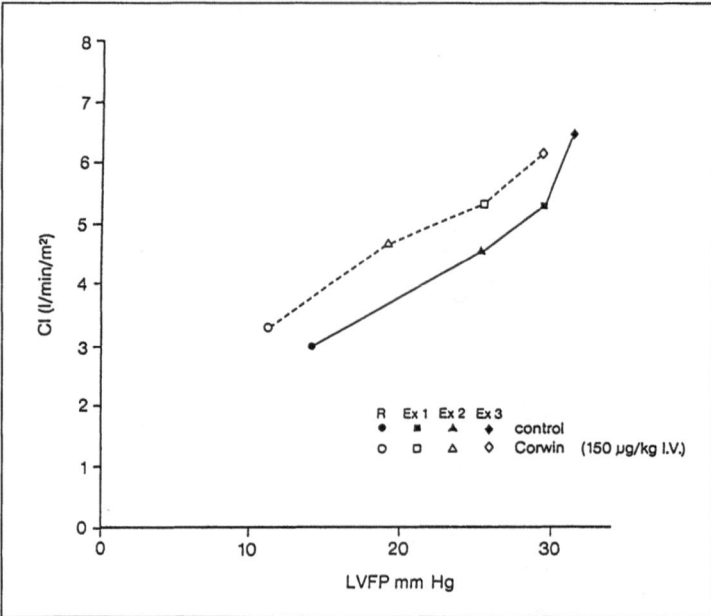

Fig. 10. The effect of xamoterol on the relation between cardiac index and pulmonary wedge pressure in 10 patients with heart failure (NYHA II/III) at rest and at three levels of supine exercise using a cycle ergometer. At all levels of activity the cardiac output is achieved with a lower pulmonary wedge pressure in the presence of xamoterol

prove the balance between energy supply and demand [10, 23]. Thus, occupation of the β_1-adrenoceptor by a partial agonist, lacking β_2-adrenoceptor agonism, leads to an alteration in the pattern of autonomic control of the heart on going from rest to exercise. This acute effect is of hemodynamic benefit to patients with left ventricular dysfunction. The imposition of such an altered pattern of cardiac activity over a period of time with associated changes in heart size and the prevention of myocardial ischaemia may cause beneficial adaptations, for example, changes in structure and β-adrenoceptors. Such changes could account for the observed improvement in the overall clinical condition of patients during long term treatment with xamoterol [15, 5].

References

1. Astrand PO, Cuddy ET, Saltin B, Stenberg J (1964) Cardiac output during submaximal and maximal work. J Appl Physiol 19:268–274
2. Bainbridge FA (1915) The influence of venous filling upon the rate of the heart. J Physiol 50:65–84
3. Barlow JJ, Main BG, Snow HM (1981) β-Adrenoceptor stimulant properties of amidoalkylamino substituted 1-anyl-2-ethanols and 1-(aryloxy)-2-propanols. J Med Chem 24:315–322
4. Barnett DB, Maguire M (1986) Comparison of the effects of chronic infusion of xamoterol and isoprenaline on rat ventricular β-adrenoceptors. Br J Pharmacol 87:223P
5. Blackwood R, Marlow HR (1988) Xamoterol in the management of patients with exertional breathlessness and fatigue due to cardiac disease. J Am Coll Cardiol 11:143A

6. Bruce RA, Hossack KF, Kusumi F, Clarke L (1979) Acute effect of oral propranolol on hemodynamic response to upright exercise. Am J Cardiol 44:132–140
7. Carlsson E, Ablad B, Brandstrom A, Carlsson B (1972) Differential blockade of the chronotropic effects of various adrenergic stimuli in the cat heart. Life Sci 11:953–958
8. Cook N, Richardson A, Barnett DB (1984) Comparison of the β_1 selective affinity of Prenalterol and Corwin demonstrated by radioligand binding. Eur J Pharmacol 98:407–412
9. Detry JR, Decoster PM, Brasseur LA (1983) Haemodynamic effects of Corwin (ICI 118587) a new cardioselective β-adrenoceptor partial agonist. Eur Heart J 4:584–591
10. Detry JR, Decoster PM, Buy J, Rousseau MF, Brasseur LA (1984) Antianginal effects of Corwin: a new β-adrenoceptor partial agonist. Am J Cardiol 53:439–443
11. Epstein SE, Robinson BF, Kahler RL, Braunwald E (1965) Effects of β-adrenergic blockade on the cardiac response to maximal and submaximal exercise in man. J Clin Invest 44:1745–1753
12. Finegan RE, Marlow AM, Harrisson DC (1972) Circulatory effect of practolol. Am J Cardiol 29:315–322
13. Furnival CM, Linden RJ, Snow JM (1971 a) The inotropic and chronotropic effects of catecholamines on the dog heart. J Physiol 214:15–28
14. Furnival CM, Linden RJ, Snow HM (1971 b) Reflex effects on the heart of stimulating left atrial receptors. J Physiol 218:447–463
15. German and Austrian Study Group (1988) Double-blind placebo-controlled comparison of Digoxin and Xamoterol in chronic heart failure. Lancet I:489–493
16. Hetherington M, Teo KK, Haennel RG, Rossall RE, Kappagoda CT (1987) Response to upright exercise after myocardial infarction. Cardiovascular Res 21:399–406
17. Higginbotham MB, Morris KG, Williams RS, McHale PA, Coleman RE, Cobb FR (1986) Regulation of stroke volume during submaximal and maximal upright exercise in normal man. Circ Res 56:281–291
18. Jennings G, Bobick A, Oddie C, Restall R (1984) Cardioselectivity, kinetics, hemodynamics and metabolic effects of xamoterol. Clin Pharmacol Ther 35:594–602
19. Levy MN, Zieske A (1969) Autonomic control of cardiac pacemaker activity and atrioventricular transmission. J Appl Physiol 27:465–470
20. Linden RJ, Snow HM (1974) The inotropic state of the heart. In: Linden RJ (ed) Recent advances in physiology. Churchill Livingstone, Edinburgh
21. Lofdahl C-G, Svedmyr N (1984) Effects of xamoterol (ICI 118587) in asthmatic patients. Br J Clin Pharmacol 18:597–601
22. McCaffrey PM, Riddell JG, Shanks RG (1986) The selectivity of the partial agonist activity of xamoterol in man measured by its effects in the presence of ICI 118551. Br J Pharmacol 89:594P
23. Molajo AO, Bennett DH, Marlow HF, Snow HM, Bastain W (1987) The effects and dose-response relationship of xamoterol in patients with ischaemic heart disease. Br J Clin Pharmacol 24:373–379
24. Nuttall A, Snow HM (1982) The cardiovascular effects of ICI 118587: a β_1-adrenoceptor partial agonist. Br J Pharmacol 77:381–388
25. Ohyagi A, Sasayama S, Nakamura Y, Lee JD, Kihara Y, Kawai C (1984) Effect of ICI 118587 on left ventricular function during gradual treadmill exercise in conscious dogs. Am J Cardiol 54:1108–1113
26. Parker JO, West RO, Di Giorgi S (1968) Hemodynamic effect of propranolol in coronary heart disease. Am J Cardiol 21:11–19
27. Pouleur H, Rousseau MF, Hanet C, Marlow HF, Charier AA (1987) Left ventricular sensitivity to β-adrenoceptor-stimulating drugs in patients with ischaemic heart disease and varying degrees of ventricular dysfunction. Circ Res 4 [Suppl 1]:91–95
28. Reybrouck T, Amery A, Billiet L (1977) Hemodynamic response to graded exercise after chronic β-adrenergic blockade. J Appl Physiol 42:133–138
29. Rousseau MF, Brasseur LA, Detry JM (1973) Haemodynamic and electrocardiographic effects of practolol during upright exercise in coronary heart disease. Cardiovascular Res 7:306–312
30. Sasayama S, Yokawa S, Akiyama M, Mikawa M, Sakai O (1986) Cardiovascular effects of ICI 118587 a new β-adrenoceptor partial agonist in man. Jpn Circ J 50:636–643

31. Sato H, Inooe M, Matsyama T, Ozaki H, Shimazu T, Takeda H, Ishida Y, Kamada T (1987) Hemodynamic effects of the β_1-adrenoceptor partial agonist xamoterol in relation to plasma norepinephrine levels during exercise in patients with left ventricular dysfunction. Circulation 75:213–220
32. Starling EH (1920) On the circulatory changes associated with exercise. J R Army M Corps 34:258–272

Author's address:

H. M. Snow, Bioscience II, ICI Pharmaceuticals, Macclesfield, Cheshire, U.K.

Dopexamine in congestive heart failure: how do the pharmacological activities translate into the clinical situation?

T. Meinertz, H. Drexler, Hj. Just

Medizinische Universitätsklinik, Innere Medizin III, Kardiologie, Universität Freiburg, FRG

Summary

Dopexamine is a newly developed sympathetic catecholamine which combines dopaminergic (DA-1) and β_2-adrenergic agonist activity with only minor β_1-adrenergic action. Thereby, this compound exerts systemic and preferential renal vasodilation, causing afterload reduction, increases in cardiac output, and improved renal perfusion in animals and normal volunteers. Short-term administration of dopexamine in congestive heart failure established benefical effects in central hemodynamics, that is, reduction of systemic and pulmonary vascular resistance, combined with a decrease in LV-filling pressures and increased renal blood flow. The effect on central hemodynamics are comparable to sodium nitroprusside. With higher doses, however, heart rate may increase substantially with dopexamine along with increased myocardial oxygen consumption. Experiences with prolonged administration of this drug are scarce and have, so far, yielded conflicting results. Thus, dopexamine appears to be a promising agent in the short-term management of congestive heart failure. However, its ultimate value for prolonged administration remains to be established.

Introduction

Treatment of severe congestive heart failure often requires the use of several compounds such as β-adrenoceptor agonists, dopamine, and pre- and/or afterload lowering agents. Since no single agent has been shown to provide adequately successful control for the hemodynamic needs in this situation, a variety of compounds has been developed to provide a more adequate therapy.

One such agent is dopexamine hydrochloride, a new synthetic catecholamine which has been developed specifically to improve the treatment of congestive heart failure (Fig. 1). In this review we describe the pharmacological profile of the compound and examine how these activities translate into the clinical situation, using data from recent clinical studies.

The development of dopexamine hydrochloride as a therapy for acute severe heart failure had two primary objectives: to produce a compound which would have the beneficial effects of existing compounds – in particular afterload reduction and selective renal vasodilatation – and, secondly, to avoid some of the deleterious effects of conventional catecholamines such as an excessive myocardial oxygen demand, the provocation of ventricular arrhythmias or loss of activity during long-term treatment.

Dopexamine is an afterload-reducing agent with additional properties of renal vasodilatation and mild positive inotropism. These hemodynamic effects are achieved by

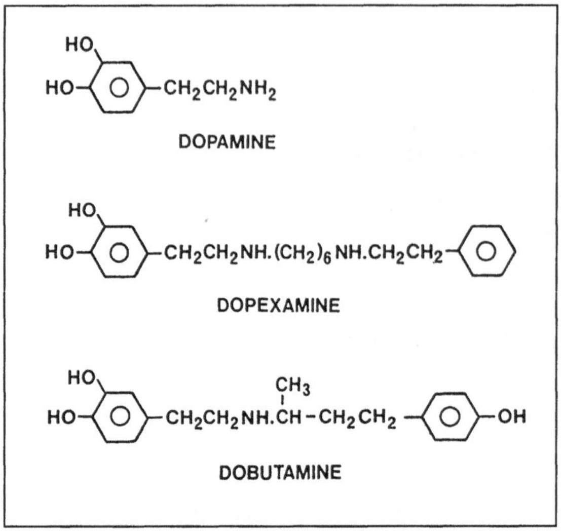

Fig. 1. Chemical structures

a unique pharmacological profile: it is primarily a combined dopaminergic (DA 1) and β_2-adrenergic agonist, and it has only a weak activity at the cardiac β_1-adrenoceptor and is devoid of α-adrenergic activity. This combination of properties gives dopexamine hydrochloride the ability to increase cardiac output secondary to afterload reduction and to improve renal blood flow. Therefore, it can be expected that myocardial oxygen consumption is not significantly altered.

Pharmacology

The activity of dopexamine at peripheral *dopamine (DA 1 and DA 2) receptors and adrenoceptors* (β_1, β_2, α) has been evaluated and compared with standard receptor agonists (Table 1). DA1 receptors occur predominantly in the arterial smooth muscle of certain organs, i.e., renal, mesenteric, coronary and cerebral vascular beds. Stimulation causes vasodilatation and, hence, increased blood flow at these sites; activation of the renal tubular DA1-receptors produces, in addition, natriuresis and diuresis. The vascular DA1-receptor activity of dopexamine was assessed in pentobarbitone-anesthetized dogs [5] in which DA2-receptors and α- and β-adrenoceptors were blocked. Dopexamine hydrochloride produced an increase in renal blood flow and a fall in renal vascular resistance, with a potency about one-third that of dopamine.

DA2-receptors are located prejunctionally on sympathetic nerve terminals; stimulation leads to a reduction in noradrenaline release. This results in a general reduction in sympathetic tone throughout the cardiovascular system. Excessive peripheral DA2-receptor activity is undesirable, since DA2-like receptors occur in the chemoreceptor trigger zone of the brain and stimulation of these receptors results in nausea and emesis. Dopexamine's DA2-agonist activity was studied in two systems. DA2-receptor stimulation produces inhibition of neurogenic vasoconstriction in the isolated perfused rabbit ear artery and inhibition of tachycardia caused by electrical stimulation of the cardiac accelerans nerve in the anesthetized cat. Dopexamine showed DA2-agonist effects in both tests, with a potency approximately one-fourth to one-sixth that of dopamine.

Table 1. Comparative agonist activities and actions of dopexamine hydrochloride, dopamine and dobutamine

Activity	Response to activity	Dopexamine hydrochloride	Dop- amine	Dobut- amine
DA_1	Renal and mesenteric vasodilatation; natriuresis, diuresis	+++	+++	NA
DA_2	Presynaptic sympathetic inhibition of noradrenaline release: generalized reduction of sympathetic tone (excessive stimulation leads to increased risk of nausea and emesis)	+	++	NA
β_1	Positive chronotropy/inotropy. Increased A-V conduction (leads to increased risk of arrhythmias)	(+)	++	+++
β_2	Peripheral vasodilatation (afterload reduction)	+++	+	++
α	Peripheral vasoconstriction. Arteriolar constriction	NA	++	++
Uptake-1 inhibition	Increase in the amount of noradrenaline in sympathetic synaptic cleft	+++	++	+

Activities: NA not active; (+) weak; + mild; ++ moderate; +++ powerful.

The β_1-*agonist activity* of dopexamine was assessed in isolated spontaneously beating guinea-pig atria [4]. Dopexamine produced only a small increase in atrial rate, acting as a very weak β_1-stimulant, and it was clearly weaker than isoprenaline and dopamine.

The agonist activity of dopexamine hydrochloride at the β_2-*receptor* was assessed in the isolated tracheal chain preparation of the guinea-pig [4]. Dopexamine was found to be a full agonist, 60-times more potent than dopamine but 32-times less potent than salbutamol (Fig. 2). Agonist activity of dopexamine hydrochloride at cardiac β_2-adrenoceptors in the human myocardium in vitro has been described by Brodde [6].

The agonist activity of dopexamine hydrochloride at α-*adrenoceptors* was investigated in the isolated saphenous vein strip of the dog [4], a preparation which contains postjunctional α_1- and α_2-adrenoceptors. Dopexamine hydrochloride was found to be devoid of direct α-adrenoceptor stimulant activity (Fig. 3).

In addition to the actions at dopaminergic and adrenergic receptors dopexamine has been investigated for *other pharmacological properties* which might be relevant to its therapeutic use in the treatment of low cardiac output states. Re-uptake into adrenergic nerve terminals (uptake-1) is a major physiological mechanism for the inactivation of neurogenic and circulating noradrenaline – and, to a lesser extent, adrenaline and dopamine. Under certain circumstances, the physiological effects of sympathetic nerve activity or circulating catecholamines may be enhanced and/or prolonged in the presence of uptake-1 blockade. There is experimental evidence that dopexamine produces an indirect sympathomimetic action mediated by inhibition of neuronal catecholamine uptake [14].

The propensity to cause or exacerbate cardiac arrhythmias is an undesirable feature of the catecholamines and many other agents used to treat heart failure. Dopexamine produced no arrhythmias at doses up to 10–50 µmol in isolated perfused guinea pig

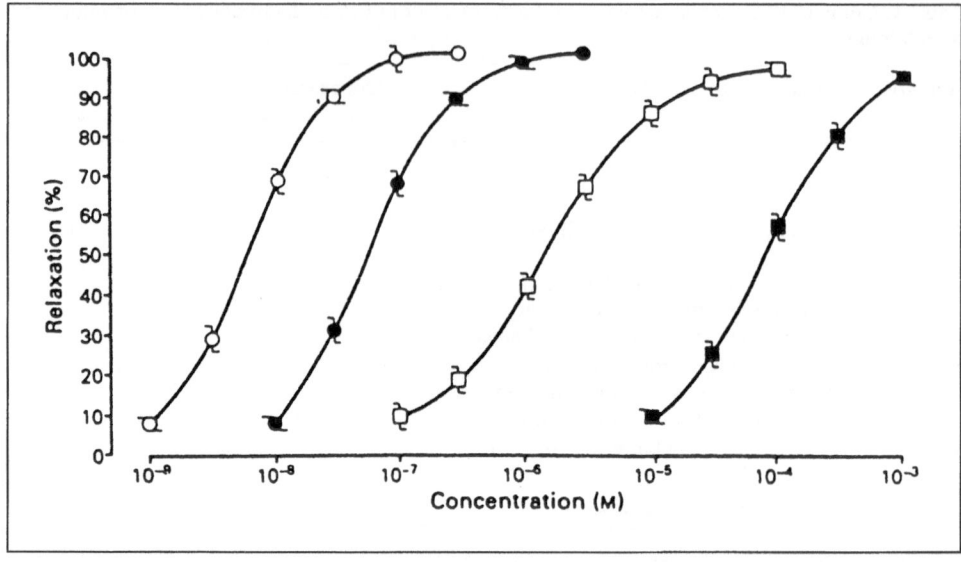

Fig. 2. Isolated tracheal chain of the guinea-pig: β_2-adrenoceptor-mediated inhibition of spontaneous contracture (mean with s.e. mean shown by bars), produced by isoprenaline (o, n = 28), salbutamol (●, n = 28), dopexamine (□, n = 24) and dopamine (■, n = 16): α- and β_1-adrenoceptors were blocked by phentolamine (10^{-5} M) and atenolol (4×10^{-6} M) respectively; (from [4], used with permission)

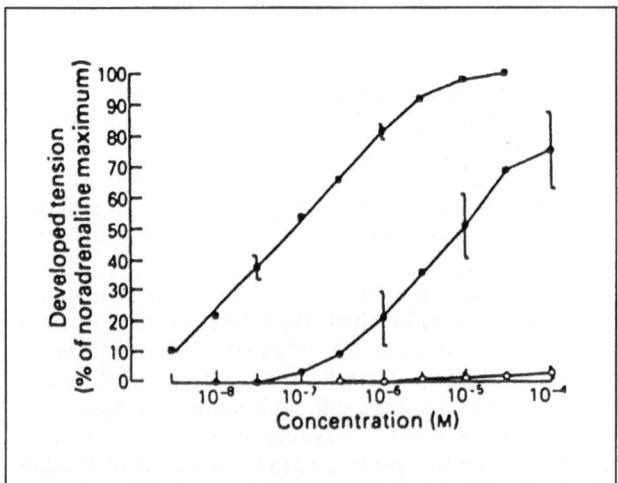

Fig. 3. Canine isolated saphenous vein strip: α-adrenoceptor-mediated contractures produced by noradrenaline (■, upper curve, n = 9) and dopamine (●, middle curve, n = 6) expressed as a percentage of the maximal noradrenaline (mean ± s.e. mean) response, contrasting with the lack of effect with dopexamine (o, bottom curve, n = 6); (from [4], used with permission)

hearts, while dopamine caused rhythm abnormalities [4]. Similarly, in the halothane-anesthetized rat dopexamine hydrochloride was not arrhythmogenic, whereas dopamine, adrenaline, and isoprenaline each caused arrhythmias.

Hemodynamic profile in animals

In pentobarbitone-anesthetized dogs, intravenous infusions of dopexamine produced dose-related falls in blood pressure and total peripheral resistance with increases in cardiac output, contractility, and heart rate. In contrast, dopamine did not significantly reduce blood pressure but increased both blood pressure and heart rate at the highest dose and produced larger dose-related increases in contractility. Both drugs caused renal and mesenteric vasodilatation [5].

Using different adrenergic and dopaminergic antagonists it could be shown that a combined stimulation of β_2-, DA1- and DA2-receptors was responsible for the fall in blood pressure and total peripheral resistance induced by dopexamine [15].

The increases in cardiac output, contractility, and heart rate produced by dopexamine hydrochloride were virtually abolished by beta-2-blockade (implicating β_2-adrenoceptor stimulation) whereas only a weak β_1-adrenoceptor-mediated cardiac stimulation could be blocked by atenolol [15].

In conscious dogs, intravenous infusions of dopexamine hydrochloride and dopamine have been compared for their hemodynamic and behavioral effects [5]. Dopexamine produced a fall in blood pressure, accompanied by increases in heart rate and cardiac contractility and renal vasodilatation with increased renal blood flow. Dopamine had similar effects on the renal bed and on cardiac contractility but did not effect blood pressure and heart rate except at the highest dose, when small increases occurred in both variables.

Hemodynamic profile and kinetics in healthy volunteers

Infusions of dopexamine (0.5–12.0 µg/kg/min, each dose-level being maintained for 12 min) produced a small increase in systolic blood pressure and a fall in diastolic pressure, and a dose-related increase in heart rate of approximately 50 beats/min at the top dose. Cardiac performance (measured by systolic time intervals and echocardiography) showed dose-related increases.

Dopexamine was administered to healthy volunteers by infusion in a stepwise manner from 1–4 µg/kg/min to give a total dose of 180 µg/kg. Following cessation of the infusion, blood concentrations fell rapidly with an initial halflife of 6–7 min. Blood clearance was high, ranging from 21–39 ml/min/kg.

Human renovascular profile

Magrini et al. [13] studied the hemodynamic and renovascular effects of dopexamine in 8 male patients with mild to moderate hypertension, who were undergoing renal vein catheterization. They found a dose related increment in renal blood flow, heart rate and systolic blood pressure. There was a slight but significant reduction in renal vascular resistance which was greater than that attributable to the increase in cardiac output or perfusion pressure alone.

Hemodynamic studies in congestive heart failure

The hemodynamic profile of dopexamine has been studied in short-term open studies in a large number of patients with congestive heart failure. Most of these patients received the drug at infusion rates of 0.5 to 6.0 µg/kg/min, each dose-level being maintained for 10–15 min. Dawson et al. [8] administered dopexamine to 10 patients with severe congestive heart failure at rates of 1, 3, and 6 µg/kg/min during diagnostic cardiac catheterization. They observed a significant dose-related increase in cardiac index, stroke volume index, and heart rate, and falls in systemic vascular resistance, pulmonary vascular resistance, and left ventricular enddiastolic pressure; aortic and pulmonary artery pressures remained unchanged. Isovolumic contraction and ejection phase indices were increased but these changes could not be attributed to improved myocardial contractility since dopexamine produced an increase in heart rate as well.

Svensson et al. [16] administered dopexamine at infusion rates ranging from 0.5 to 4.0 µg/kg/min to eight patients with congestive heart failure. A dose of 1 µg/kg/min produced a significant increase in cardiac index and a decrease in systemic vascular resistance. Stroke volume index and heart rate increased as well.

Dopexamine has been studied in patients with congestive heart failure up to a maximum rate of 6 µg/kg/min by Colardyn et al. [7]. They infused a dose that produced an "optimal improvement" for up to 48 h. The beneficial hemodynamic effects which were found during the titration period (increase in cardiac index and decrease in systemic vascular resistance) persisted during the 48-h-observation period.

Recently the hemodynamic effects of dopexamine were studied in a placebo controlled randomized double blind study [12]. The central and regional hemodynamic effects and responses in renal function were measured in 12 patients with low output congestive heart failure. Dopexamine significantly increased cardiac output and stroke volume index at a dose of more than 0.25 µg/kg/min. Similar to other studies dopexamine lowered significantly the systemic and pulmonary vascular resistance. Right and left ventricular filling pressures decreased and indices of left ventricular performance were improved. For the first time it could be convincingly shown in patients with congestive heart failure that dopexamine preferentially increased visceral (renal, hepatic-splanchnic) blood flow. Correspondingly, urine volume and sodium excretion were slightly increased by dopexamine.

The improvement of ventricular performance produced by dopexamine in this study could not be explained by the afterload reduction alone. The magnitude of the changes in cardiac performance was far greater than those observed after administration of agents with predominantly after- and/or preload reducing properties. The authors concluded that a portion of the dopexamine-induced improvement in ventricular performance was related to a positive inotropic response, probably mediated by the stimulation of ventricular β_2-receptors and a mild β_1-receptor stimulation as well.

The important question of whether the beneficial effects of dopexamine on ventricular performance are only secondary to afterload reduction and reflex cardiac stimulation, or whether dopexamine has a substantial positive inotropic effect was discussed recently [9]. Based on the data od Jaski et al. [10] and Tan et al. [18] there is also some evidence that the dopexamine-induced improvement of cardiac performance is at least in part due to a direct positive inotropic action of the drug.

The direct positive inotropic action of dopexamine might be more pronounced in patients with congestive heart failure than in animals with normal heart function. First there is some evidence that β_2-receptors have a greater importance in hearts of patients with endstage heart failure in whom the number of β_1-receptor has been found to be reduced

[10]. Secondly, patients with congestive heart failure have high levels of circulating cate-cholamines, and prevention of uptake of noradrenaline by dopexamine might increase the noradrenaline concentration at the β_1-receptor site. However, it should be kept in mind, that down-regulation of adrenoceptors and attenuated hemodynamic response has been observed following prolonged administration of β-agonists [7a]. The development of tolerance occurred both to β_1-agonists such as dobutamine and β-agonists stimulating predominantly β_2-receptors such as pirbuterol.

Data on prolonged administration of dopexamine in severe heart failure are scarce, however, in a recent report, a rapid decline of the therapeutic effect was observed after prolonged infusion of dopexamine [14a]. In contrast, others reported a sustained hemody-namic effect of dopexamine [7, 1]. Thus, data on prolonged administration are conflicting and therefore the ultimate value of this new β-adrenoceptor agonist during prolonged in-fusion remains to be established.

In this context, also the positive chronotropic effect of dopexamine merits consider-ation. In fact, the positive chronotropic effect observed with dopexamine in most clinical studies may limit its use in patients with coronary artery disease in whom the relation be-tween myocardial supply and demand may be precarious [8].

Comparative hemodynamic studies

Bonnier [2] compared the hemodynamic effects of dopexamine and dobutamine in a ran-domized cross-over study in 12 patients with congestive heart failure. A similar increase in cardiac index was seen with infusion rates of dopexamine of 2 µg/kg/min compared to 10 µg/kg/min of dobutamine. Reductions of pulmonary artery capillary wedge pres-sure were similar with both drugs, whereas the decrease in systemic vascular resistance was more pronounced with dopexamine.

Jackson et al. [11] reported that dopexamine was more effective on a molar basis than dopamine. They found that infusion rates of 0.5, 1, and 2 µg/kg/min of dopexamine pro-duced an increase in cardiac index similar to that with dopamine infusion rates of 2.5, 5, and 10 µg/kg/min.

Baumann et al. [1] compared the acute hemodynamic effects of dopexamine, dobut-amine, and sodium nitroprusside in 12 patients with dilated cardiomyopathy and chronic heart failure. All treatments significantly increased cardiac output but hemodynamic pro-files differed. Dobutamine caused a large increase in cardiac output, a small rise in sys-temic blood pressure, and a small decrease in pulmonary capillary wedge pressure, with little change in heart rate. Dopexamine had similar effects on cardiac output and heart rate but the decrease in pulmonary and systemic vascular resistance was clearly more pro-nounced, indicating a greater degree of vasodilatation. Nitroprusside had less effect on cardiac output than the other treatments but the marked reduction in blood pressure caused a large increase in heart rate. At doses producing an increase of 30% in cardiac output dobutamine and dopexamine caused this effect almost entirely through an im-provement in stroke volume.

For nitroprusside a significant contribution came from the increased heart rate. Sys-temic vascular resistance was reduced most by nitroprusside and least by dobutamine. Dopexamine and nitroprusside each produced a 40% reduction in pulmonary vascular resistance, double that seen with dobutamine. In summary, according to the data of Bau-mann et al. [1], dopexamine produced acute hemodynamic effects intermediate between those of dobutamine and nitroprusside, when vascular resistance, systemic blood pres-sure, and stroke volume are compared. Such a drug with a combined vasodilatory and inotropic action might have advantages over a pure β_1-receptor stimulator or a pure vasodilator in the treatment of congestive heart failure.

Side effects

Most of the adverse effects reported with dopexamine are short-lived and rapidly reversed by down-titration or discontinuation of the infusion. Of 200 patients receiving dopexamine two deaths have occurred during dopexamine infusion and one additional patient died during the post-infusion period. These deaths were not considered to be related to dopexamine infusion but rather due to the severity of the underlying diseases.

Cardiovascular events, such as myocardial infarction, ventricular fibrillation, angina pectoris, third-degree AV-block, atrial fibrillation, and ventricular arrhythmias were seen in a few patients treated with dopexamine. The relationship between these events and the dopexamine infusion remains questionable in most of these cases. There is however some evidence that higher infusion rates of dopexamine (6 µg/kg/min) might produce angina pectoris in some patients with severe coronary artery disease.

Nausea and vomiting are the commonest side-effects seen with dopexamine, the former being more common than the latter. Both are considered to be due to the DA2-agonist property of the drug with stimulation of the chemoreceptor trigger zone. In the majority of these cases infusion rates above the current recommended maximum were being administered.

Open questions

Despite the promicity pharmacological profile of dopexamine important questions remain to be answered:

1) Is dopexamine less arrhythmogenic compared to other catecholamines?

2) What happens during long-term administration with the β_2-receptors? Is there a significant β_2-receptor down-regulation as recently described and discussed with β_1-receptors [3]?

3) Which substances might be combined with dopexamine in the clinical daily work? Which are the substances to achieve additional effects (e.g., phosphodiesterase inhibitors)?

4) Are there beneficial effects in acute myocardial infarction?

5) Is there a rationale to combine an α-agonist with a β_2-agonist in septic conditions?

6) Which are the treatment effects in patients on long-term β-blocker medication?

7) Is there a rationale for using the drug in pretransplant patients?

Conclusion

The hemodynamic results from more than 400 patients have shown that dopexamine doses in the range of 0.5–4 µg/kg/min are effective and well tolerated in the majority of patients. For most patients the 4 µg/kg/min dose is the maximum dose required to achieve the optimal hemodynamic effect. A minority of patients – mainly those who have received recent β-blockade – may benefit from increasing the infusion rate to 6 µg/kg/min.

Most of the adverse events reported with dopexamine are predictable from the knowledge of the drug's pharmacology. These events are short-lived and rapidly reversed by down-titration or discontinuation of the infusion. In the disease areas being treated it is important to remember that a background of concurrent events is normal and it is therefore almost impossible to attempt to associate cause with event.

The data so far generated on dopexamine would suggest that this compound has many properties which are considered to be desirable in the acute management of severe heart failure. The profile of preferential renal vasodilatation with afterload reduction and mild cardiac stimulation should prove useful in the short-term management of low cardiac output states.

References

1. Baumann G, Gutting M, Pfafferoth C, Ningel K, Klein G (1988) Comparison of acute haemodynamic effects of dopexamine hydrochloride, dobutamine and sodium nitroprusside in chronic heart failure. Eur Heart J 9:503–512
2. Bonnier JJRM (1986) Dopexamine hydrochloride, haemodynamic effects in haemochronic cardiac failure, a comparison with dobutamine. Abstracts of X World Congress of Cardiology, Washington DC, September, no. 1076, p 188
3. Bristow MR, Ginsburg R, Umans V, Fowler M, Minobe W, Rasmussen R, Zera P, Menlove R, Shah P, Jamieson S, Stinson EB (1986) Beta-1- and beta-2-adrenergic receptor subpopulations in nonfailing and failing human ventricular myocardium: coupling of both receptor subtypes to muscle contraction and selective beta-1-receptor down-regulation in heart failure. Circ Res 59(3):297–309
4. Brown RA, Dixon J, Farmer JB, Hall JC, Humphries RG, Ince F, O'Connor SE, Simpson WT, Smith GW (1985a) Dopexamine: a novel agonist at peripheral dopamine receptors and $beta_2$-adrenoceptors. Br J Pharmacol 85(3):599–608
5. Brown RA, Farmer JB, Hall JC, Humphries RG, O'Connor SE, Smith GW (1985b) The effects of dopexamine on the cardiovascular system of the dog. Br J Pharmacol 85 (3):609–619
6. Brodde OE (1988) The functional importance of $beta_1$ and $beta_2$ adrenoceptors in the human heart. Am J Cardiol 62:24c–29c
7. Colardyn F, Clement DL (1986) Acute and long-term haemodynamic effects of dopexamine hydrochloride in patients with chronic cardiac failure. Abstracts of X World Congress of Cardiology, Washington, DC, September, no. 1980, p 346
7a. Colucci WS, Wright RF, Braunwald E (1986) New positive inotropic agents in the treatment of congestive heart failure. N Engl J Med 314:290–299
8. Dawson JR, Thompson DS, Signy M, Juul SM, Turnbull P, Jenkins BS, Webb-Peploe MM (1985) Acute heamodynamic and metabolic effects of dopexamine, a new dopaminergic receptor agonist, in patients with chronic heart failure. Br Heart J 54(3):313–320
9. Goldberg LI (1988) Dopamine agonists in intensive care medicine: from receptors to clinical application. In: Vincent JL (ed) Update in intensive care and emergency medicine, vol 5. Springer, Berlin Heidelberg New York, pp 687–689
10. Jaski BE, Wijns W, Foulds R, Serruys PW (1986) The haemodynamic and myocardial effects of dopexamine: a new $beta_2$-adrenoceptor and dopaminergic agonist. Br J Clin Pharmacol 21(4):393–400
11. Jackson N, Frais M, Sharma SK, Reynolds G, Taylor SH (1986) A dose response haemodynamic study of dopexamine vs dopamine in ischaemic left ventricular failure. Abstract X World Congress of Cardiology, Washington, DC, September, no. 2302, p 400
12. Leier CV, Binkley PF, Carpenter J, Randolph PH, Unverferth DV (1988) Cardiovascular pharmacology of dopexamine in low output congestive heart failure. Am J Cardiol 62(1):94–99
13. Magrini F, Foulds R, Roberts N, Macchi G, Mondadori C, Zanchetti A (1987) Human renovascular effects of dopexamine hydrochloride: a novel agonist of peripheral dopamine and $beta_2$-adreno-receptors. Eur J Clin Pharmacol 32(1):1–4

14. Mitchell PD, Smith GW, Wells E, West PA (1987) Inhibition of Uptake1 by dopexamine hydro-
    chloride in vitro. Br J Pharmacol 92(2):265–270
14a. Murphy JJ, Hampton JR (1988) Failure of dopexamine to maintain haemodynamic
    improvemennt in patients with chronic heart failure. Br Heart J 60(1):45–49
15. Smith GW, Hall JC, Farmer JB, Simpson WT (1987) The cardiovascular actions of dopexamine
    hydrochloride, an agonist at dopamine receptors and beta$_2$-adrenoceptors in the dog. J Pharm
    Pharmacol 39(8):636–641
16. Svensson G, Sjogren A, Erhardt L (1986) Short-term haemodynamic effects of dopexamine in
    patients with chronic congestive heart failure. Eur Heart J 7(8):697–703
17. Tan LB, Smith SA, Littler WA, Murray RG (1987) Haemodynamic and renal effects of dopex-
    amine in low output heart failure. Br Heart J 57(1):81
18. Tan LB, Littler WA, Murray RG (1987) Beneficial haemodynamic effects of intravenous dopex-
    amine in patients with low-output heart failure. J Cardiovasc Pharmacol 10(3):280–286

Author's address:

Prof. Dr. T. Meinertz, Medizinische Universitätsklinik, Innere Medizin III, Kardiologie,
Universität Freiburg, Hugstetter Strasse 55,D-7800 Freiburg, FRG

The use of levodopa, an oral dopamine precursor, in congestive heart failure

G. Broderick, S. I. Rajfer

Committee on Clinical Pharmacology,
Departments of Medicine (Section of Cardiology) and
Pharmacological and Physiological Sciences, The University of Chicago,
Chicago, Illinois, USA

Summary

The successful treatment of congestive heart failure with intravenous dopamine in the acute setting has prompted investigation into the development and use of oral dopamine analogs. The administration of dopamine can lead to an improvement in myocardial pump performance through a combination of afterload reduction and augmented contractile state. The ingestion of levodopa, an oral dopamine precursor, is associated with sustained hemodynamic and clinical improvement in patients with congestive heart failure. Improvements in hemodynamic performance correlated with the generation of substantial amounts of dopamine. Current research efforts are directed at developing oral dopamine analogs that exhibit improved bioavailability and do not traverse the blood-brain barrier.

Introduction

Congestive heart failure, which affects approximately four million Americans, is a devastating disease with an extremely poor prognosis. Therapeutic strategies aimed at improving the derangements in loading conditions and contractile state of the failing left ventricle may improve the clinical status of patients with heart failure. Digitalis glycosides are currently the only oral positive inotropic agents approved for the treatment of heart failure in the United States. The beneficial effects of intravenous dopamine, a potent positive inotropic drug, in patients with congestive heart failure have prompted the investigation and development of orally available dopamine analogs. The ingestion of one such agent, levodopa, an oral dopamine precursor, has been associated with significant clinical and hemodynamic improvement in patients with heart failure. An understanding of the mechanisms by which dopamine produces an improvement in cardiac performance is essential in order to further refine oral dopamine analogs for use in congestive heart failure.

Effects of dopamine in heart failure

The pharmacological actions of dopamine can be attributed to the activation of dopamine receptors and β_1- and α-adrenoceptors. Two pharmacologic subtypes of the dopamine receptor have been defined: DA_1 and DA_2. DA_1 receptors are located postsynaptically and cause vasodilation primarily in the renal, mesenteric, coronary, and cerebral blood vessels [5]. DA_2 receptors are found on postganglionic sympathetic nerves and autonomic ganglia; DA_2 activation causes inhibition of norepinephrine release. DA_2 receptors are also found in the chemoreceptor trigger zone and induce emesis when activated. At low infusion rates (usually 0.5 to 2 µg/kg/min) dopamine acts predominantly on DA receptors resulting in a decrease in peripheral vascular resistance, and elevation in renal blood flow, urine volume, and sodium excretion [6]. Significant stimulation of β_1-adrenoceptors becomes apparent at infusion rates of 2 µg/kg/min and above; an increase in myocardial contractility and heart rate occurs as a result of this action of dopamine. The recruitment of α-adrenoceptors is observed at infusion rates above 4–6 µg/kg/min and causes progressive vasoconstriction which may decrease myocardial performance. The undesired vasoconstriction can be blocked by α-adrenoceptor antagonists or countered with other vasodilating agents.

Recent studies have used more sophisticated measures of myocardial mechanics to highlight the effects of β_1-adrenoceptor and DA receptor activation on ventricular performance [2, 11]. Using a combined cardiac catheterization and echocardiographic protocol, the relative contributions of afterload reduction and augmented contractility on ventricular performance in heart failure patients given dopamine or dobutamine were examined. Measurements of left ventricular end-systolic wall stress [7, 8] and the rate-corrected velocity of left ventricular fiber shortening were determined during graded infusions of dopamine and dobutamine. The relationship between these two parameters has been used as a load-independent measure of contractility in man [1]. While both agents caused a significant increase in myocardial contractility, dopamine also produced a significant decrease in left ventricular end-systolic wall stress (an index of left ventricular afterload); with dobutamine, end-systolic wall stress either remained unchanged or increased [11]. The decline in left ventricular afterload seen with dopamine was ascribed to the activation of DA receptors.

Levodopa

Levodopa is an oral dopamine precursor which has been used for years in patients with Parkinson's disease. Levodopa, which is pharmacologically inert, is converted to dopamine endogenously by aromatic amino acid decarboxylase. The dopamine thus generated is degraded to homovanillic acid as its major metabolite. Accordingly, it was hypothesized that levodopa given to patients with congestive heart failure would result in hemodynamic responses similar to those seen with dopamine.

Patients with congestive heart failure were studied hemodynamically after acute ingestion of 1.5–2 g of levodopa following a brief titration period [9, 10]. Serial hemodynamic measurements along with plasma dopamine and norepinephrine levels were obtained. Levodopa ingestion caused a significant increase in cardiac and stroke-volume index accompanied by a significant reduction in systemic vascular resistance. Mean arterial blood pressure, heart rate, and pulmonary capillary wedge pressure changed minimally. A generous production of dopamine was observed and correlated with the increase in cardiac index. These hemodynamic responses to orally administered levodopa resembled those seen with dopamine infusions of 1–4 µg/kg/min, and the dopamine levels measured after levodopa ingestion are similar to levels obtained in other studies at these infusion rates.

After long-term treatment with levodopa (3–14 months), patients with heart failure exhibited a sustained improvement in cardiac performance [10]. Plasma dopamine levels were markedly elevated after ingestion of levodopa and, again, correlated with improved pump function. These beneficial hemodynamic effects of levodopa in patients with heart failure can be attributed to the activation of myocardial β_1-adrenoceptors and DA receptors by the generated dopamine.

The acute oral administration of levodopa to patients with heart failure is associated with an increase in plasma norepinephrine levels, which did not correlate with the improvement in cardiac index. Interestinly, after chronic therapy the plasma norepinephrine concentrations return to baseline levels [10]. The acute increase in norepinephrine levels may be due to its release from neuronal storage sites at the synaptic cleft, which over the long-term can decline due to stimulation of prejunctional α_2-adrenoceptors [4]. It would appear that the DA_2 activity of dopamine is not responsible for the improvement in pump performance observed with levodopa.

The most common adverse effects seen with levodopa therapy are nausea and vomiting. These side effects can be frequently controlled with gradual titration of dosage. Most standard anti-emetic agents are DA_1 and DA_2 antagonists, and their use in these patients could lessen the beneficial hemodynamic effects of levodopa. Low doses of metoclopramide, however, may selectively block DA_2 receptors on the chemoreceptor trigger zone and can be given to patients with levodopa without clinical deterioration. The importance of gradual drug titration has also been underscored by the observations of Shah et al. [13] who noted clinical deterioration in a patient with congestive heart failure given levodopa manifested by a rise in pulmonary capillary wedge pressure and systemic vascular resistance. Levodopa ingestion may have generated enough dopamine to cause activation of α-adrenoceptors in this instance, leading to the observed hemodynamic changes.

Since levodopa can have complicated hemodynamic effects and randomized, placebo-controlled trials have not yet been performed, its widespread use in congestive heart failure cannot be recommended at this time. No data are available on its effects on survival in congestive heart failure. However, early studies show that its augmentation of myocardial performance can result in sustained hemodynamic and clinical improvement in heart failure. A search for orally effective dopamine analogs that do not traverse the blood-brain barrier is currently underway.

References

1. Borow KM, Neumann A, Lang RM (1986) Milrinone vs dobutamine: contribution of altered myocardial mechanics and augmented inotropic state to improved left ventricular performance. Circulation 73 (Suppl III):153–160
2. Colan SD, Borow KM, Neumann A (1984) Left ventricular end-systolic wall stress-velocity of fiber shortening relation: a load-independent index of myocardial contractility. J Am Coll Cardiol 4:715–724
3. Colucci WS, Alexander RW, Williams GH, Rude R, Holman B, Konstam M, Wynne J, Mudge G, Braunwald E (1981) Decreased lymphocyte β-adrenergic receptor density in patients with heart failure and tolerance to the β-adrenergic agonist pributerol. New Engl J Med 305:185–190
4. Cousineau D, Govesky CA, Rose CP (1986) Decreased basal cardiac interstitial norepinephrine release after neuronal uptake inhibition in dogs. Circ Res 58:859–866
5. Goldberg LI, Rajfer SI (1985) Dopamine receptors: applications in clinical cardiology. Circulation 72:245–248
6. Goldberg LI, Hsieh YY, Resnekov L (1977) Newer catecholamines for treatment of heart failure and shock: an update on dopamine and a first look at dobutamine. Prog Cardiovasc Dis 4:327–340

7. Laskey WK, St John Sutton M, Zeevie G, Hirshfeld JW, Reichek N (1984) Left ventricular mechanics in dilated cardiomyopathy. Am J Cardiol 54:620–625
8. Lang RM, Borow KM, Neumann A, Janzen D (1986) Systemic vascular resistance: an unreliable index of left ventricular afterload. Circulation 74:1114–1123
9. Rajfer SI, Anton AH, Rossen JD, Goldberg LI (1984) Beneficial effects of oral levodopa in heart failure. N Engl J Med 310:1357–1362
10. Rajfer SI, Rossen JD, Nemanich JW, Douglas F, Davis F, Osinski J (1987) Sustained hemodynamic improvement during long-term therapy with levodopa in heart failure: role of plasa catecholamines. J Am Coll Cardiol 10:1286–1293
11. Rajfer SI, Borow KM, Lang RM, Neumann A, Carroll JD (1988) Effects of dopamine on left ventricular afterload and contractile state in heart failure: relation to the activation of β_1-adrenoceptors and dopamine receptors. J Am Coll Cardiol 12:498–506
12. Rajfer SI, Rossen JD, Douglas FL, Goldberg LI, Karrison T (1986) Effects of long-term therapy with oral ibopamine on resting hemodynamics and exercise capacity in patients with heart failure: relationship to the generation of N-methyldopamine and to plasma norepinephrine levels. Circulation 73:740–748
13. Shah PK, Amin DK, Horn E (1985) Adverse clinical and hemodynamic effects of oral levodopa in chronic congestive heart failure. Am Heart J 110:488–489

Author's address:

Dr. George Broderick, Committee on Clinical Pharmacology, Departments of Medicine (Section of Cardiology) and Pharmacological and Physiological Sciences, The University of Chicago, 947 East 58th Street, Chicago, IL 60637, USA

Clinical relevance of long-term therapy with levodopa and orally active dopamine analogues in patients with chronic congestive heart failure

G. Hasenfuss, Hj. Just

Medizinische Universitätsklinik, Innere Medizin III, Kardiologie,
Universität Freiburg, FRG

Summary

Beneficial effects of long-term treatment with dopamine analogues in patients with congestive heart failure may result from their vasodilating properties, in particular from renal artery vasodilation. Oral application of levodopa results in increased dopamine plasma levels and can improve cardiac performance and renal function in patients with congestive heart failure. A daily levodopa dosage of at least 4 g appears to a prerequisite for long-term response to the drug. Because of frequent side effects including nausea, vomiting, and dyskinesia at this dosage, the clinical usefulness of levodopa seems to be limited to a minority of patients. Ventricular arrhythmias have been shown to increase significantly during long-term levodopa therapy, probably due to stimulation of myocardial beta receptors. Increased ventricular arrhythmias or significant central nervous side effects have not been observed after administration of ibopamine and fenoldopam, which are orally active analogues of dopamine. Both agents exhibit potent arterial vasodilating properties and have been shown to increase cardiac performance in patients with congestive heart failure after short-term administration. The long-term beneficial effects of ibopamine and fenoldopam in the treatment of congestive heart failure have not yet been clarified. However, available results are encouraging and warrant further clinical evaluation of these agents, as well as the development of new analogues of dopamine, in particular of potent vascular dopamine$_1$ agonists.

Introduction

Intravenous administration of dopamine is well established for the short-term treatment of patients with congestive heart failure. Beneficial hemodynamic effects result from inotropic stimulation by activation of beta-receptors and from vasodilation by activation of dopaminergic receptors. The degree to which the different receptors are activated depends on the dosage applied [12, 23]. Because of the acute hemodynamic effects of dopamine, interest has been directed towards orally active dopamine analogues for the long-term treatment of congestive heart failure. The beneficial effects of long-term inotropic therapy with beta-receptor agonists or phosphodiesterase inhibitors, both of which increase cyclic AMP is questionable. Some studies even indicate detrimental effects of those agents on survival of patients with congestive heart failure [10, 20]. In contrast, long-term therapy with vasodilators can improve clinical symptoms and prognosis in patients with congestive heart failure [6, 7, 9]. Therefore, dopamine analogues with potent vasodilating actions may be favorable in the long-term treatment of those patients. In particular, dopamine analogues with potent effects on the renal vasculature may be of advantage. Renal hypoperfusion in patients with congestive heart failure activates the renin-angiotensin system which results in constriction of the efferent arteriole. This mechanism maintains glomerular filtration rate by increasing glomerular hydraulic pressure and filtration fraction despite reduced renal plasma flow [22]. The disadvantage of increased filtration fraction is avid tubular reabsorption of sodium and water and reduced efficiency of loop diuretics [2, 14, 30]. Dopamine analogues, by increasing renal plasma flow, counteract avid tubular sodium and water reabsorption and may restore the efficiency of diuretics [1]. Moreover, increased renal blood flow may prevent the development of renal insufficiency frequently associated with angiotensin converting enzyme inhibitor treatment [21, 22]. Therefore, dopamine analogues may be especially beneficial in combination with angiotensin converting enzyme inhibitors. Several studies have evaluated the usefulness of levodopa and other orally available analogs of dopamine like ibopamine or fenoldopam for acute and chronic treatment in patients with congestive heart failure.

Effects of levodopa, ibopamine, and fenoldopam in patients with congestive heart failure

Levodopa has been used for the treatment of Parkinson's disease for many years. Finlay et al. and Whittsett and Goldberg et al. demonstrated that acute administration of levodopa increases renal blood flow and cardiac performance in patients with Parkinson's disease [11, 32]. The clinical usefulness of levodopa for treatment of patients with congestive heart failure was suggested by Rajfer et al. This group demonstrated that short-term oral administration of 1.5–2 g levodopa results in increased dopamine plasma levels, increased cardiac index, and reduced systemic vascular resistance and that similar hemodynamic effects are present after long-term treatment with the drug for 12 weeks [3, 24, 25]. The beneficial effects of levodopa obtained in these studies seem to result predominantly from arterial vasodilation [25].

To gain more information about the clinical relevance of levodopa for the treatment of congestive heart failure we conducted a long-term study in which symptoms (NYHA classes), hemodynamics (right heart catheterization), arrhythmias (24-h Holter electrocardiogram), and renal plasma flow (I-hippuran clearance) were evaluated in patients with severe congestive heart failure (NYHA III–IV) [15]. The study was designed as a withdrawal study, comparing evaluations after an average period of 30 days on levodopa therapy with those obtained after withdrawal of levodopa. Patients who showed both clinical (NYHA classes) as well as hemodynamic deterioration after withdrawal of levodopa were defined as long-term responders to the drug. Levodopa was added to the previous treatment which included digitalis and diuretics. The levodopa dosage has been titrated beginning with 4×250 mg per day to the maximum permitted dosage or up to 4×1 g per day. Of the 17 patients who entered the study, 11 completed the protocol. Two patients died during levodopa therapy. In one patient levodopa was discontinued because of progressive deterioration of the clinical symptoms, and in three patients levodopa was discontinued because of side effects. During levodopa therapy, pulmonary capillary wedge pressure and mean right atrial pressure were significantly lower and renal plasma flow was significantly higher compared to measurements after withdrawal of levodopa. Of the 11 patients who completed the study, seven improved by one NYHA class during levodopa therapy. After withdrawal of levodopa, four patients deteriorated again by one NYHA class (Fig. 1). Three patients in whom the deterioration in NYHA classification was associated with impaired hemodynamic parameters after withdrawal of levodopa were classified as long-term responders to the drug. All responders were treated with

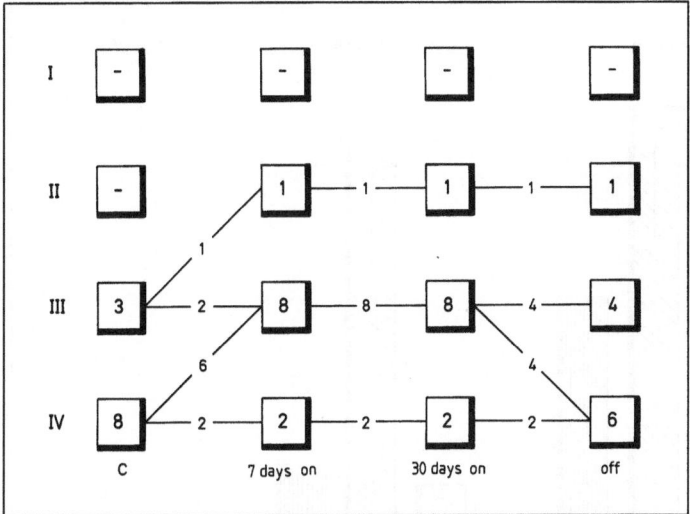

Fig. 1. Clinical status according to New York Heart Association functional classes of 11 patients with chronic congestive heart failure. Before levodopa therapy (C), 7 days after initiation of levodopa therapy (7 days on), 30 ± 1 days after initiation of levodopa therapy (30 days on), and after withdrawal (off) of levodopa. (Reproduced with permission from [15])

4×1 g levodopa per day and had higher dopamine plasma levels than the nonresponders. Dopamine plasma levels of the responders were comparable to those obtained during continuous infusion of dopamine at 2 µg/kg/min, a dosage which predominantly causes arterial vasodilation.

A daily dosage of 4×1 g of levodopa, therefore, seems to be a prerequisite for a potential long-term response to levodopa in patients with congestive heart failure. Because of side effects, a daily dosage of 4×1 g was permitted in only six of the patients. This is in contrast to the studies of Rajfer et al. in which patients were treated with levodopa dosages between 6 to 8 g levodopa per day [3, 24, 25]. The reason that higher dosages could be used without significant side effects in those studies may be related to differences in the patient populations or in the dose titration protocols used compared to our study.

The incidence of side effects was considerable in our patients. Three of 17 patients who entered the study dropped out because of intolerable side effects (dyskinesia, nausea and vomiting) even at lower dosages of levodopa. Five of the 11 patients who completed the study exhibited side effects which could be controlled by reducing the levodopa dosages or by additional treatment with metoclopramide. Ventricular arrhythmias were significantly increased during levodopa therapy (Fig. 2) in both responders and nonresponders to the drug.

From this data we conclude that long-term oral levodopa therapy may be useful in only a minority of patients with congestive heart failure who tolerate a daily dosage of levodopa of 4×1 g without significant side effects.

Ibopamine is the diisobutyric ester of epinine (N-methyl-dopamine) which after ingestion is hydrolyzed to the dopamine analog epinine [27]. Like dopamine, epinine activates dopaminergic, β- and α-receptors in a dose-dependent fashion [16, 26]. In patients with congestive heart failure, acute oral administration of ibopamine [100-350 mg) resulted in increased cardiac output and decreased systemic vascular resistance [8, 26, 28,

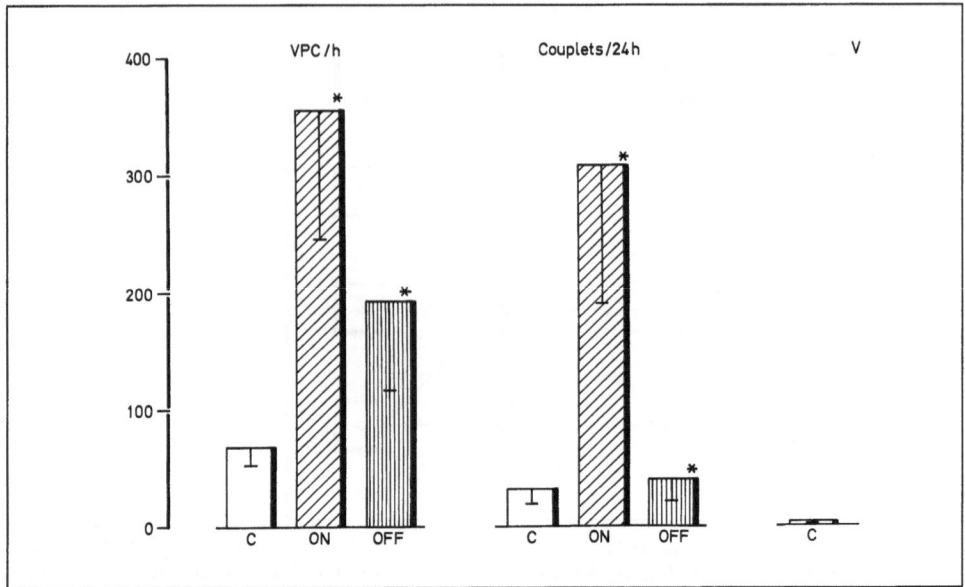

Fig. 2. Cardiac arrhythmias before (C), during (on), and after withdrawal of levodopa therapy (off). VPC = ventricular premature contractions; VT = nonsustained ventricular tachycardia (three or more VPC, duration less than 30 s); *=$p<0.05$. VPC are given as events/h; couplets and VT as events/24 h. (Reproduced with permission from [15])

29]. To a large extent, these effects appear to be mediated through vasodilatory properties rather than direct positive inotropy [29]. Furthermore, ibopamine was shown to increase diuresis and natriuresis in patients with congestive heart failure [18]. In contrast to levodopa, ibopamine does not seem to be associated with significant side effects from the central nervous system [4, 8, 19, 26, 28]. Holter electrocardiogram recordings over 24 h showed no increase in ventricular arrhythmias with ibopamine [4]. The long-term effects of ibopamine in patients with congestive heart failure have not yet been clarified. Rajfer et al. observed attenuation of the initial hemodynamic response to the drug after 8 weeks of treatment. Exercise tolerance did not improve during ibopamine therapy in these patients [26]. In contrast, in two placebo-controlled trials, Caponnetto et al. [5] and Cadel et al. [4] observed improved exercise capacity after 10 days and 1 month of ibopamine treatment, respectively.

Fenoldopam is a benzazepine derivative that appears to act primarily by stimulation of dopamine$_1$ receptors, thus being a potent vasodilator of the renal vasculature [13]. In healthy humans fenoldopam resulted in increased renal blood flow and fractional excretion of sodium [31]. Short-term administration of fenoldopam (50–200 mg) in patients with congestive heart failure increased cardiac index and reduced systemic vascular resistance; no significant side effects were observed [33]. The preliminary analysis of a double blind, placebo-controlled trial shows increased exercise tolerance and improved clinical symptoms after 8 weeks of continuous treatment with fenoldopam (200–300 mg per day) [17].

In conclusion, potential beneficial effects of long-term treatment of patients with congestive heart failure with dopamine analogs may result from their vasodilating properties, in particular from renal artery vasodilation. Short-term oral administration of levodopa, ibopamine, and fenoldopam resulted in beneficial hemodynamic effects in patients with congestive heart failure. The long-term effects of these agents and other dopamine analogues in the treatment of congestive heart failure warrant further investigations.

References

1. Beregovich J, Bianchi C, Rubler S, Lomnitz E, Cagin N, Levitt B (1974) Dose-related hemodynamic and renal effects of dopamine in congestive heart failure. Am Heart J 87:550–557
2. Brater DC (1981) Resistance to diuretics: emphasis on a pharmacological perspective. Drugs 22:477–494
3. Broderick G, Rajfer SI (1989) The use of levodopa, an oral dopamine precursor in congestive heart failure (this book)
4. Cadel A, Brusoni B, Pirelli P, Osculati G, Rovati A, Fanciulli R, Carati L, Valagussa F (1986) Effects of digoxin, placebo and ibopamine on exercise tolerance and cardiac rhythm of patients with chronic post-infarct left ventricular failure. Arzneim-Forsch/Drug Res 36(I):376–379
5. Caponnetto S, Allegro A, Bellotti G, et al. (1986) Positive inotropic effects of ibopamine in patients with congestive heart failure. Arzneim-Forsch/Drug Res 36(I):386–390
6. The Captopril-Digoxin Multicenter Research Group (1988) Comparative effects of therapy with captopril and digoxin in patients with mild to moderate heart failure. JAMA 259:539–544
7. Cohn JN, Archibald DG, Ziesche S, et al. (1986) Effects of vasodilator therapy on mortality in chronic congestive heart failure. Results of a Veterans Administration Cooperative Study (V-HeFT). N Engl J Med 314:1547–1552
8. Col J, Mievis E, Reynaert M (1983) Ibopamine in very severe congestive heart failure: pilot hemodynamic invasive assessments. Eur J Clin Pharmacol 24:297–300
9. The CONSENSUS Trial Study Group (1987) Effects of enalapril on mortality in severe congestive heart failure. Results of the Cooperative North Scandinavian Enalapril Survival Study (CONSENSUS). N Engl J Med 314:1547–1552
10. Dies F, Krell MJ, Whitlow P, et al. (1986) Intermittend dobutamine in ambulatory outpatients with chronic cardiac failure. Circulation 74 (suppl II):38
11. Finlay GD, Whittsett TL, Cucinell E, Goldberg LI (1971) Augmentation of sodium and potassium excretion, glomerular filtration rate and renal plasma flow by levodopa. N Engl J Med 284:865–870
12. Goldberg LI (1972) Cardiovascular and renal actions of dopamine: potent clinical applications. Pharmacol Rev 24:1–29
13. Hahn RA, Wardell JR, Sarau HM, Ridley PT (1982) Characterization of the peripheral and central effects of SK&F 82526, a novel dopamine receptor agonist. J Pharmacol Exp Ther 223:305–313
14. Hasenfuss G, Holubarsch C, Herzog C, Knauf H, Spahn H, Mutschler E, Just H (1987) Influence of cardiac function on the diuretic and hemodynamic effects of the loop diuretic piretanide. Clin Cardiol 10:83–88
15. Hasenfuss G, Kasper W, Meinertz Th, Busch W, Lehmann M, Krause Th, Hofmann Th, Revenaugh M, Holubarsch Ch, Just H (1987) Evaluation of long-term oral levodopa therapy in chronic congestive heart failure. Klin Wochenschr 65:1087–1094
16. Itoh H, Kohli ID, Rajfer SI, Goldberg LI (1985) Comparison of the cardiovascular actions of dopamine and epinine in dogs. J Pharmacol Exp Ther 207:16–22
17. King BD, Beck TR, Snyder MM, Musser Th (1987) Efficacy of fenoldopam (SK&F 82526) in chronic congestive heart failure. Circulation 76 (suppl IV):71
18. Melloni GF, Minoja GM, Scarazzati G, Bauer R, Brusoni B, Ghirardi P (1981) Renal effects of SB 7505: a douple blind study. Eur J Clin Pharmacol 21:369–371
19. Melloni GF, Melloni R, Minoja GM, Scarazzati G, Bruni GC, Loreti P, Bauer R (1981) Clinical tolerability of ibopamine hydrochloride (SB 7505). Eur J Clin Pharmacol 19:409–411

20. Packer M, Medina N, Yushak M (1984) Hemodynamic and clinical limitations of long-term inotropic therapy with amrinone in patients with severe chronic heart failure. Circulation 70:1038–1047
21. Packer M, Lee WH, Kessler PD (1986) Preservation of glomerular filtration rate in human heart failure by activation of the renin-angiotensin system. Circulation 74:766–774
22. Packer M (1988) Neurohumoral interactions and adaptations in congestive heart failure. Circulation 77:721–730
23. Rajfer SI, Goldberg LI (1982) Dopamine in the treatment of heart failure. Eur Heart J 3 (suppl D):103–106
24. Rajfer SI, Anton AH, Rossen JD, Goldberg LI (1984) Beneficial hemodynamic effects of oral levodopa in heart failure. Relation to the generation of dopamine. N Engl J Med 310:1357–1362
25. Rajfer SI, Rossen JD, Nemanich JW, Douglas FL, Davis F, Osinski J (1986) Sustained hemodynamic improvement during long-term therapy with levodopa in heart failure: role of plasma catecholamines. J Am Col Cardiol 10:1286–1293
26. Rajfer SI, Rossen JD, Douglas FL, Goldberg LI, Karrison TH (1986) Effects of long-term therapy with oral ibopamine on resting hemodynamics and exercise capacity in patients with heart failure: relationship to the generation of N-methyldopamine and to plasma norepinephrine levels. Circulation 73:740–748
27. Randolph WC, Swagzdis JE, Joseph GL, Gifford R (1983) Circulating levels of epinine and epinine conjugates in rat, dog, and man following oral administration of ibopamine. Pharmacologist 25:117
28. Reffo GC, Turrin M, Gabellini A, Forattini C (1984) Hemodynamic evaluation of ibopamine in severe congestive heart failure. Eur J Pharmacol 26:19–22
29. Ren JH, Unverferth DV, Leier CV (1984) The dopamine congener, ibopamine, in congestive heart failure. J Cardiovasc Pharmacol 6:748–755
30. Skorecki KL, Brenner BM (1982) Body fluid homeostasis in congestive heart failure and cirrhosis with ascites. Am J Med 72:323–338
31. Stote RM, Dubb JW, Familar RG, Erb BB, Alexander F (1983) A new oral renal vasodilator, fenoldopam. Clin Pharmacol Ther 34:309–315
32. Whittsett TL, Goldberg LI (1972) Effects of levodopa on systolic preejection period, blood pressure, and heart rate during acute and chronic treatment of Parkinson's disease. Circulation 45:97–106
33. Young JB, Leon CA, Pratt CM, Suarez JM, Aronoff RD, Roberts R (1985) Hemodynamic effects of an oral dopamine receptor agonist (fenoldopam) in patients with congestive heart failure. J Am Col Cardiol 6:792–796

Author's address:

Gerd Hasenfuss, Medizinische Universitätsklinik, Innere Medizin III, Kardiologie, Universität Freiburg, Hugstetter Strasse 55, D-7800 Freiburg, FRG

Cardiovascular effects of forskolin and phosphodiesterase-III inhibitors

M. Schlepper, J. Thormann, V. Mitrovic

Kerckhoff-Klinik of the Max-Planck Society, Bad Nauheim, FRG

Summary

In the first part of this presentation, data is reported on the hemodynamic effects of forskolin given to patients with dilated cardiomyopathy in a concentration of 3 µg/kg/min and 4 µg/kg/min. At the lower dosage, forskolin had no effect on dP/dt_{max}, cardiac index, ejection fraction, or myocardial oxygen consumption. With small dosages of dobutamine, however, an increase of all four parameters has been observed in the same group of patients. Systemic vascular resistance and left ventricular enddiastolic pressure fell with forskolin given at the lower concentration. Forskolin administered at a dosage of 4 µg/kg/min induced an increase in dP/dt_{max} by 19% and a 16% rise in heart rate. However, these changes were associated with symptomatic flush syndromes. Therefore, forskolin may serve as a vasodilating substance in lower concentrations, but cannot be used as a positive inotropic compound because of the subjective symptoms.

In the second part, a study is reported in which an anti-ischemic effect of the phosphodiesterase inhibitor enoximone was observed in patients with proven significant coronary heart disease. With respect to the hemodynamic parameters, the most striking findings were the decreases in left ventricular enddiastolic pressure and systemic vascular resistance. Furthermore, when left ventricular stroke work index was plotted as a function of the left ventricular enddiastolic pressure, enoximone shifted the left ventricular function curve to the left. Therefore, the anti-ischemic effect of enoximone may not only be due to a reduction in preload and afterload but may rather reflect an effect on diastolic compliance. Studies with intracoronary injections of enoximone and animal experiments support this hypothesis.

Introduction

Forskolin, a diterpene derivative of an Indian plant (*coleus forskohlii*), and the phosphodiesterase (PDE) inhibitors currently under investigation are chemically not related and bear no structural similarities. Forskolin is classified as an adenylate cyclase activator, while PDE inhibitors have a direct effect on cyclic AMP [1–3]. The final pathway of effectiveness ultimately leads to changes in the mobilization of Ca^{2+}-ions. However, there are more common features justifying the joint presentation of both substances.

First, the target site of action is not confined to the myocardium, but is rather ubiquitous and wide-spread. Second, based on these prerequisites, there are general effects influencing the cardiocirculatory system. Thus, affecting both the vasculature and the myocardium, vasodilation and positive inotropic stimulation are to be expected when the compounds are administered systemically to healthy volunteers or patients.

Last but not least, the mode of action of both compounds is independent of β-receptors, and therefore cannot be abolished by β-receptor blockers. On the other hand, blockade of their effects by Ca^{2+}-antagonists can be demonstrated. At least for the PDE in-

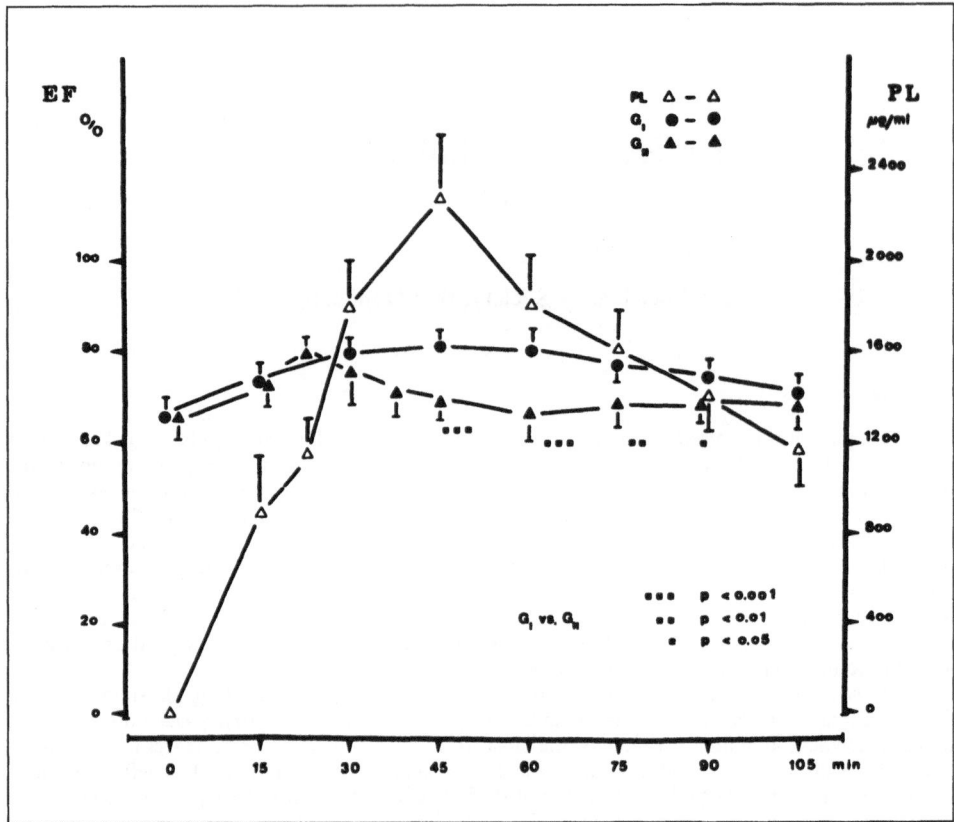

Fig. 1. The graph illustrates sulmazole-induced alterations of ejection fraction (EF) in healthy volunteers and the average serum levels (PL) following infusion of sulmazole 1) without verapamil (group I = G_I) and 2) with the additional application of verapamil (group II = G_{II}).

The conclusions are: 1) In group I alterations of left ventricular performance correspond with the increase of serum levels up to 1800 ng/ml; further elevations do not result in an additional increase in efficiency, i.e., beyond a certain serum level of sulmazole a dosage/efficacy relationship no longer exists; 2) In group II the sulmazole-induced increases in EF are abolished with the influence of verapamil setting in: sulmazole effects are blocked by verapamil

hibitors such a counteraction – induced by verapamil – has been observed when obtaining cardiac parameters after sulmazole infusion at increasing dosage in healthy volunteers [4, 5]. The distinct increase in fractional shortening and ejection fraction and the decrease in the preload- and afterload-independent but rate corrected electromechanical systole (QS_2I) could be abolished when verapamil was added. Looking at the serum levels of sulmazole which were identical in both courses of infusion, it is apparent that the three parameters observed remained at a plateau, although the serum level was still increasing (Fig. 1). Is this a feature inherent to all PDE inhibitors, and can it be assumed that at a certain serum level "target sites" have been totally activated or occupied?

When compounds which exhibit both positive inotropic effects and vasodilating properties are tested clinically, the assessment of inotropy faces several methodological difficulties. Methods applied should deliver data sufficient to discriminate between pre- and afterload-induced changes of overall cardiac pump function and those caused by true positive inotropic effects, which by definition are load-independent. Most of the parameters which inotropy is judged by in clinical investigations, however, are to some extent, load-dependent, even the so-called contractility parameters. The QS_2 index is load-independent provided heart rate is identical. Pressure/volume diagrams or the endsystolic pressure/volume relationship are able to discriminate between load dependency and inotropic alterations, when the extent of controlled load alterations is used as a measure.

It can be taken as an axiom that with augmentation of inotropy a concomitant increase in oxygen consumption occurs. However, accompanying vasodilation may mask the increase in oxygen consumption so that the net effect may not reflect oxygen-demanding augmentation caused by positive inotropy. Clinical investigations trying to establish positive inotropic effects must therefore be interpreted with caution.

Forskolin

In 1983 Bristow et al. [6] reported on positive inotropic effects of forskolin in membrane preparations derived from failing and normal functioning human left ventricles. There was a dose-related activation of adenylate cyclase, which was 4.38-fold as compared to that caused by maximal isoproterenol application in the normal myocardium and 6.02-fold in the failing heart muscle. The important message from this study, however, was that in a failing myocardium β-receptor density was diminished and that the action of forskolin was independent of this density and could not be abolished by β-receptor blockers. The same investigators [7] also elicited positive inotropic effects in isolated right ventricular papillary muscles from seven failing human hearts and compared these hemodynamic properties of forskolin to isoproterenol in dog experiments. Forskolin increased muscle tension to a lesser degree than isoproterenol, while, in the dog experiments, the effects of forskolin were qualitatively similar to isoproterenol, namely increase in heart rate, cardiac output and dp/dt_{max}, and a decrease in mean arterial pressure, peripheral resistance, and left ventricular filling pressure. However, forskolin proved to be 10- to 100-times less potent on a molar basis, and it appeared to produce relatively greater effects on left ventricular filling pressure. By critically reviewing these investigations, the vasodilator effects were certainly dominating and prevailing over the inotropic effects of forskolin.

In experiments carried out on isolated guinea pig hearts, inotropic action could be studied without interfering with vasodilator effects. It was shown that with increasing doses of forskolin an augmentation of contraction became apparent, while heart rate changes were only minimal. There was an increase in coronary flow and oxygen consumption, although to a much lesser degree than oxygen supplies, pointing to an additional vasodilator effect of this drug on the coronary circulation [8].

There are only few publications dealing with the clinical application of forskolin. Lele et al. [9] reported on observations in which the effects of forskolin were studied by means of a nuclear stethoscope. The cohort of patients in that study was inhomogenous as far as the etiology was concerned. Furthermore, the dosis ranged from 1 µg to 16 µg/kg BW. Following administration of forskolin, there was a statistically significant increase in left ventricular ejection fraction, ejection rate and peak filling rate, and an approximate rise of 15% over the basal values was observed after infusion of 8 µg/kg and 16 µg/kg. Heart rate did not change, cardiac output was elevated and so was stroke volume (in dose-dependent fashion) which amounted to a maximum of 70% following the dosis of 16 µg/kg.

The observed changes showed a large scatter, but it might be suggested that, in addition to an increase in overall pump function, there could have been positive lusitropic effects involved, as indicated by an increase in peak filling rate. As heart rate also increased and no data are given of changes in total peripheral resistance and in arterial pressure, these measurements are not sufficient to discriminate between positive inotropic and vasodilating effects.

Linderer and Biamino [10] likewise studied a group of patients which was etiologically inhomogeneous. In patients with congestive heart failure, mostly due to dilated cardiomyopathy, a 16% increase of dp/dt_{max} was found, systemic vascular resistance was decreased, and coronary blood flow measured by the argon method was either uneffective or even declined in some patients inspite of the increase in dp/dt_{max}. The most striking feature was the drop in left ventricular enddiastolic pressure, which in the author's opinion may explain the reduction in oxygen consumption and the unaltered coronary blood flow due to an improved diastolic compliance of the ventricle.

In our own study, the results were obtained from 15 patients with dilated cardiomyopathy, and the effects of forskolin were compared to those of dobutamine at a rather small dosis of 10 µg/kg infused over 10 min [11]. Since dobutamine has a rather short half-life of only 2.5 min, data after dobutamine application were obtained first and were then followed by a second course of investigation during which forskolin was given at a dosage of 3 µg/kg/min over an infusion period of 10 min. Before and at the end of each infusion period heart rate was maintained constant by atrial pacing. With the dosages applied, heart rate changes were less than 10% and insignificant after both drugs. Left ventricular systolic pressure showed a tendency to fall with forskolin and a slight increase with dobutamine; both changes were insignificant. When patients were divided into two subgroups (Group A with normal resting dp/dt_{max}, and Group B consisting of patients with decreased dp/dt_{max} at rest), dobutamine elicited a significant rise in dp/dt_{max}, while forskolin failed to show a significant increase of this contractility parameter in both subgroups. Cardiac index and EF did not show any increase with either drug. The missing inotropic effect could also be seen by determining myocardial oxygen consumption according to the Bretschneider formula [12, 13]. The slight positive inotropic effect of dobutamine was accompanied by a significant increase in oxygen consumption, whereas forskolin did not induce any alterations in oxygen consumption at all. Left ventricular enddiastolic pressure decreased significantly with both drugs during the infusion period and regained normal values as soon as dobutamine was discontinued. Systemic vascular resistance was decreased with both drugs, but the reduction was more marked with forskolin than with dobutamine. Using the simple compliance index according to Gaasch et al. [14], both drugs exhibited a decrease in this compliance index accompanying the fall in left ventricular enddiastolic pressure. Whether or not this is a specific action of the drug remains speculation (Fig. 2).

To further establish genuine and therefore load-independent positive inotropic effects, left ventricular pressure/volume loops were recorded in all patients. Volume was determined by scintigraphic methods and left ventricular pressure by tip manometer [11]. For volume determinations 250–300 heart beats were mandatory; both signals were digitized and computer-assisted pressure/volume curves were obtained. With dobutamine the endsystolic pressure/volume relationship is shifted to the left and downwards. Lack of slope-k deviations proved that the alterations appearing after the administration of forskolin were due to changes in pre- and afterload exclusively and positive inotropic effects were lacking.

To provoke a more pronounced inotropic stimulation, the dosage of forskolin was increased to 4 µg/kg/min in a pilot study. This resulted in an average increase of dp/dt_{max}

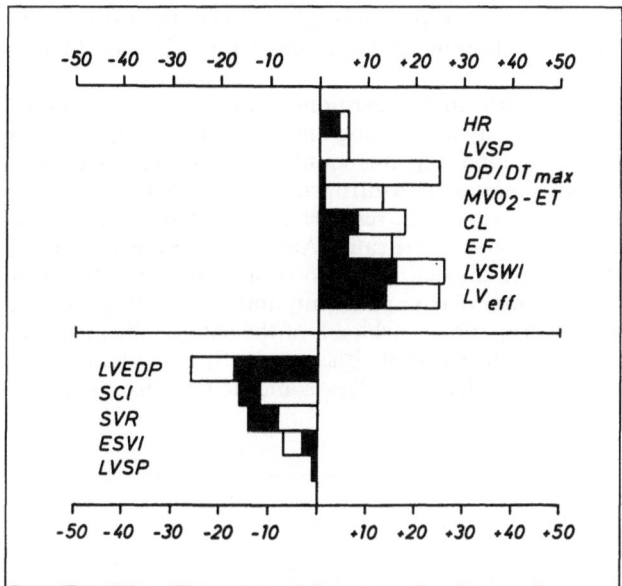

Fig. 2. Maximal percentile parameters alterations of left ventricular function during forskolin intervention (black columns) and dobutamine (white columns).

Abbreviations: HR = heart rate, LVSP = left ventricular systolic pressure, MVO_2-ET = myocardial oxygen consumption (indirectly assessed), CI = cardiac index, EF = ejection fraction, LVSWI = stroke work index, LV_{eff} = left ventricular efficiency, LVEDP = left ventricular enddiastolic pressure, SCI = compliance index according to GAASCH, SVR = systemic vascular resistance, ESVI = endsystolic volume index

by 19%, but was accompanied by a 20% decrease in arterial pressure and a 16% rise in heart rate. Clinically, these hemodynamic changes were associated with symptomatic flush snydromes. For these reasons, further investigations with an increased dosage were delayed. As far as preliminary conclusions can be drawn from our results, we comment that the main effect of forskolin is vasodilation and that, in order to achieve positive inotropic effects, the drug would have to be administered in dosages which are not tolerable for the patient, leading to a fall in blood pressure and systemic vascular resistance and concomitantly to undue subjective symptoms.

To our present knowledge, no further clinical investigations have been carried out in patients with failing hearts. Although the mode of action is sound by principle and certainly valed in theory as regards treatment of the failing heart, we are still in search of a cardiospecific adenylate cyclase activator whch can be used safely and with inotropy as prevailing effect.

PDE Inhibitors

With the exception of amrinone, all PDE inhibitors are still on clinical trial. While favorable effects with short- and long-term treatment of patients with congestive heart failure have been published [15–23], there have been conflicting reports as to whether the beneficial effects were due to vasodilatory or true positive inotropic action of these drugs [24–27]. Wilmshurst et al. deny any positive action of amrinone and claim only vasodilating properties [28].

In a study carried out in patients with dilated cardiomyopathy, pressure-volume curves were obtained to distinguish between vasodilating and positive inotropic effects [29]. Using the conductance catheter technique [30], the required rapid load changes were achieved by either pharmacological intervention with nitroprusside and phenylephrine or by temporary balloon occlusion of the inferior caval vein. Atrial pacing was used to obtain pressure-volume diagrams at comparable heart rates. According to the severity of left ventricular impairment, patients were divided into two subgroups. In patients of Group A with a minor degree of left ventricular impairment, amrinone was given at a low i.v. dosage of 0.5 mg/kg BW. The drug-induced slope k of the endsystolic pressure-volume relationship did not deviate from the slope of either controls, vasodilation with nitroprusside or vasoconstriction with phenylephrine. The isometric maxima all appear

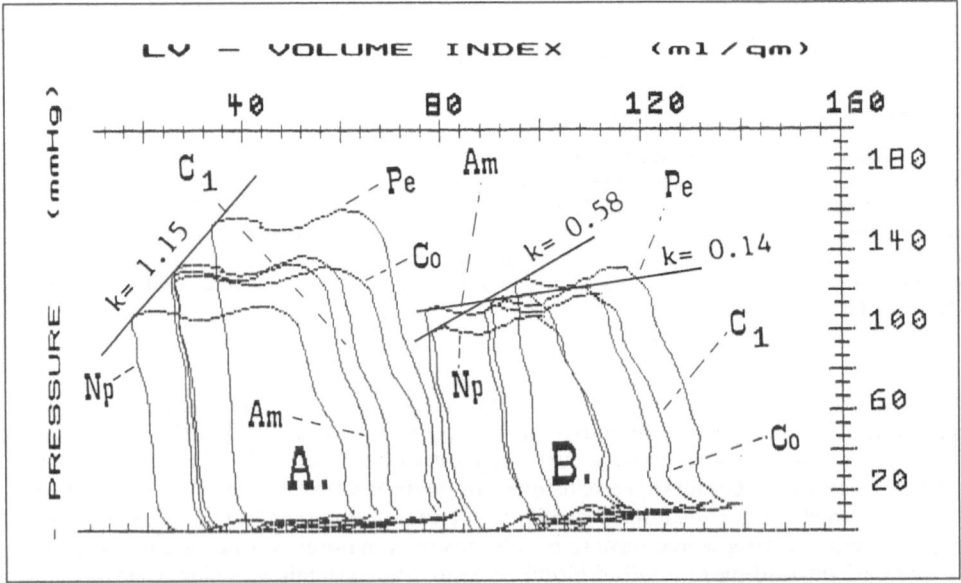

Fig. 3. Amrinone effects as analyzed by endsystolic pressure-volume relationships (ESPVR). Rapid load changes were induced pharmacologically by 1) nitroprussid (Np) and 2) phenylephrine (Pe); pacing for constant heart rates was kept up during the entire investigation. The computer print out demonstrates the ESPVR of the five representative investigational phases, displayed as summation loops for 1) group A with smaller ventricular volumes and 2) group B with larger volumes. Am marks the phase of maximal amrinone effects at a dosage of 1.5 mg/kg and C1 the control during pacing.

Thus, improved LV-pump function in group A was achieved by load changes and not by inotropy: the Am-induced leftward motion of the loop of the ESPVR occurs along the endsystolic interventional line C1-Am which is identical with the endsystolic load line Np-Pe; slope k is 1.15 mm Hg/ml.

In group B patients with larger ventricles a minor (not significant) Am-induced change in inotropy can be assumed since the leftward motion (representing the interventional line C-1-Am) shows a noticeable deviation form the loading line Np-Pe on one line; vasoactivity of the drug is indicated, but no change in contractility. In the patients of Group B with severely impaired left ventricular function characterized by larger volumes and dp/dt_{max} reduced to a range of 1000 mm Hg/s, amrinone shifted the loop of the endsystolic pressure/volume relationship slightly to the left with a minor deviation of slope k, indicating tendencies towards positive inotropic action induced by amrinone at a low i.v. dosage of 1.5 mg/kg (Fig. 3).

This becomes more apparent, when, in a different cohort of patients with equally impaired left ventricular function, the dosage of amrinone was increased to 2.5 mg/kg, and the required rapid load changes were achieved by temporary balloon occlusion of the vena cava inferior. Without medication the slope k was 0.52 mm Hg/ml, and deviated with amrinone increasing to 0.69 mmHg/ml, while the summarized loop shifted leftwards. When dobutamine (10 μg/kg/min) was infused additionally at this point, slope k deviated to 0.80 mm Hg/ml, thus contributing to the mild inotropic effects of amrinone [29]. A mild, but distinct inotropic action of amrinone can, therefore, be expected. However, it apparently exhibits stronger effects on the more severely impaired cardiac function (Fig. 4).

The same behavior, namely a stronger cardiotonic effect, was obtained when enoximone was studied. Of 27 patients with dilated cardiomyopathy and a cardiac index of about or below 2 l/min/m^2, enoximone was administered in 13 patients at a dosage of 0.5–1 mg/kg, and in 14 patients at a dosage of 1.5–2.0 mg/kg. There was a dose-related

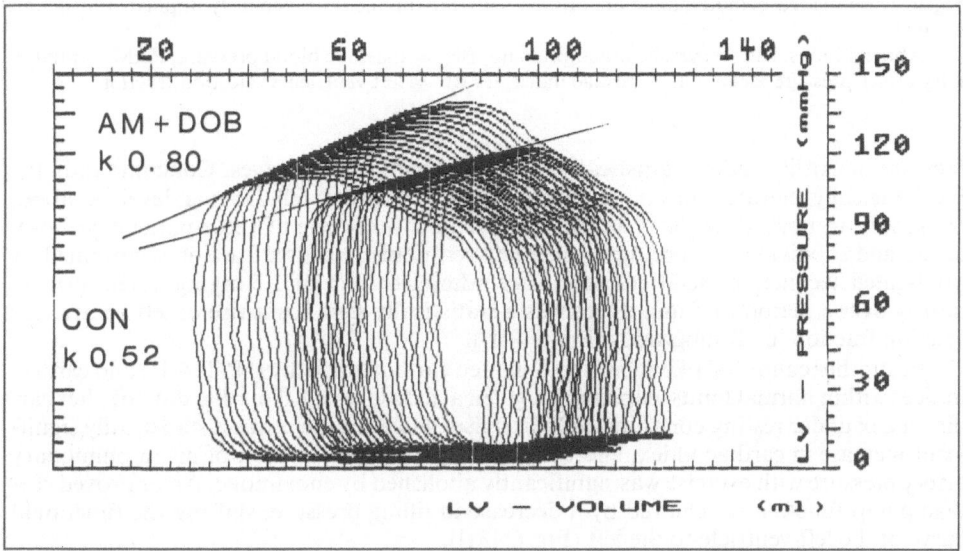

Fig. 4. Depicted are amrinone effects as analyzed by endsystolic pressure-volume relationships (ESPVR) in patients with ventricular impairment. The required load changes were (in contrast to those in Fig. 3) induced by temporary balloon occlusion of the vena cava inferior.

The computer print out demonstrates the ESPVR, displayed as summation loop-series, before and after amrinone (AM), 2.5 mg/kg plus dobutamine (DOB). Under the influence of both drugs a leftward shift of the loops was achieved, with the slope "k" of the ESPVR increasing from 0.52 mmHg/ml to 0.8 mmHg/ml. This indicates an amrinone-induced increase of inotropy which can be enhanced by applying additionally dobutamine

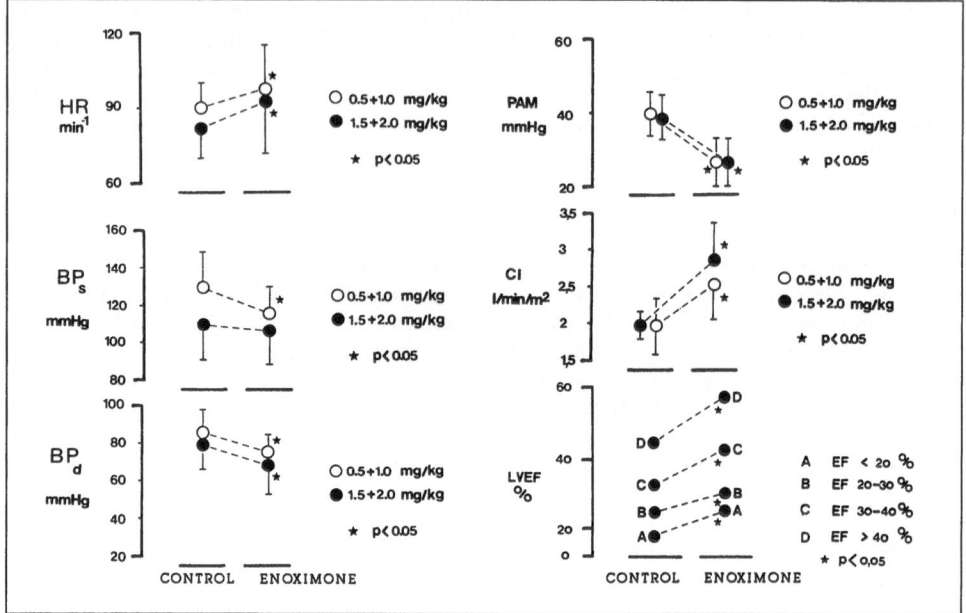

Fig. 5. Dose-related hemodynamic alterations with enoximone, intravenously applied. For details see text.

Abbreviations: BP_s = systolic blood pressure, BP_d = diastolic blood pressure, PAM = pulmonary artery pressure mean, CI = cardiac index, LVEF = left ventricular ejection fraction

increase in cardiac index, statistically significant with both dosages. Concomitantly the pathologically elevated mean pulmonary artery pressure fell from a level of about 40 mm Hg to normal levels of about 25 mm Hg. Twenty-one of these patients were selected and subdivided according to their left ventricular ejection fraction determined by pool-gated technetium scintigraphy. After administration of 1.5 mg/kg given intravenously over a period of 5 min, there was a statistically significant rise of left ventricular ejection fraction in all subgroups (Fig. 5; [21]).

In another cohort of 14 patients with dilated cardiomyopathy (NYHA II) and cardiac indices within normal limits, enoximone, given at a dosage of 1.0 mg/kg, did not alter cardiac index under resting conditions. On exercise, however, there was a statistically significant increase in cardiac index, and the pathological augmentation of mean pulmonary artery pressure with exercise was significantly abolished by enoximone. An improved cardiac pump function is achieved by a decrease in filling pressure, shifting the functional curve of the left ventricle to the left (Fig. 6; [31]).

Again, these data do not allow discrimination between vasodilatory and positive inotropic effects. Whether or not this relationship is in the same range for all PDE inhibitors under investigation has not been established. Pharmacological research has furnished evidence for a compound-specific profile of PDE inhibitors as far as the relationship between vasodilation and inotropy is concerned. The investigations were carried out in anesthetized dogs in which PDE inhibitors were applied by bolus injection. Inotropy was graded by the increase of dp/dt/P; total peripheral resistance (TPR) was calculated and cardiac output (CO) determined by thermodilution.

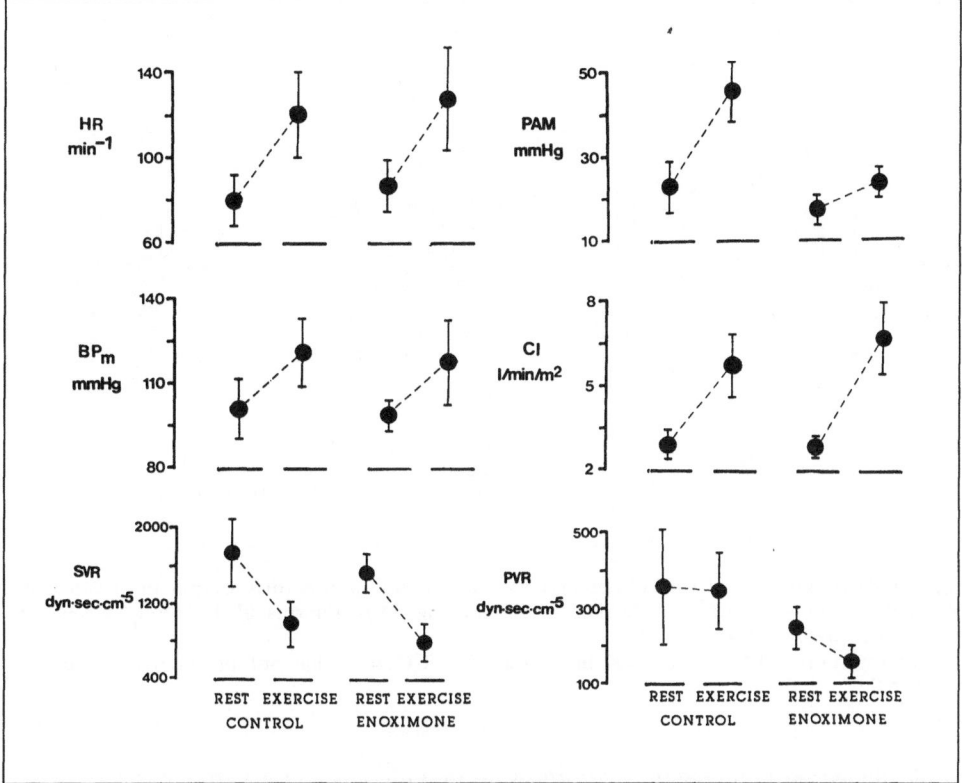

Fig. 6. Enoximone (1.0 mg/kg)-induced hemodynamic changes in 14 patients with dilated cardiomyopathy, under resting conditions and during exercise. For details see text.

Abbreviations: HR = heart rate, BP_m = blood pressure mean, PAM = pulmonary artery pressure mean, CI = cardiac index, SVR = systemic vascular resistance, PVR = pulmonary vascular resistance

When dp/dt/P was measured at a 10% fall of TPR, and TPR and cardiac output at a level of 60% increase of dp/dt/P, a compound-specific profile in the vasodilator to inotropic effects could be established, ultimately leading to a classification of PDE inhibitors into three classes:

Class I drugs in which the increase of dp/dt/P was predominating, while vasodilator effects were only weak;

Class II drugs showed a profile with only slightly prevailing positive inotropy, and

Class III drugs with a balanced ratio of cardiotonic and vasodilating effects or slightly accentuated vascular activity (Fig. 7; [32]).

Since positive inotropy feeds on increased myocardial oxygen consumption, it has to be questioned whether these drugs might provoke ischemia in patients with coronary artery disease.

In a recent publication, patients have been reported to experience angina pectoris when enoximone was applied. This did not only occur in patients with proven coronary artery disease, but, given at relatively high dosage, also in patients suffering from idio-

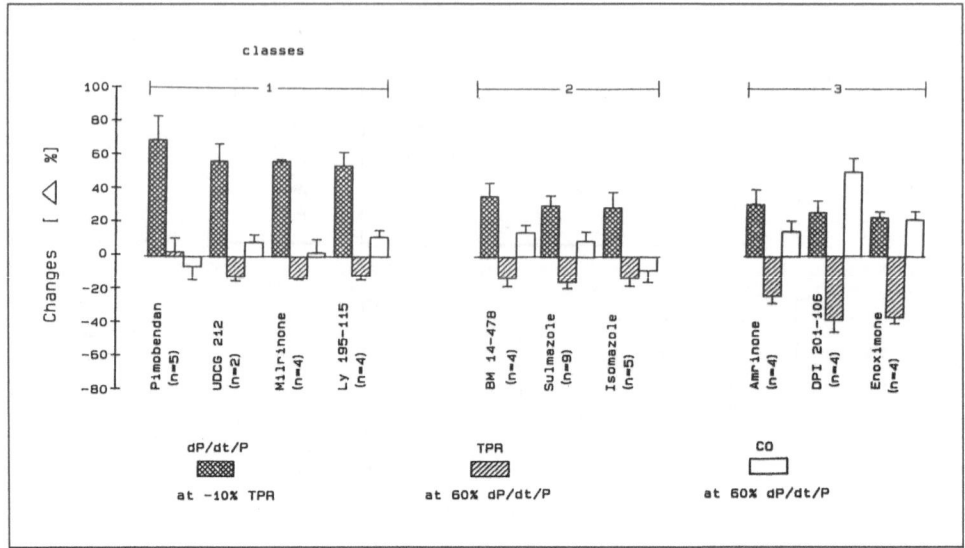

Fig. 7. Relative changes of cardiotonic activity, vasodilation and cardiac output in anesthetized beagle dogs (courtesy Drs. H. J. Schliep and J. Harting, Dept Pharmacol, E Merck/Darmstadt, FRG). For details see text.

Abbreviations: TPR = total peripheral resistance, CO = cardiac output, dp/dt/P = index of contractility

pathic dilated cardiomyopathy [33]. When we first investigated the effects of sulmazole, the drug was applied in 17 patients with angiographically proven 2– or 3-vessel disease in stable angina. Ischemia could be provoked by ventricular pacing as evidenced by lactate production and loss of regional wall motion. The patients received 2 mg/kg BW of sulmazole intravenously over 5 min, and the acute effects were studied at rest and after 9, 14, and 19 min after administration of the compound. Heart rate slightly increased by 19%, left ventricular systolic pressure fell by 13%, and ventricular enddiastolic pressure by 42%. There was a rise in dp/dt$_{max}$ by 27%, systemic vascular resistance was reduced by 24% and cardiac index increased by 30%. Coronary sinus flow was augmented by 39%, coronary vascular resistance declined by 37%, and myocardial oxygen consumption, as calculated by the Bretschneider formula [12, 13], was enhanced by 35%. None of the patients, however, experienced angina pectoris [34].

When ischemia was induced by ventricular pacing (stimulation rate 170 over 75 s) and the procedure repeated 10 min after i.v. administration of 2.5 mg/kg BW of sulmazole, no ischemic patterns could be elicited by the pacing maneuver. An increased pump function was observed and, with the abolition of pacing-induced ischemia, ventricular wall motion considerably improved [35].

Because of these apparently anti-anginal effects of PDE inhibitors, enoximone was investigated with regard to its anti-ischemic effects. In these consented studies, patients with proven significant coronary stenoses and stable angina were investigated. In all, angina could be provoked by either exercise or ventricular pacing. Prior to the investigation, all anti-anginal drugs, with the exception of nitrates, were withdrawn for a period of 5 half-lives, and nitrates were withheld 8–12 h before examinations.

Seventeen patients were exercised on a bicylce ergometer in supine position applying a modified Bruce protocol. A five-lead ECG was recorded, pulmonary-artery and PC-pressures and cardiac output were measured by Swan Ganz catheter. Systemic blood pressure was obtained by the Riva-Rocci method, and resistances were calculated. In addition, lactate, potassium and pH were determined in blood samples taken from the pulmonary artery. Exercise load was individually adjusted and terminated when at least two of the following three criteria were fulfilled: 1) occurrence of angina, 2) ST-segment depression in at least two leads exceeding 0.2 mV, and 3) an elevated mean pulmonary artery pressure above 30 mm Hg. Immediately after termination of exercise, measurements were taken and the patients were allowed to rest for 30 min. Then 0.75 mg of enoximone/kg BW were infused over 10 min; after another 15 min measurements were carried out at rest and after exercise which was repeated for exactly the same time at an identical work load. At the end, all parameters were again taken and calculated. The alterations in pH, lactate and potassium were only slightly altered by enoximone. The lesser decrease in pH and potassium seen after exercise and enoximone can probably be attributed to an increased peripheral blood flow due to the vasodilatory properties.

In summary, it can be assumed that the work load compared by these rough parameters was almost identical. However, systolic and diastolic blood pressure did not change; heart rate increased slightly, but significantly, at rest, but did not do so during exercise. Calculating myocardial oxygen consumption by the double product and the pressure work volume index (Rooke and Feigl [36]), there was only a slight increase in the double product at rest, while the values as calculated by both formulas were not altered on exercise. Cardiac output was increased during exercise and after enoximone. But the most striking feature was a pronounced fall in mean pulmonary pressure and pulmonary capillary pressure, accordingly lowering pulmonary vascular resistance. At an identical workload and with probably identical oxygen consumption, mean pulmonary artery was kept at normal levels compared to pathological levels of about 40 mm Hg during exercise without enoximone. ST-segment depression as a sum of all ST-segment depressions and elevations in the five leads was almost completely abolished, while it was pronounced under exercise without the drug (Fig. 8; [37]). To explain these findings, one could reason that a marked reduction of preload caused by enoximone, which became more apparent under exercise, is responsible for the elevation of the anginal threshold, since all of these patients experienced angina without enoximone and none did so after the drug had been adminsitered. Have we been dealing with just a better nitrate preparation?

Almost identical anti-ischemic effects were seen in another cohort of patients in whom enoximone was given as a single oral dose of 150 mg. The maximal effects were found 2 h after application of the drug showing a strong correlation between plasma levels, hemodynamic improvement at rest and after exercise, and anti-ischemic action [38].

To further evaluate the possible role of intracavitary pressure and reduction of ventricular size and volume, another cohort of 12 patients was studied during routine cardiac catheterization [39]. The same protocol as outlined with sulmazole [35] was applied to these patients, i.e., ventricular pressure recordings were taken at rest and after pacing with and without enoximone and three biplane ventriculograms were performed under the same conditions. On the average, there was an increase of dp/dt_{max} of 20% after post-pacing enoximone as compared to post-pacing without the drug. The oxygen consumption measured on-line by the Bretschneider formula and the Rooke and Feigl formula [12, 13, 36] was identical with and without enoximone. Left ventricular effectiveness was increased by 17%. There were only insignificant changes in heart rate, but, again, the most striking feature was the decrease in left ventricular enddiastolic pressure by 37% and the fall in systemic vascular resistance by 31%. Ejection fraction, which fell under

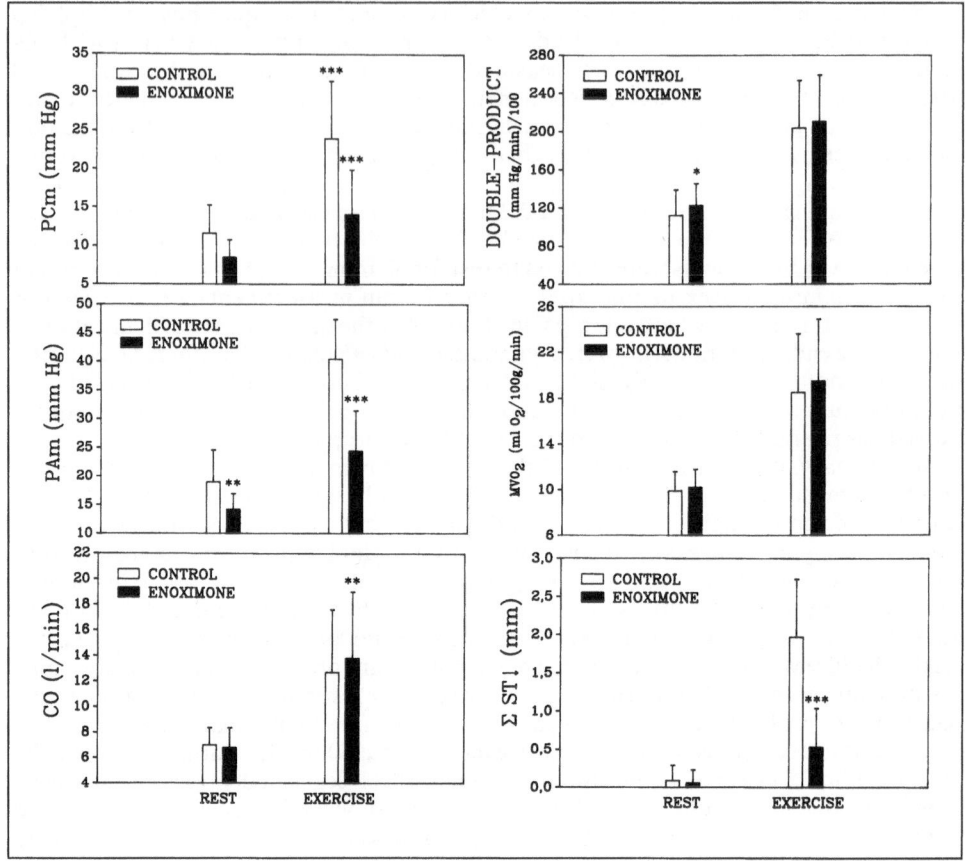

Fig. 8. Enoximone-induced hemodynamic changes in 17 patients with ischemic coronary artery disease at rest and during exercise. Note that cardiac output increased significantly with enoximone only during exercise and that exercise-induced ischemia was abolished by the influence of enoximone. For further details see text. (Abbreviations as in Figs. 2, 5, and 6)

the pacing procedure without enoximone, returned to control values, when the drug had been administered. It is certainly unconventional to average regional wall motion in patients with a different pattern of stenotic lesions in the coronary system. But doing so, it turned out that the loss in regional wall motion which was seen in the five segments in RAO- and LAO-projections and which occurred after the pacing procedure without enoximone, was completely compensated when the pacing procedure was repeated after application of enoximone (Fig. 9; [39]).

When left ventricular stroke work index is plotted against left ventricular enddiastolic pressure, it becomes apparent that less cardiac work is performed at higher left ventricular enddiastolic pressure during pacing-induced ischemia, and that after enoximone left ventricular stroke work index is back to resting values, but this was now achieved at a lower enddiastolic pressure, shifting the functional curve of the left ventricle to the left (Fig. 10).

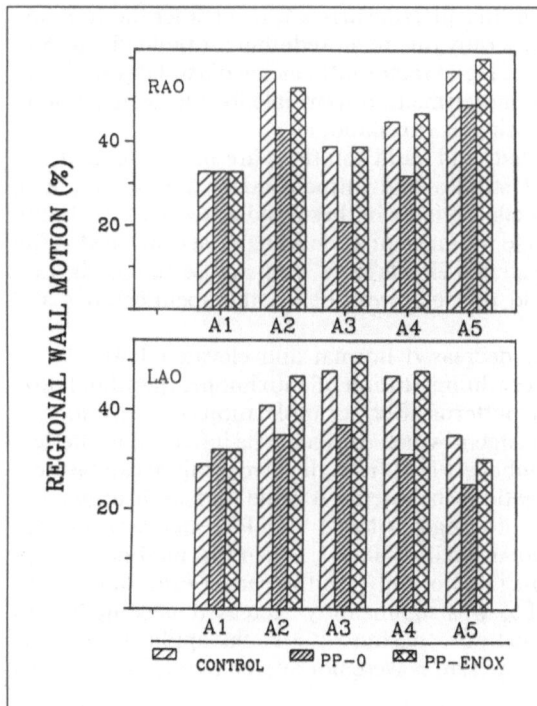

Fig. 9. Group-percentile changes in regional wall motion (RWM) for the 5 segments in both angiographic positions, comparing the postpacing phase with (PP-ENOX) and without enoximone (PP-O). Although this type of demonstration ignores the non-uniformity of individual coronary lesions, the general impression is that with the exception of the septal and posterobasal area, RWM is greatly reduced in PP-O and RWM is improved in PP-ENOX up to and/or above resting levels

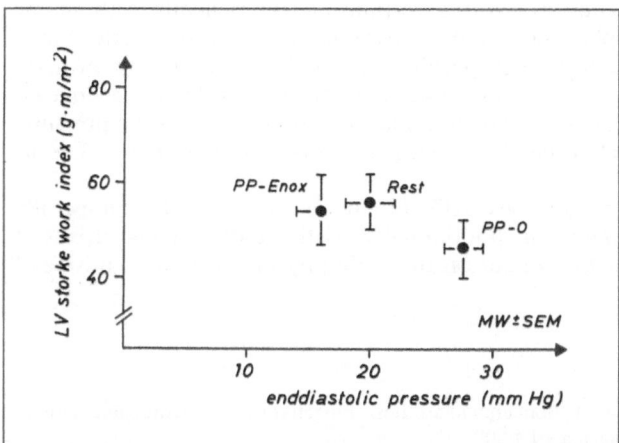

Fig. 10. For 12 patients (same group as in Fig. 9) the changes in left ventricular pump function are defined by the relation of left ventricular stroke work index (LVSWI) versus enddiastolic pressure (LVEDP): In the postpacing phase without medication (PP-O) ischemia was produced, so that LVSWI declined while LVEDP rose (rightward motion down). In the enoximone-medicated postpacing phase (PP-ENOX), LVSWI improved by 18% while LVEDP decreased by 37% (leftward motion, up) and myocardial ischemia was no longer present (PP-ENOX vs PP-O: $p < 0.05$)

Again, when interpreting these findings the question arises as to whether the abolition of the ischemic patterns after enoximone is only due to a predominant preload and possible afterload reduction; or is there a myocardial factor influencing diastolic compliance and, hence, extracoronary coronary flow impairment, responsible for the decrease of intraventricular pressure and elevation of ischemic threshold?

These problems have to be investigated, and valid solutions are not at hand at this time. It could be demonstrated in pilot studies that intracoronary administration of enoximone had similar anti-ischemic effects. Patients in these studies received a 10-min-infusion of 0.075 mg/kg BW of enoximone into a stenosed coronary vascular bed. With these dosages given intravenously, no systemic effects were seen, and serum levels after intracoronary administration of the said dosages were below therapeutic levels and hardly detectable.

Intracoronary application, however, decreased normal and elevated LVEDP. As compared to ventricular pacing before i.c. administration of enoximone, the stimulation procedures were unable to elicit ischemic patterns after i.c. application of the drug.

With a newly developed echo contrast agent which can safely be infused into the coronary arteries, the washout time of the bubbles ($T\frac{1}{2}$) from the myocardium can be measured. In animal experiments and investigations in patients, it could be demonstrated that $T\frac{1}{2}$ is flow-dependent and is markedly prolonged when LVEDP is increased in consequence to eliciting ischemia by ventricular pacing [40, 41]. When this method was applied in studying the effects of enoximone i.c., not only did the pacing procedure fail to elevate LVED or to cause angina, but $T\frac{1}{2}$ was significantly decreased as compared to control values and measurements obtained after pacing without the application of the drug. LVEDP was decreased, but arterial pressures were not altered at all, i.e., identical perfusion pressures.

In animal experiments, Taira [42] demonstrated vasodilation of the epicardial coronary arteries after amrinone i.c. In interpreting our results, a similar mechanism of enoximone can be assumed. But this assumption does not explain the marked drop in LVEDP or the lacking increase after pacing procedures, responsible for ischemia without the drug. Preliminary echocardiographic and Doppler investigations point to a better filling property of the ventricle and a positive lusitropic effect of enoximone after intracoronary application has to be considered, possibly contributing to the beneficial anti-ischemic effects of PDE inhibitors. This ultimately leads to a marked fall in enddiastolic pressure, and this fall, in turn, takes off the burden of an extravascular flow impairment in the coronary circulation.

The findings so far obtained need to be clarified by further studies, which hopefully may lead to a better understanding of the undoubtedly existing anti-ischemic effects of PDE inhibitors, but may also shed new light on the pathophysiological mechanisms of ischemic events.

References

1. Kariya T, Wille LJ, Dage RC (1982) Biochemical studies of mechanism of cardiotonic activity of MDL 17,043. J Cardiovasc Pharmacol 4:509
2. Endoh M, Yamashita S, Taira N (1981) Positive inotropic effect of amrinone in relation to cyclic nucleotide metabolism in the canine ventricular muscle. J Pharm Exp Ther 216:220–224
3. Honerjager P, Schafer-Korting M, Reiter M (1981) Involvement of cyclic AMP in the direct inotropic action of amrinone: biochemical and functional evidence. Naunyn Schmiedeberg's Arch Pharamcol 318:112–120
4. Kramer W, Thormann J, Schlepper M (1981) Effects of the new cardiotonic agent AR-L115 in normal hearts and loss of its effectivity by the Ca^{++} antagonist verapamil. Eur Heart J 2 (Suppl A):56

5. Kramer W, Thormann J, Schlepper M, Bittner C, Zrenner E (1981) Aktivitätsprofil von AR-L 115 BS bei therapierefraktärer kongestiver Kardiomyopathie (CC) und Herzgesunden (HG): Wirkungsverlust durch Ca^{2+}-Antagonisten. Verh Dtsch Ges Inn Med 87:446–450
6. Bristow MR, Strosberg AM, Ginsburg R (1983) Forskolin activation of human myocardial adenylate cyclase. Circulation 69 (Suppl III):60
7. Ginsburg R, Strosberg AM, Bristow MR, Montgomery W (1983) Inotropic potential of forskolin in failing human hearts. Circulation 69 (Suppl III):373
8. Linderer E, Metzger H (1985) The positive inotropic and smooth muscle relaxing effects of forskolin by direct activation of adenylate cyclase. In: Rupp RH, de Souza NJ, Dohadwalla AN (eds) Proceedings of the International Symposium on Forskolin: Its chemical biological and medical potential. Bombay, January, pp 83–101
9. Lele RD, Kamdar NB, Popat N, Nair KG (1985) Study of forskolin with the nuclear stethoscope. In: Rupp RH, de Souza NJ, Dohadwalla AN (eds) Proceedings of the International Symposium on Forskolin: Its chemical biological and medical potential. Bombay, January, pp 115–127
10. Linderer T, Biamino G (1985) Hemodynamic and cardiac metabolic effects of forskolin. In: Rupp RH, de Souza NJ, Dohadwalla AN (eds) Proceedings of the International Symposium on Forskolin: Its chemical biological and medical potential. Bombay, January, pp 103–115
11. Kramer W, Thormann J, Kindler M, Schlepper M (1987) Forskolin's effects on left ventricular function in dilated cardiomyopathy. Arzneim-Forschg/Drug Res 37 (I)/3:364–367
12. Bretschneider HJ, Cott La, Hellige G, Hensel I, Kettler D, Martl J (1971) A new hemodynamic parameter consisting of 5 additive determinants of estimation of the O_2-consumption of the left ventricle. Proceedings of the International Congress of Physiological Scienes, pp 98–111
13. Baller D, Bretschneider HJ, Hellige G (1981) A critical look at currently used indirect indices of myocardial oxygen consumption. Basic Res Cardiol 76:163–181
14. Gaasch WH, Batlle WE, Oboler AA, Banas JS, Levine HJ (1972) Left ventricular stress and compliance in man. Circulation 45:746–751
15. Amin DK, Shah PK, Hulse S, Shellock FG, Swan HJC (1984) Myocardial metabolic and haemodynamic effects of intravenous MDL 17,043, a new cardiotonic drug, in patients with chronic severe heart failure. Am Heart J 108:1285
16. Arbogast R, Brandt C, Haegele KD, Fincker JL, Schechter PJ (1983) Haemodynamic effects of MDL 17,043, a new cardiotonic agent in patients with congestive heart failure: comparison with sodium nitroprusside. J Cardiovasc Pharmacol 5:998
17. Benotti J, Grossman W, Braunwald E et al. (1980) Effects of amrinone on myocardial energy metabolism and hemodynamics in patients with severe congestive heart failure due to coronary artery disease. Circulation 62:28
18. Grosse R, Strain M, Greenberg M et al. (1986) Systemic and coronary effects of intravenous milrinone and dobutamine in congestive heart failure. J Am Coll Cardiol 7:1107
19. Jentzer J, LeJemtel T, Sonnenblick E, Kirk E (1981) Beneficial effect of amrinone on myocardial oxygen consumption during acute left ventricular failure in dogs. Am J Cardiol 48:75–83
20. Likoff MJ, Martin JL, Andrews V, Weber KT (1983) Long-term therapy with the cardiotonic agent MDL 17,043, in chronic cardiac failure. Circulation 68 (Suppl III):III-373
21. Neuzner J, Mitrovic V, Kornecki P, Schlepper M (1987) Enoximone (MDL 17,043): Hämodynamische Effekte bei dilatativer Cardiomyopathie (DCM). Z Kardiol 76 (Suppl):69
22. Uretsky BF, Generalovich T, Reddy RS, Spangenberg RB, Follansbee WP (1983) The acute haemodynamic effects of a new agent, MDL 17,043, in the treatment of congestive heart failure. Circulation 67:823
23. Weber KT, Jain MC, Janicki JS (1985) Enoximone (MDL 17,043) in the long-term treatment of chronic cardiac failure of mild to moderate severity. Circulation 72 (Suppl III):III-202
24. Cardenas LM, Vidaurri DA (1979) Estudio de los efectos hemodynamicos de differentes dosis de un nuevo intropico: la amrinona. Arch Inst Cardiol Mex 49:961–968
25. Firth BG, Ratner AV, Grassman ED, Winniford MD, Nocod P, Hillis LD (1984) Assessment of the inotropic and vasodilator effects of amrinone versus isoproterenol. Am J Cardiol 54:1331–1336
26. Hermiller JB, Leithe ME, Magorien RD, Unverferth DV, Leier CV (1984) Amrinone in severe congestive heart failure: another look at an intriguing new cardiotonic drug. J Pharmac Exp Ther 228:319–326

27. Wilmshurst PT, Thompson DS, Jenkins BS, Coltart DJ, Webb-Peploe MM (1983) Hemodynamic effects of intravenous amrinone in patients with impaired left ventricular function. Br Heart J 49:77–82
28. Wilmshurst PT, Thompson DS, Juul SM, Jenkins BS, Coltart DJ, Webb-Peploe MM (1984) Comparison of the effect of amrinone and sodium nitroprusside on hemodynamics, contractility and myocardial metabolism in patients with cardiac failure due to coronary artery disease and dilated cardiomyopathy. Br Heart J 52:38–48
29. Thormann J, Kramer W, Kindler M, Kremer P, Schlepper M (1987) Bestimmung der Wirkkomponenten von Amrinon durch kontinuierliche Analyse der Druck/Volumen-Beziehungen; Anwendung der Conductance (Volumen-) Katheter-Technik und der schnellen Laständerung durch Ballon-Okklusion der Vena cava inferior. Z Kardiol 76:530–540
30. Baan J, Van der Velde ET, De Bruin HG, Smeenk GJ, Koops J, Van Duk AD, Temmerman D, Senden J, Buis B (1984) Continuous measurement of left ventricular volume in animals and humans by conductance catheter. Circulation 70:812–823
31. Mitrovic V, Neuzner J, Dieterich HA, Schlepper M (1988) Hemodynamic effects of an i.v.- and p.o. single dose of enoximone in patients with impaired LV-function. Eur Heart J 9 (Suppl 1)
32. Schliep HJ, Harting I (1986) Effect of ismazole and reference compounds on hemodynamic parameters in anesthetized dogs. Naunyn-Schmiedeberg's Arch Pharmacol [Suppl] 334:R36
33. Herrmann HC, Ruddy TD, Dec W, Strauss HW, Boucher CA, Fifer MA (1987) Inotropic effect of enoximone in patients with severe heart failure: demonstration by left ventricular end-systolic pressure-volume analyis. J Am Coll Cardiol 9:1117–1123
34. Thormann J, Kramer W, Schlepper M (1982) Hemodynamic and myocardial energetic changes induced by the new cardiotonic agent, AR-L 115, in patients with coronary artery disease. Am Heart J 104:1294–1302
35. Thormann J, Schlepper M, Kramer W, Gottwik M, Kindler M (1983) Effects of AR-L 115 BS (Sulmazol), a new cardiotonic agent, in coronary artery disease: improved ventricular wall motion, increased pump function and abolition of pacing-induced ischemia. J Am Coll Cardiol 2:332–337
36. Rooke GA, Feigl EO (1982) Work as a correlate of canine left ventricular oxygen consumption, and the problem of catecholamine oxygen wasting. Circul Res 50:273
37. Mitrovic V, Schlepper M, Neuzner J, Bahavar H, Volz M, Dieterich HA (1988) Hämodynamische, antiischämische, metabolische und neurohumorale Effekte von Enoximon (MDL 17,043) bei Patienten mit koronarer Herzkrankheit. Z Kardiol 77:660–667
38. Mitrovic V, Bahaver H, Neuzner J, Dieterich HA, Schlepper M (1989) Antiischämische Wirksamkeit nach Einzelgabe von 150 mg Enoximon bei Patienten mit koronarer Herzkrankheit. Z Kardiol (in press)
39. Thormann J, Kremer P, Mitrovic V, Neuzner J, Bahavar H, Schlepper M (1989) Effects of enoximone in coronary artery disease: increased pump function, improved ventricular wall motion, and abolition of pacing-induced myocardial ischemia. J Appl Cardiol 4:152–167
40. Berwing K, Schlepper M, Kremer P, Bahavar H (1986) Assessment of myocardial perfusion abnormalities in patients with coronary heart disease by intracoronary injection of a new echo contrast agent. Circulation 74:(Suppl II)475
41. Berwing K, Schlepper M, Kremer P, Bahavar H (1989) Beurteilung myokardialer Perfusionsstörungen bei Patienten mit koronarer Herzkrankheit mittels intrakoronarer Injektion eines neuen Echo-Kontrastmittels. In: Grube E (Hrsg) Farb-Doppler- und Kontrast-Echokardiographie. Thieme, Stuttgart New York, S 323–335
42. Taira N (1983) Amrinone. Nippon Rinsho 41:15

Author's address:

Prof. Dr. M. Schlepper, Kerckhoff-Klinik, Benekestrasse 2–8, D-6350 Bad Nauheim, FRG

Phosphodiesterase inhibitors: alterations in systemic and coronary hemodynamics

K. Chatterjee

University of California, San Francisco, California, USA

Summary

This paper reviews the effects on myocardial contractility, left ventricular afterload and left ventricular distensibility induced by the following phosphodiesterase inhibitors: Enoximone, piroximone, RO 13-6438, amrinone and milrinone. For all these compounds, direct positive inotropic effects have been shown in experimental studies. For amrinone and milrinone, a direct stimulating effect on myocardial contractility has been demonstrated by an increase in dP/dt_{max} when intracoronary applications of the compounds were performed. A direct stimulating effect on the myocardium was also demonstrated for enoximone and piroximone by analyzing the systolic pressure versus end-systolic volume ratio.

For all of the phosphodiesterase inhibitors, a marked decrease of systemic vascular resistance has been observed indicating direct peripheral vasodilation. Although it has been demonstrated that phosphodiesterase inhibition increases left ventricular distensibility, the nature of this effect is not clear. For most of the phosphodiesterase inhibitors an increase in myocardial oxygen requirements was demonstrated due to overall contractility increase. However, these phosphodiesterase inhibitors induce increased coronary blood flow in excess so that a direct effect of these compounds on the coronary vasculature has been postulated. The clinical significance of such changes, however, remains unclear.

Introduction

Introdcution of vasodilator therapy has been an important milestone in the management of acute and chronic heart failure and many patients appear to benefit from such therapy. However, symptomatic and hemodynamic deterioration, despite vasodilator therapy, occurs rather frequently, particularly in patients with severe unstable heart failure. Furthermore, some patients either do not tolerate or do not respond to vasodilators or angiotensin converting enzyme inhibitors. Thus, the search continues to discover other therapeutic modalities for the management of heart failure. As decreased contractile function is the primary intrinsic pathophysiologic mechanism of heart failure in the vast majority of patients, it is not surprising that recently there has been an increasing interest to identify effective positive inotropic drugs which can be useful in the management of heart failure. Indeed, research during the last decade has resulted in the discovery of a fairly large number of pharmacologic agents with positive inotropic effects which can be potentially clinically useful. Among these newer inotropic drugs, beta-adrenoceptor agonists, dopamine receptor agonists, and phosphodiesterase inhibitors have drawn the most attention. At the University of California, San Francisco, we had the opportunity to evaluate the systemic hemodynamic effects of a few of the phosphodiesterase inhibitors in patients with chronic heart failure and this review summarizes our experience.

Phosphodiesterase inhibitors

A number of structurally different phosphodiesterase inhibitors are currently under investigation (Table 1) and the list is increasing rapidly. One common mechanism for enhanced contractility, however, is an increase in intracellular cyclic AMP concentration,
due to inhibition of FIII-phosphodiesterase which is required for degradation of cyclic
AMP [1, 2]. Increased cyclic AMP promotes cellular calcium influx and calcium-depen-
dent calcium release and increased calcium delivery to contractile units – the principal
mechanism for the positive inotropic effect of these agents. A few of these agents have
been thought to enhance sensitivity of myofilaments to calcium. It should be noted, however, that despite differences in the chemical structures and possibly in the mechanisms
of action, changes in systemic hemodynamics and ventricular performance with these
agents are remarkably similar. It is not surprising that the physiologic mechanisms by
which these agents improve cardiac performance are very similar. We studied the systemic hemodynamic effects of milrinone, enoximone, piroximone, and RO13-6438, in
different patient populations with severe chronic heart failure, and the systemic hemodynamic effects of these agents were both qualitatively and quantitatively similar. The acute
systemic hemodynamic effects of oral enoximone – an imidazole derivative in 38 patients
with chronic heart failure are summarized in Table 2 [3]. There was a substantial increase
in cardiac index, stroke volume index, and stroke work index along with a significant decrease in right atrial and pulmonary capillary wedge pressures, as well as systemic and
pulmonary vascular resistance. A slight increase in heart rate and a modest decrease in
mean arterial pressure was also observed.

Piroximone, another imidazole derivative phosphodiesterase inhibitor, produces
similar hemodynamic effects in patients with chronic heart failure. After intravenous administration of piroximone, the hemodynamic response was observed within 10 min and
lasted approximately 4 h. Cardiac stroke volume and stroke work indices increased significantly along with a reduction in right atrial and pulmonary capillary wedge pressures
and systemic and pulmonary vascular resistance (Table 3) [4]. Mean pulmonary and systemic arterial pressures and heart rate remained unchanged. Oral piroximone produced
similar hemodynamic changes in these patients. RO13-6438, a nonglycoside, noncate-

Table 1. Phosphodiesterase inhibitors

Agents
  Amrinone
  Milrinone
  Enoximone
  Piroximone
  RO 13-6438
  Berferine
  CI-914
  CD-OG115
  DPI-2010106

Potential Mechanisms of Action
  Inhibition of Phosphodiesterase F-III
  Adenosine inhibition
  Inhibition of calcium reuptake by sarcoplasmic reticulum
  Increased sensitivity of myofilaments to calcium
  Acivation of myocardial α-adrenergic receptors

Table 2. Hemodynamic effects of oral MDL 17043 in patients with congestive heart failure

	Control	12.5 min	30 min	1 h	2 h	3 h	4 h
Heart rate (beats/min)	91 ± 26	90 ± 15	91 ± 15	91 ± 14	93 ± 14	93 ± 14	92 ± 16
MAP (mmHg)	79 ± 11	78 ± 13	74 ± 10***	74 ± 9	71 ± 9*	72 ± 10*	74 ± 11***
PAP (mmHg)	38 ± 9	32 ± 8*	32 ± 8*	30 ± 8*	30 ± 8*	30 ± 9*	31 ± 10*
PCW (mmHg)	26 ± 7	19 ± 7*	18 ± 8*	18 ± 9*	17 ± 8*	18 ± 8*	18 ± 9*
RAP (mmHg)	13 ± 8	9 ± 6*	8 ± 7*	8 ± 7*	8 ± 6*	8 ± 6*	8 ± 7*
SVR (dynes·s·cm^{-5})	1405 ± 430	1061 ± 331*	997 ± 336*	996 ± 351*	930 ± 92*	915 ± 260*	1023 ± 260*
PVR (dynes·s·cm^{-5})	262 ± 152	206 ± 98**	208 ± 120***	191 ± 135	200 ± 132**	183 ± 34	209 ± 35***
SVI (ml/m²)	25 ± 7	34 ± 9*	35 ± 8*	34 ± 8*	35 ± 7*	35 ± 8*	33 ± 8*
CI (l/min per m²)	2.2. ± 0.5	3 ± 0.6*	3.1 ± 0.7*	3.1 ± 0.7*	3.2 ± 0.6*	3.2 ± 0.6*	3 ± 0.6*
SWI (g·m/m²)	19 ± 9	28 ± 11*	27 ± 10*	27 ± 11*	26 ± 9*	26 ± 10*	26 ± 11*

* $p < 001$ vs control; ** $p < 0.01$ vs control; *** $p < 0.05$ vs control. CI = cardiac index; MAP = mean arterial pressure; PAP = mean pulmonary arterial pressure; PCW = mean pulmonary capillary wedge pressure; PVR = pulmonary vascular resistance; RAP = mean right atrial pressure; SVI = stroke volume index; SVR = systemic vascular resistance; SWI = stroke work index.

Table 3. Changes in hemodynamic variables in 15 patients after intravenous piroximone

	Control	Peak effect	p value
HR (beats/min)	90 ± 17	92 ± 16	NS
RAP (mmHg)	13 ± 6	7 ± 5	<0.005
PAM (mmHg)	35 ± 8	29 ± 10	NS
LVFP (mmHg)	25 ± 7	19 ± 7	<0.001
MAP (mmHg)	73 ± 6	73 ± 8	NS
CI (l/min per m^2)	1.7+ 0.3	2.5± 0.5	<0.001
SVR (dynes·s·cm^{-5})	1633 ±394	1183 ±278	<0.001
PVR (dynes·s·cm^{-5})	304 ±164	192 ±111	<0.005
SVI (ml/m^2)	20 ± 6	30 ± 8	<0.001
SWI (g·m/m^2)	13 ± 4	22 ± 7	<0.005

CI = cardiac index; HR = heart rate; LVFP = left ventricular filling pressure; MAP = mean arterial pressure; PAP = mean pulmonary arterial pressure; PCW = mean pulmonary capillary wedge pressure; PVR = pulmonary vascular resistance; RAP = mean right atrial pressure; SVI = stroke volume index; SVR = systemic vascular resistance; SWI = stroke work index.

Table 4. Changes in hemodynamics at rest with RO13-6438 (n = 12)

	Baseline (mean ± SD)	Peak (mean ± SD	Difference (mean ± SD)	Range	p
HR (beats/min)	91 ± 12	95 ± 13	+ 4 ± 7	− 11 to + 18	NS
SBP (mmHg)	113 ± 11	108 ± 11	− 5 ± 9	− 20 to + 10	NS
DBP (mmHg)	63 ± 7	54 ± 10	− 9 ± 5	− 23 to − 3	<0.01
Mean BP (mmHg)	80 ± 6	72 ± 8	− 8 ± 6	− 19 to + 0	<0.01
PCWP (mmHg)	26 ± 7	16 ± 8	− 10 ± 5	− 19 to + 5	<0.01
RAP (mmHg)	11 ± 6	6 ± 8	− 5 ± 3	− 13 to + 0	<0.01
CI (l/min/m^2)	2.09± 0.45	3.30± 0.73	+ 1.21± 0.7	+ 0.08 to + 2.40	<0.01
SVI (ml/m^2)	23.4 ± 6.8	35.8 ± 11.4	+ 12.4 ± 7.7	+ 0.85 to + 2.51	<0.01
SWI (g·m/beat/ m^2)	22.8 ± 10.8	35.9 ± 13.5	+ 13.1 ± 6.6	+ 0.38 to + 21.6	<0.01
SVR (dynes·s·cm^{-5})	1450 ±330	910 ±310	−540 ±290	−980 to −130	<0.01

BP = blood pressure; CI = cardiac index; DBP = diastolic blood pressure; HR = heart rate; NS = not significant; PCWP = pulmonary capillary wedge pressure; RAP = right atrial pressure; SBP = systolic blood pressure; SD = standard deviation; SVI = stroke volume index; SVR = systemic vascular resistance; SWI = stroke work index.

cholamine, imidazoquinazolinone derivative has been shown in animal studies to possess both positive inotropic and vasodilating properties. The systemic hemodynamic effects of this agent are also similar to those of enoximone or piroximone. The hemodynamic effects were determined in 12 patients with severe chronic heart failure unresponsive to conventional and vasodilator therapy (Table 4) [5]. There was a marked increase in cardiac, stroke volume, and stroke work indexes and a concomitant fall in right atrial and pulmonary capillary wedge pressures and systemic vascular resistance. Although heart rate and systolic blood pressure did not change, diastolic blood pressure fell significantly.

Table 5. Peak hemodynamic effects of oral milrinone in 37 patients with severe congestive heart failure

	Baseline	Peak effect	P value
Heart rate (beats/min)	88 ± 18	95 ± 16	0.0001
MAP (mmHg)	80 ± 11	69 ± 11	0.0001
RAP (mmHg)	13 ± 8	7 ± 5	0.0001
PCW (mmHg)	28 ± 9	18 ± 8	0.0001
SVR (dynes·s·cm^{-5})	1694 ±485	1139 ±287	0.0001
CI (l/min per m²)	1.9± 0.5	2.4± 0.5	0.0001
SWI (gm/m²)	15.6± 8.0	22.4± 9.6	0.0001
SVI	21.8± 6.6	31.4± 9.5	0.0001

Values are mean ± standard deviation. CI = cardiac index; MAP = mean arterial pressure; PCW = pulmonary capillary wedge pressure; RAP = right arterial pressure; SVI = stroke volume index; SVR = systemic vascular resistance; SWI = stroke work index.

Milrinone, a bipyridine derivative, causes a substantial increase in cardiac and stroke volume indexes, along with a decrease in right atrial and pulmonary capillary wedge pressures, and pulmonary and systemic vascular resistances (Table 5) [6]. Heart rate and mean arterial pressure usually remain unchanged. In uncontrolled studies, sustained beneficial hemodynamic effects of milrinone have been reported in some patients with severe heart failure.

It is apparent that the systemic hemodynamic effects of most phosphodiesterase inhibitors are very similar, despite differences in their chemical structures. An improvement in left ventricular pump function occurs even in patients with severe heart failure.

Mechanisms for Improvement in Ventricular Function

The major determinants of ventricular performance are prelaod, afterload, and contractility. Enhanced contractility is associated with increased cardiac output and decreased ventricular filling pressure resulting in an upward shift of the ventricular function curve. Similar shifts in ventricular function curve also occur when resistance to ventricular ejection declines [7]. It is not widely appreciated that changes in compliance also influence ventricular performance. Decreased distensibility implies a greater increase in filling pressure for any given change in enddiastolic volume; such a change is associated with a downward shift of the ventricular function curve, similar to that associated with decreased contractility or increased afterload. Conversely, increased distensibility or compliance results in an upward shift of the ventricular function curve as observed with enhanced contractility or decreased afterload. Concomitant changes in preload or end-diastolic volume with changes in contractility or afterload influence the net changes in stroke volume. The expected increase in stroke volume from increased contractility or decreased afterload may be curtailed when there is also a reduction in end-diastolic volume. Similarily, increased end-diastolic volume may cause a greater increase in stroke volume with a similar increase in contractility, or a decrease in afterload.

Phosphodiesterase inhibitors exert positive inotropic effects and enhance contractile function. A direct positive inotropic effect of enoximone, piroximone, RO13-648, and milrinone has been demonstrated in in vitro studies [8–10]. Absolute force and its rate of development at muscle isolength increase. In ejecting ventricles the end-systolic stress-endsystolic volume ration increases [11, 12]. In patients with heart failure it is more difficult to assess changes in contractility. However, in a number of studies a substantial increase in left ventricular dp/dt with little or no change in heart rate or blood pressure and with a fall in left ventricular filling pressure has been reported in response to phosphodiesterase inhibitors [13–15]. These hemodynamic changes provide evidence for increased contractility. Intracoronary administration of amrinone and milrinone increases left ventricular dp/dt without causing a substantial systemic hemodynamic change, which also suggests a direct positive inotropic effect of these agents. Changes in the contractile function have been assessed by determining the ratio of peak systolic pressure/end-systolic volume relations and a reasonable correlation has been reported between the peak systolic pressure to end-systolic volume ratio and the end-systolic pressure to end-systolic volume ratio. In response to enoximone and piroximone, the ratio of peak systolic pressure/end-systolic volume tends to increase in many patients with chronic heart failure [16]. Similar changes have been observed in response to bipyridine derivatives. A reduction in left ventricular ejection time, despite a marked increase in left ventricular forward stroke volume, has been observed in response to enoximone in patients with chronic heart failure [16]. When the increase in stroke volume primarily results from decreased afterload or increased preload, left ventricular ejection time prolongs or remains unchanged [17, 18]. This shortening of ejection time does indicate enhanced contractile function. Although these studies suggest that these phosphodiesterase inhibitors exert positive inotropic effects, even in patients with severe heart failure, the relative contribution of increased contractility to the improvement of left ventricular function is difficult to estimate in these patients. Nevertheless, enhanced contractility appears to be one of the mechanisms for improvement in cardiac performance with phosphodiesterase inhibitors.

Phosphodiesterase inhibitors are also potent vasodilators. In the perfused dog hind limb preparations, systemic administration of enoximone produced a dose-related decrease in hind limb perfusion pressures [19]. Vasodilation in these preparations were not altered by prior adrenergic or histaminergic receptor blockade. Although enoximone has both alpha adrenergic and histamine receptor blocking properties, these effects are very weak and do not appear to have clinical relevance. The intravenous or oral administration of enoximone causes a significant reduction of systemic vascular resistance. There is also a slight reduction in mean arterial pressure.

Piroximone also exerts vasodilatory effects and systemic vascular resistance declines substantially in most patients in response to intravenous or oral piroximone. A similar reduction in systemic vascular resistance has been observed with RO13-6438 [5]. The bipyridine derivatives, amrinone and milrinone also cause a substantial reduction in systemic vascular resistance following their oral or intravenous administration. Milrinone increases limb blood flow, decreases limb vascular resistance, and raises limb venous capacitance. The direct vasodilatory effect of milrinone, however, appears to be less pronounced than the direct-acting vasodilator agent nitroprusside [20]. Nevertheless, a direct peripheral vasodilatation appears to be a physiologic effect common to all phosphodiesterase inhibitors, although the magnitude of vasodilatation in response to the individual agent may vary. Peripheral vasodilatation, particulary, decreased systemic vascular resistance, is an important mechanism for the improvement in left ventricular pump function. Left ventricular ejection impedance is inversely related to its stroke volume; thus, decreased left ventricular outflow resistance results in increased stroke volume and cardiac output, along with a reduction in left ventricular filling pressure, suggesting improved left ventricular function.

In addition to enhanced contractility and decreased afterload, phosphodiesterase inhibitors may also improve left ventricular function by improving left ventricular diastolic compliance. In a group of patients with severe left ventricular failure and secondary right heart failure, left ventricular end-diastolic volume was determined by measuring left ventricular ejection fraction (computerized nuclear probe) and forward stroke volume before and after enoximone (thermodilution technique). In the group as a whole, end-diastolic volume index did not change significantly at the time of peak systemic hemodynamic effects following intravenous enoximone, although pulmonary capillary wedge pressure decreased markedly, indicating the possibility of improved left ventricular diastolic distensibility [16]. In individual patients there was also an increase in end-diastolic volume, along with a marked reduction in pulmonary capillary wedge pressure. In other patients, however, pulmonary capillary wedge pressure decreased along with a decrease in end-diastolic volume. Axelrod et al. [4] also determined changes in left ventricular filling pressure and end-diastolic volume following intravenous piroximone in 10 patients with severe chronic left ventricular failure. The group mean left ventricular filling pressure decreased (26 ± 8 to 19 ± 9 mmHg P < 0.01), whereas end-diastolic volume index tended to increase (106 ± 42 to 132 ± 60 ml/m^2, P = 0.07). In eight of 10 patients a decrease in left ventricular filling pressure was associated with either no change or an increase in end-diastolic volume index. These findings suggested that piroximone improved overall left ventricular distensibility in these patients. Monrad et al. [21] also reported an improvement in indexes of diastolic performance in patients with congestive heart failure treated with milrinone, another phosphodiesterase inhibitor. In this study, left ventricular volumes were measured by radionuclide ventriculography. In six of the nine patients studied, left ventricular diastolic pressure volume curve shifted downward after milrinone therapy. Indeed, in the majority of patients, there was an increase in end-diastolic volume after milrinone, concomitant with a decrease in left ventricular end-diastolic pressure. These findings also suggested improved left ventricular distensibility in response to milrinone.

Improved left ventricular diastolic compliance in response to phosphodiesterase inhibitors however, is not a uniform finding. Herman et al. [22] assessed changes in left ventricular diastolic volume and pressure in 11 patients with chronic heart failure due to dilated cardiomyopathy. In only three of these 11 patients was there a downward shift of the left ventricular diastolic pressure volume curve. Nevertheless, these studies demonstrate that phosphodiesterase inhibitors can potentially improve diastolic compliance in some patients with chronic left ventricular failure.

The mechanism for improved left ventricular distensibility with phosphodiesterase inhibitors in these patients, however, has not been fully clarified and it is probable that multiple mechanisms are contributory. Milrinone has been shown to increase left ventricular peak filling rate, decrease the time constant for the rate of fall in pressure and isovolumic relaxation time [21]. These findings suggest that phosphodiesterase inhibitors can improve left ventricular relaxation and early filling properties. It has been suggested that the downward shift of the left ventricular diastolic pressure volume curve may result from changes in ventricular interaction and can be accounted for due to changes in intrapericardial pressures. Herman et al. [22] assessed the relationship between changes in estimated transmural pressure (left ventricular end-diastolic pressure – right atrial pressure) and left ventricular diastolic volume following enoximone, and observed no downward shift of the transmural pressure-diastolic volume curve, although there was a downward shift of the left ventricular diastolic-pressure-volume curve. These findings suggested that there was only an apparent shift of left ventricular diastolic pressure-volume relation. Axelrod et al. [4] however, could not demonstrate that changes in intrapericardial pressure could always account for the observed changes in left ventricular diastolic-pressure volume relation. In 10 patients with chronic heart failure, they determined left ventricular diastolic volume by concurrent determination of ejection fraction (radionuclide scintigraphy) and forward stroke volume (thermodilution technique). Following piroximone, in eight of 10 patients, there was a decrease in left ventricular filling pressure associated with either no change or an increase in end-diastolic volume index. Transmural pressure was again estimated by subtracting mean right atrial pressure from pulmonary capillary wedge pressure. When the changes in estimated transmural pressure were related to changes in end-diastolic volume, following piroximone, no consistent correlation was found. In five patients an increase in end-diastolic volume was associated with a decrase in estimated transmural pressure. Thus, in these patients, the observed improvement in distensibility persists even after accounting for the reduction in pericardial pressure. It is apparent from these studies that changes in left ventricular distensibility following phosphodiesterase inhibitors may result from a number of interacting factors which may be influenced by the hemodynamic changes.

Coronary hemodynamics and myocardial energetics

Coronary blood flow and myocardial perfusion are determined by changes in myocardial metabolic demand, coronary artery driving pressure, transmyocardial pressure gradient, and coronary vascular resistance. The major determinants of myocardial oxygen demand are heart rate, contractility, and left ventricular wall stress; and wall stress is related to left ventricular volume, pressure and wall thickness. In patients with obstructive coronary artery disease, the degree of coronary artery stenosis and the collateral circulation are also important determinants of coronary blood flow and myocardial perfusion.

Phosphodiesterase inhibitors can potentially influence both determinants of myocardial oxygen demand and myocardial perfusion. Changes in heart rate, arterial pressure and contractility are common to all phosphodiesterase inhibitors, although the magnitude of change in these determinants of myocardial oxygen requirements vary according to the specific type and the dose of the phosphodiesterase inhibitors. Phosphodiesterase inhibitors can also induce coronary vasodilatation and thus increase coronary blood flow. Thus, the net effect on coronary hemodynamics and myocardial energetics in response to the phosphodiesterase inhibitors is regulated by their relative effects on the factors that influence myocardial metabolic demand and perfusion.

The changes in coronary hemodynamics in 17 patients with severe chronic heart failure following enoximone are summarized in Table 7 [23]. In the group as a whole there was a significant increase in coronary sinus blood flow. The double product, i.e., the product of the systolic blood pressure and the heart rate tended to be higher, but the changes were not statistically significant. There was a significant increase in myocardial oxygen delivery and consumption. However, myocardial oxygen extraction also decreased significantly. Transmyocardial lactate extraction in most patients remained unchanged.

Decreased myocardial oxygen extraction suggests that coronary blood flow was in excess of metabolic demand, which probably results from coronary vasodilatation following enoximone. Coronary vasodilatation is further supported by the observation that the coronary sinus oxygen content tended to be higher following enoximone, in the majority of patients. Although these findings suggest that enoximone can cause coronary vasodilatation and potentially increase coronary blood flow, the mechanism for the decreased coronary vascular resistance remains unclear: both primary vasodilatation or secondary

Table 6. Changes in coronary hemodynamics in 17 patients with severe heart failure after MDL 17043

	Baseline	Peak	P value
Myocardial oxygen extraction (vol-%)	11.7± 2.0	10.1± 3.3	<0.05
Myocardial oxygen delivery (ml/min)	12.0± 4.3	16.4± 9.6	<0.05
Myocardial oxygen consumption (ml/min)	8.9± 3.3	11.8± 5.4	<0.05
Lactate extraction (%)	27 ±19	21 ±19	NS
Coronary sinus flow	75 ±29	111 ±51	<0.01
HR × SBP × 10^{-3} (mmHg/min)	8.9± 1.5	10.4± 1.9	NS

Data are reported as mean ± standard deviation. HR = heart rate; NS = not significant; SBP = systolic blood pressure.

Table 7. Changes in coronary hemodynamics following RO 13-6438 in chronic heart failure (n = 10)

	Control	RO 13-6438	P
CSBF (ml/min)	139 ± 69	138 ± 61	NS
CSO$_2$ (vol-%)	3.2± 0.8	4.3± 1.5	<0.05
ACSO$_2$ (vol-%)	11.7± 2.1	10.4± 1.9	<0.05
MVO$_2$ (ml/min)	15.3± 6.8	14.9± 6.8	NS
MW/MVO$_2$ (gm/ml O$_2$)	299 ±143	580 ±367	<0.05

CSBF = coronary sinus blood flow; CSO$_2$ = coronary sinus oxygen; ACSO$_2$ = arterial-coronary sinus venous oxygen difference; MVO$_2$ = myocardial oxygen consumption; MW/MVO$_2$ = minute work per myocardial oxygen consumption.

to increased metabolic demand may decrease coronary vascular resistance. In this study, enoximone also increased myocardial oxygen consumption which indicates that there must have also been an increase in myocardial oxygen requirements. The product of heart rate and blood pressure did not change significantly; however, the enhanced contractility with enoximone might have contributed to augmented myocardial oxygen consumption and coronary blood flow. It should be noted, however, that the magnitude of increase in coronary blood flow (+44%) was in excess of an increase in oxygen consumption (+20%), indicating that in addition to metabolically mediated reduction in coronary vascular resistance, enoximone exerts a direct vasodilatory effect on the coronary vascular bed.

RO13-6438 also appears to induce coronary vasodilatation in patients with chronic heart failure due to dilated cardiomyopathy. Daly et al. [5] assessed the effects of RO13-6438 on coronary hemodynamics in a group of patients with chronic severe heart failure (Table 7). The overall coronary blood flow or myocardial oxygen consumption remained unchanged. These findings suggest that improved left ventricular function wiht RO13-6438 is not usually associated with increased metabolic cost. Although stroke work or minute work forms only a portion of total mechanical work (and therefore correlates poorly with MVO$_2$), they reflect changes in pump function. Thus, minute work/MVO$_2$ provides a clinical estimate of left ventricular efficiency which appears to improve following RO13-6438. Myocardial oxygen consumption is usually controlled by changes in coronary blood flow rather than changes in myocardial oxygen extraction, i.e., arterial-coronary sinus venous oxygen difference. In response to Ro13-6438, however, arterial-coronary sinus oxygen difference decreased and coronary sinus oxygen increased, suggesting a direct coronary vasodilatory effect. In this study, the relative changes in coronary blood flow and myocardial oxygen consumption were similar, indicating that myocardial oxygen supply/demand ratio was unaffected by RO13-6438.

Amrinone, milrinone, and other phosphodiesterase inhibitors also appear to produce similar changes in coronary hemodynamics [24, 25]. Intravenous amrinone in patients with chronic heart failure did not cause any significant change in myocardial oxygen consumption or coronary sinus blood flow. Transmyocardial lactate extraction also remained unchanged in most patients. Intravenous milrinone however, increased coronary blood flow significantly but myocardial oxygen extraction decreased. Thus, overall myocardial oxygen consumption remained unchanged. Heart rate blood pressure product also remained unchanged. The relative increase in coronary blood flow was considerably greater than the change in myocardial oxygen requirements and myocardial oxygen supply/demand ratio was higher following milrinone. Decreased myocardial oxygen extraction and a substantial increase in coronary blood flow suggest a primary decrease in coronary vascular resistance and coronary vasodilatation.

Newer phosphodiesterase inhibitors, therefore, appear to affect both the determinants of oxygen requirements and myocardial perfusion. Enhanced contractility, a slight increase in heart rate and occasionally increased diastolic volume, enhance myocardial oxygen consumption. Higher coronary sinus oxygen content and decreased myocardial oxygen extraction indicate a primary reduction in coronary vascular resistance which contributes to maintaining or increasing coronary blood flow for the changes in metabolic demand. Thus, in response to these phosphodiesterase inhibitors, myocardial oxygen supply/demand ratio increases or remains unchanged. The clinical significance of the excess oxygen delivery for metabolic demand and the alterations in the regulatory mechanisms of coronary blood flow with phosphodiesterase inhibitors, however, remain unclear. Myocardial oxygen delivery is primarily determined by coronary blood flow as arterial-venous oxygen difference, i.e., myocardial oxygen extraction is near maximum even during basal metabolic state. In the presence of intact coronary vascular autoregulation, coronary blood flow increases with increased metabolic demand without any appreciable change in myocardial oxygen extraction. Precise estimations of changes in the determinants of myocardial oxygen requirements in response to phosphodiesterase inhibitors have been difficult. Although heart rate-blood pressure product usually do not change, rather the overall contractility increases; ventricular volume may also increase in some patients. Thus, it is very likely that the overall myocardial oxygen requirements increased following phosphodiesterase inhibitors. Despite increased metabolic demand, myocardial oxygen extraction usually declines and coronary sinus venous PO2 rises. As perfusion pressure did not change significantly in most patients, the increased coronary blood flow must have resulted, at least partly, from a primary decrease in coronary vascular resistance. The coronary vasodilating effect may provide improved coronary reserve, particularly in patients with coronary artery disease.

In summary, phosphodiesterase inhibitors have the potential to improve left ventricular function even in patients with severe refractory heart failure. Enhanced contractility, decreased left ventricular afterload and possibly increased left ventricular distensibility contribute to improved left ventricular function. Phosphodiesterase inhibitors exert coronary vasodilatory effects and usually coronary blood flow and oxygen supply matches or exceeds changes in myocardial oxygen demand. The clinical significance of such changes, however, remain unclear.

References

1. Colucci WS, Wright RF, Braunwald E (1986) New positive inotropic agents in the treatment of congestive heart failure. Mechanisms of action and recent clinical developments. N Engl J Med 314:290
2. Colucci WS, Wright RF, Braunwald E (1986) New positive inotropic agents in the treatment of congestive heart failure. Mechanism of action and recent clinical developments. N Engl J Med 314:3498
3. Kereiakes D, Chatterjee K, Parmley WW et al. (1984) Intravenous and oral DML 17043 (a new inotrope-vasodilator agent) in congestive heart failure: hemodynamic and clinical evaluation in 38 patients. J Am Coll Cardiol 4:884
4. Axelrod RJ, DeMarco T, Dae M, Botvinick E, Chatterjee K (1987) Hemodynamic and clinical evaluation of piroximone, a new inotrope-vasodilator agent, in severe congestive heart failure. J Am Coll Cardiol 9:1124
5. Daly PA, Chatterjee K, Viquerat CE et al. (1983) RO13-6438, a new inotrope-vasodilator: systemic and coronary hemodynamic effects in congestive heart failure. Chest 84:408
6. Simonton CA, Chatterjee K, Cody RJ et al. (1985) Milrinone in congestive heart failure: Acute and chronic hemodynamic and clinical evaluation. J Am Coll Cardiol 6:453
7. Chatterjee K, Parmley WW (1977) The role of vasodilator therapy in heart failure. Prog Cardiovasc Dis 19:301–325
8. Alousi AA, Stankus GP, Stuart JC, Walton LH (1983) Characterization of the cardiotonic effects of milrinone, a new and potent cardiac bipyridine, on isolated tissues from several animal species. J Cardiovasc Pharmacol 5:804
9. Kariya T, Willia LJ, Dage RC (1982) Biochemical studies on the mechanism of cardiotonic activity of RMI 17043 (1,3-dihydro-4-methyl-5-4(methylthio)-(benzoyl)-2H-imidizone-2-one) (abstr). Fed Proc 41:1310
10. Dage RC, Hsieh CP, Kariya T, Schnettler RA (1985) Cardiotonic activity on MDL 19,205 (4-ethyl-1,3-dihydro-5-(4-pyridinyl-carbonyl)-2H-imidazol-2-one) (abstr). Fed Proc 42:1131
11. Janicki JS, Shroff SG, Weber KT (1987) Physiologic response to the inotropic and vasodilator properties of enoximone. Am J Cardiol 60:15C–20C
12. Suga H, Sagawa K, Shoukas AA (1973) Load independence of the instantaneous pressure-volume ratio of the canine left ventricle and effects of epinephrine and heart rate on the ratio. Circ Res 32:314
13. Crawford MH, Sorensen SG, Richards KL, Sedums M (1983) Demonstrations of combined vasodilator-inotropic effects of RMI 17,043 in patients with reduced left ventricular heart failure. Circulation 67:823
14. Mancini D, LeJemtel T, Sonnenblick EP (1985) Intravenous use of amrinone for the treatment of the failing heart. Am J Cardiol 56:8B
15. Benotti JR, Grossman W, Braunwald E et al. (1978) Hemodynamic assessment of amrinone: A new inotropic agent. N Engl J Med 299:1373
16. Kereiakes D, Viquerat C, Lanzer P, Botvinick E, Spangenberg R, Buckingham M, Parmley WW, Chatterjee K (1984) Mechanisms of improved left ventricular function following intravenous MDL 17,043 in patients with severe chronic heart failure. American Heart Journal 108:1278
17. Salzman SH, Wolfson S, Jackson B, Schechter E (1971) Epinephrine infusion in man. Standardization, normal response, and abnormal response in idiopathic hypertrophic subaortic stenosis. Circulation 43:137
18. Horshen RJ, Cuddy TE (1975) Dose-responsiveness relation between therapeutic levels of serum digoxin and systolic time intervals (abstr). Am J Cardiol 35:469
19. Roebel LE, Dage RC, Cheng HC, Woodward J (1982) Characterization of the cardiovascular activities of a new cardiotonic agents, MDL 17,043 (1,3-dihydro-4-methyl-5E40(methylthio)-benzoyl)-2H-imidazol-2-one. J Cardiovasc Pharmacol 4:721
20. Cody RG, Muller FB, Kubo SH et al. (1986) Identification of the direct vasodilator effect of milrinone with an isolated limb preparation in patients with chronic congestive heart failure. Cirulation 73:124

21. Monrad ES, McKay RG, Baim DS, Colucci WS, Fifer MA, Heller GV, Royal HD, Grossman W (1984) Improvement in indexes of diastolic performance in patients with congestive heart failure treated with milrinone. Circulation 70:1030
22. Herman HC, Ruddy TD, Dec GW, Strauss HW, Boucher CA, Fifer MA (1987) Diastolic function in patients with severe heart failure: comparison of the effects of enoximone and nitroprusside. Circulation 75:1214
23. Viquerat C, Kereiakes D, Morris L, Daly P, Wexman M, Frank P, Parmley WW, Chatterjee K (1985) Alterations in left ventricular function, coronary hemodynamics and myocardial catecholamine balance with MDL 17043, a New Inotropic Vasodilator Agent, in Patients with Severe Heart Failure. J Am Coll Cardiol 5:326
24. Benotti JR, Grossman W, Braunwald E, Carabello BA (1980) Effects of amrinone on myocardial energy metabolism and hemodynamics in patients with severe congestive heart failure due to coronary artery disease. Circulation 62:28
25. Monrad ES, Baim DS, Smith HS, Lanoue A, Branwald E, Grossman W (1985) Effects of milrinone on coronary hemodynamics and myocardial energetics in patients with congestive heart failure. Circulation 71:972

Author's address:

K. Chatterjee, University of California, San Francisco, California, USA

Influence of the calcium-sensitizer UDCG-115 on hemodynamics and myocardial energetics in patients with idiopathic dilated cardiomyopathy. Comparison with nitroprusside

G. Hasenfuss, C. Holubarsch, H. W. Heiss, B. Rattert, Hj. Just

Medizinische Universitätsklinik, Innere Medizin III, Kardiologie, Universität Freiburg, FRG

Summary

UDCG-115 is a new cardiotonic agent which in vitro increases the sensitivity of the contractile proteins to calcium ions, inhibits the activity of phosphodiesterase, and prolongs the duration of the action potential. The influence of UDCG-115 (i.v.) on hemodynamics and myocardial energetics was investigated in patients with idiopathic dilated cardiomyopathy (NYHA II–III) and compared to the effects of the pure vasodilator nitroprusside. UDCG-115 increased cardiac index from 3.2 ± 0.4 to 4.2 ± 0.8 l/min/m² ($p < 0.01$) and decreased left ventricular end-diastolic wall stress (preload) from 52 ± 21 to 28 ± 18 10^3 dyn/cm² ($p < 0.01$) and end-systolic wall stress (afterload) from 201 ± 61 to 129 ± 43 10^3 dyn/cm² ($p < 0.01$) compared to control conditions. Compared to nitroprusside, for a similar decrease in preload and afterload, UDCG-115 increased cardiac index by 40% ($p < 0.01$), stroke volume index by 37% ($p < 0.01$) and maximum rate of left ventricular pressure rise by 23% ($p < 0.05$). Heart rate did not significantly change with either drug. Myocardial oxygen consumption per beat decreased by 33% ($p < 0.05$) with UDCG-115 and by 30% ($p < 0.01$) with nitroprusside. With both drugs, the decrease of myocardial oxygen consumption correlated significantly with the decrease of left ventricular systolic stress-time integral. The slopes of the respective linear regression lines were not significantly different.

Thus, UDCG-115 given intravenously in patients with idiopathic dilated cardiomyopathy and moderate congestive heart failure exhibits significant inotropic and vasodilating properties. The systemic hemodynamic actions are associated with favorable effects on myocardial energetics.

Introduction

During the last decade numerous inotropic agents have been developed that increase the contractile force of the myocardium by elevating the amount of calcium ions available to the contractile proteins [2, 30]. Alternatively, contractile force can be increased at a given calcium concentration by elevating the sensitivity of the contractile proteins to calcium ions [24, 25]. Several inotropic agents which increase the calcium sensitivity, as indicated by a leftward shift of the force-calcium relationship of skinned myocardial fibers, are currently under investigation. Most of these so-called calcium-sensitizers exhibit additional pharmacological properties (Table 1).

UDCG-115 has been extensively investigated in animal experiments. Its influence on calcium-sensitivity of the contractile proteins has been shown in chemically skinned cardiac muscle fibers as well as in isolated troponin C [4, 9, 16, 23, 24, 31]. Two additional

Table 1. Cardiotonic agents which increase the sensitivity of skinned myocardial fibers to calcium ions

Agents	References
Sulmazole	[6]
UDCG-115	[23, 24]
APP 201-533	[26]
DPI 201-106	[28]
Adibendan	[19]
Isomazole	[14]
MCI-154	[12]

mechanisms seem to contribute to the inotropic effect of UDCG-115. Firstly, UDCG-115 prolongs the duration of the action potential, which may increase the calcium influx through calcium channels [8]. Secondly, UDCG-115 inhibits phosphodiesterases with the subsequent increases in cyclic adenosine monophosphate concentration [1, 8, 13, 29]. The extent to which each of these potential inotropic mechanisms is relevant under in vivo conditions is unclear, in particular, since UDCG-115 is rapidly metabolized to its o-de-methyl metabolite UDCG-212 which is also pharmacologically active [3, 13, 16, 29]. Besides its inotropic action, UDCG-115 exhibits pronounced vasodilating properties [3, 15].

Because of its pharmacological profile in animal experiments, UDCG-115 might be a promising drug for the treatment of congestive heart failure. In the present study, the influence of UDCG-115 on hemodynamics and myocardial energetics was investigated in patients with idiopathic dilated cardiomyopathy and moderate congestive heart failure. The influence of UDCG-115 on myocardial energetics is of particular interest, since an increase in myocardial energy demand due to inotropic stimulation may limit the clinical use of cardiotonic agents [5, 11, 20]. In order to differentiate between the influence of the inotropic and the vasodilating effect on hemodynamics and myocardial energetics the actions of UDCG-115 were compared to those of the pure vasodilator nitroprusside.

Methods

Patients

The investigations were performed in 16 patients with idiopathic dilated cardiomyopathy undergoing routine diagnostic cardiac catheterization. Fourteen patients were male and two female (mean age 46 ± 9 years); 11 were in NYHA class III and five in class II. All patients were in sinus rhythm. Idiopathic dilated cardiomyopathy was defined by increased left ventricular end-diastolic volume (> 220 ml) and reduced left ventricular ejection fraction ($< 55\%$) in the absence of coronary or valvular heart disease, or a history of arterial hypertension.

Measurements of hemodynamics and myocardial oxygen consumption

The patients were evaluated in the fasting state. Previous medications were withheld at least 48 h before the investigation. Each patient underwent coronary angiography as well as right and left heart catheterization. Left ventriculography was performed by power injection of 40 ml nonionic contrast solution at 50 frames per s while pressure was measured simultaneously using Millar microtip catheter pressure transducers. The projection was a 10° caudally angulated right anterior oblique view. Maximum rate of left ventricular pressure rise was obtained from continuous differentiation of left ventricular pressure tracings. During one cardiac cycle, left ventricular volumes were calculated from each cine frame at intervals of 20 ms using the Sandler and Dodge method [27]. Left ventricular wall thickness and muscle mass were determined according to a modification of the method of Rackley et al. [21]. Instantaneous circumferential wall stress values were obtained using the ellipsoid model of Mirsky [17]. End-diastolic wall stress (index of preload) was defined as the wall stress value at end-diastolic pressure. End-systolic wall stress (index of afterload) was defined as the stress value at the time of the initial appearance of the smallest left ventricular volume. The systolic stress-time integral was calculated by integrating instantaneous stress values from end-diastolic wall stress to the stress value derived from the cine frame before left ventricular volume increased. Cardiac output was measured by the thermodilution technique. Cardiac index was calculated as cardiac output/body surface area, and stroke volume index as cardiac index/heart rate. Heart rate was taken from the continuously recorded electrocardiogram. Systemic vascular resistance (SVR) was calculated from the formula: SVR = (mean aortic pressure – mean right atrial pressure) 80/cardiac output. Ejection fraction was calculated as left ventricular end-diastolic volume minus left ventricular end-systolic volume divided by left ventricular end-diastolic volume.

Myocardial oxygen consumption per min was measured by means of the argon method [5, 22]. Myocardial oxygen consumption per beat was obtained from myocardial oxygen consumption per min and heart rate.

Drug administration

Upon completion of basal measurements, the intravenous infusion of nitroprusside was begun at a rate of 25 µg/min. The infusion rate was increased by 12.5 or 25 µg/min until either mean aortic pressure had decreased by more than 15% or was below 75 mm Hg. The average dosage was 63 ± 40 µg/min. After the infusion rate had been stable for 10 min the measurements were repeated.

UDCG-115 was administered intravenously at 1.25 mg/h for a period of 3–5 h until the measurements were repeated.

Hemodynamics and myocardial oxygen consumption were measured during control conditions in all 16 patients. Comparative hemodynamic data (nitroprusside and UDCG-115) were obtained in nine patients. Both hemodynamics and myocardial oxygen consumption were measured following the application of nitroprusside in 10 patients and following the application of UDCG-115 in nine patients.

Statistical analysis

Values are expressed as mean $\pm$ standard deviation. Statistical analysis involving comparisons of control measurements with data obtained during infusions of nitroprusside and UDCG-115 were accomplished by using the t-test for paired samples and the Bonferroni correction for multiple comparisons [32]. When only measurements before and after the administration of pimobendan or nitroprusside were compared, the t-test for paired samples was used. A p value <0.05 was accepted as significant.

Results

Effects of UDCG-115 and nitroprusside on hemodynamic variables

The application of UDCG-115 and nitroprusside resulted in a significant decrease of left ventricular systolic pressure, end-diastolic pressure, and end-diastolic volume (Table 2, Fig. 1). Left ventricular end-diastolic wall stress (preload) decreased by 46% with UDCG-115 and by 52% with nitroprusside. Left ventricular end-systolic wall stress (afterload) decreased by 36% with UDCG-115 and by 34% with nitroprusside. For a similar decrease in preload and afterload, cardiac index increased by 40%, stroke volume index by 37% and maximum rate of left ventricular pressure rise by 23% with UDCG-115 compared to nitroprusside. The heart rate did not significantly change with either drug.

Effects of UDCG-115 and nitroprusside on myocardial energetics

Following the application of UDCG-115 there was a significant decrease of myocardial oxygen consumption per min and per beat from 14.3 ± 5.1 to 10.6 ± 3.8 ml/min/100 g ($p < 0.05$) and from 165 ± 47 to 111 ± 32 µl/beat/100 g ($p < 0.05$). Following the application of nitroprusside, myocardial oxygen consumption per min decreased from 13.3 ± 5.0 to 9.8 ± 2.2 ml/min/100 g ($p < 0.05$) and myocardial oxygen consumption per beat from 159 ± 44 to 112 ± 23 µl/beat/100 g ($p < 0.01$) (Fig. 2). With both drugs, the decrease of myocardial oxygen consumption per beat correlated significantly with the decrease of left ventricular systolic stress-time integral. The slopes of the respective linear regression lines were not statistically different (Fig. 3).

Table 2. Comparative effects of nitroprusside and UDCG-115 on hemodynamic variables

	Control	Nitroprusside	UDCG-115
HR (beats/min)	82 ± 13	88 ± 13	91 ± 16
MAP (mmHg)	96 ± 10	76 ± 9**	78 ± 12**
MPP (mmHg)	22 ± 11	9 ± 5**	14 ± 6*
CI (l//min/m²)	3.2± 0.4	3.0± 0.6	4.2± 0.8**##
SVI (ml/m²)	40 ± 10	35 ± 11	48 ± 14##
SVR dyn·s·cm⁻⁵)	1247 ±313	1089 ±273*	832 ±300**#
LVEDP (mmHg)	18 ± 8	9 ± 4**	10 ± 5**
LVEDV (ml)	353 ± 92	293 ±107**	311 ± 97**
S_{ed} (10³ dyn/cm²)	52 ± 21	25 ± 13**	28 ± 18**
S_{es} (10³ dyn/cm²)	201 ± 61	133 ± 44**	129 ± 43**
dp/dt$_{max}$ (mmHg/s)	940 ±251	862 ±207	1063 ±323#
EF (%)	34 ± 14	40 ± 17*	44 ± 18*

* =p<0.05, ** =p<0.01 vs control; # =p<0.05, ## = <0.01 vs nitroprusside. CI=cardiac index; dp/dt$_{max}$=maximum rate of left ventricular pressure rise; EF=ejection fraction; HR=heart rate; LVEDP=left ventricular end-diastolic pressure; LVEDV=left ventricular end-diastolic volume; MAP=mean aortic pressure; MPP=mean pulmonary artery pressure; S_{ed}=left ventricular enddiastolic wall stress; S_{es}=left ventricular end-systolic wall stress; SVI=stroke volume index; SVR=systemic vascular resistance

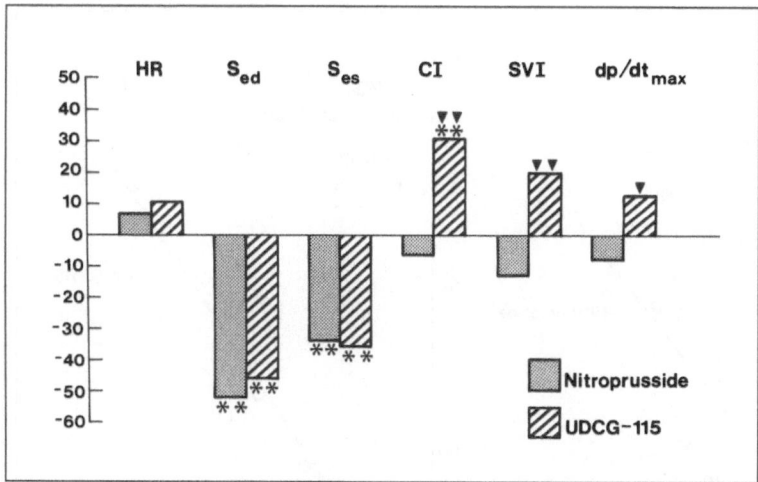

Fig. 1. Percent changes of hemodynamic variables following the application of nitroprusside and UDCG-115 in nine patients with idiopathic dilated cardiomyopathy. CI = cardiac index; dp/dt_{max} = maximum rate of left ventricular pressure rise; HR = heart rate; S_{ed} = left ventricular end-diastolic wall stress (preload); S_{es} = left ventricular end-systolic wall stress (afterload); SVI = stroke volume incex; * = $p < 0.05$, ** = $p < 0.01$ vs control; ▼ = $p < 0.05$ ▼ ▼ = $p < 0.01$ vs nitroprusside

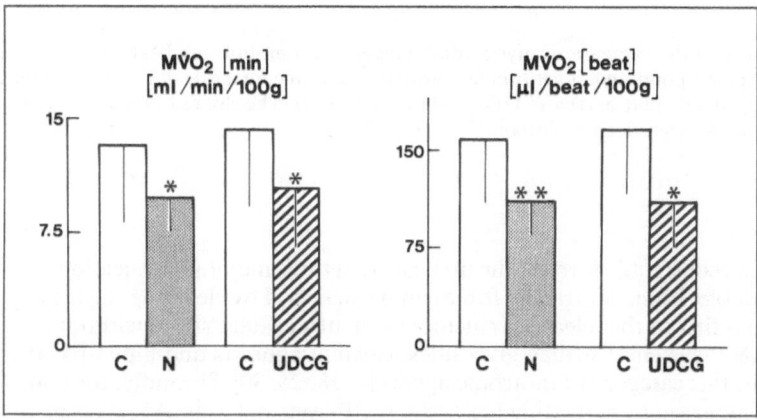

Fig. 2. Influence of nitroprusside (N; 10 patients) and UDCG-115; nine patients) on myocardial oxygen consumption per min ($M\dot{V}O_2$[min]) and myocardial oxygen consumption per beat ($M\dot{V}O_2$[beat]). C = control; * = $p < 0.05$, ** = $p < 0.01$ vs control

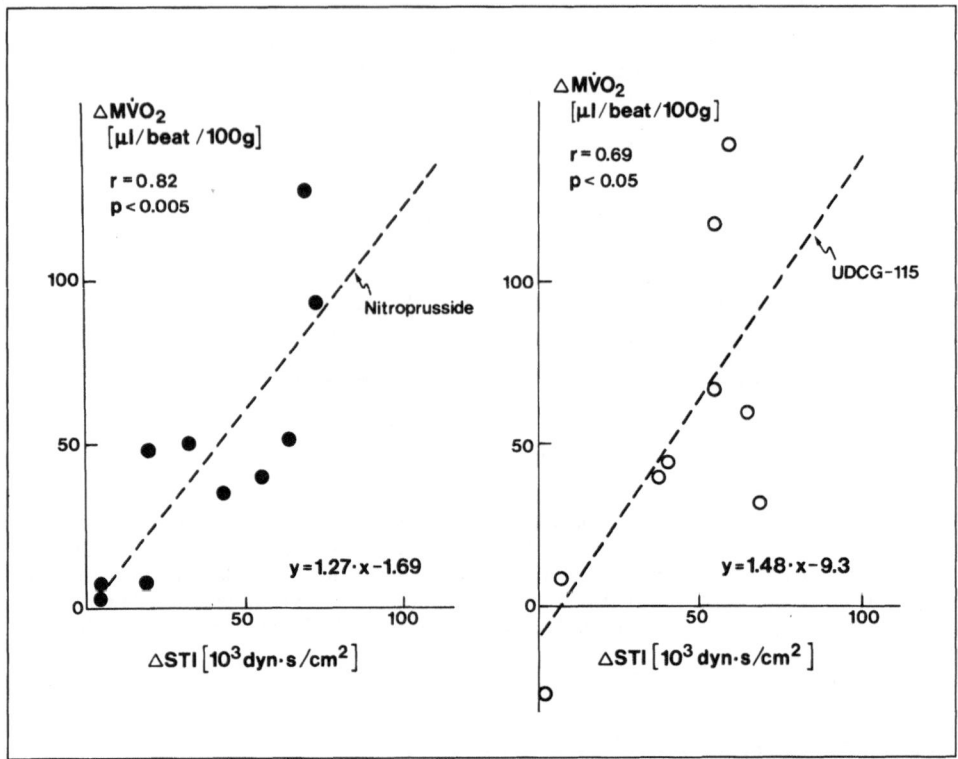

Fig. 3. Relationship between the decrease of myocardial oxygen consumption per beat (Δ M$\dot{V}O_2$) and the decrease of the corresponding left ventricular systolic stress-time integral (Δ STI) following the application of nitroprusside (left part) and UDCG-115 (right part). The slopes of the respective linear regression lines are not significantly different

Discussion

In order to increase the contractile force of the myocardium three major pharmacological mechanisms are available. First, contractile force can be increased by elevating the trans-sarcolemmal calcium influx or the release of calcium from intracellular stores without activating the cyclic AMP-system. Cardiac glycosides, calcium-agonists and alpha-receptor-agonists belong to this category of inotropic agents [2, 18, 25, 30]. Secondly, the contractile force can be increased by activating the cyclic AMP-system. Cyclic AMP via protein phosphorylation increases calcium influx through calcium channels, accelerates calcium reuptake by the sarcoplasmic reticulum and decreases the sensitivity of troponin C to calcium ions [2, 10, 25, 30]. In addition, cyclic AMP may directly influence the actomyosin interaction, increasing the cycling rate of the crossbridges [5, 7]. Beta-receptor-agonists and phosphodiesterase inhibitors are representatives of this category of inotropic agents [2, 10, 13, 29, 30]. Thirdly, contractile force can be increased by increasing the affinity of the contractile proteins to calcium ions. Substances belonging to this group of inotropic agents are listed in Table 1. All calcium-sensitizers available to date exhibit additional pharmacological properties [6, 12, 14, 19, 26, 28]. Because an increase in intracellular cyclic AMP or calcium concentration may be harmful with respect to increased myocardial energy demand or arrhythmogenic side effects calcium-sensitizers may be advantageous compared to other inotropic agents [11, 20].

In the present investigations, the influence of the calcium-sensitizer UDCG-115 was evaluated in patients with moderate congestive heart failure due to idiopathic dilative cardiomyopathy. In these patients, UDCG-115 resulted in a pronounced decrease of left ventricular preload (end-diastolic wall stress) and afterload (end-systolic wall stress) indicating considerable vasodilating properties on both the venous and the arterial bed. Since there was a similar reduction in left ventricular preload and afterload following the application of the vasodilator nitroprusside, the significant increase in cardiac index, stroke volume index and maximum rate of left ventricular pressure rise obtained with UDCG-115 in comparison with nitroprusside clearly demonstrates the inotropic effect of UDCG-115. Moreover, cardiac index, stroke volume index and maximum rate of left ventricular pressure rise showed the tendency to decrease following the application of nitroprusside. This may indicate that the pronounced reduction of preload reduced left ventricular performance according to the Frank-Starling mechanism in these patients. Analyzing the hemodynamic actions of UDCG-115 without comparison with a vasodilator, therefore, may underestimate the inotropic effect of pimobendan. Since UDCG-115 has inotropic and vasodilating effects, both may influence myocardial oxygen demand. As was shown recently, vasodilation reduces myocardial oxygen demand proportional to the reduction of left ventricular systolic stress-time integral in patients with dilative cardiomyopathy [5]. If the inotropic effect of UDCG-115 would result in an additional energy demand a given reduction of stress-time integral should reduce myocardial oxygen consumption to a lesser degree compared to the pure vasodilator nitroprusside.

Following the application of UDCG-115 as well as of nitroprusside in these patients there was a significant linear correlation between the decrease of myocardial oxygen consumption and systolic stress-time integral. The slope of the respective linear regression lines were not significantly different. In other words, the vasodilating effect of UDCG-115 reduced myocardial energy demand due to reduced left ventricular load, and the inotropic effect of UDCG-115 does not result in significant extra energy expenditure of the myocardium. When the same type of analysis was applied to evaluate the energetic consequences of the phosphodiesterase-inhibitor enoximone, for a given decrease of stress-time integral, the phosphodiesterase inhibitor reduced myocardial oxygen consumption significantly less than nitroprusside [5]. The different effects of enoximone and nitroprusside on myocardial energy consumption were interpreted in terms of additional energy cost due to increased cyclic AMP subsequent to phosphodiesterase inhibition [5]. The favorable effects of UDCG-115 on myocardial energetics, therefore, may indicate that other pharmacological mechanisms besides the cyclic-AMP system contribute to the inotropic effect of UDCG-115 in patients with congestive heart failure.

Conclusions

UDCG-115, when given as a intravenous infusion, exhibits significant inotropic and vasodilating effects in patients with idiopathic dilated cardiomyopathy and moderate congestive heart failure. UDCG-115 reduces myocardial oxygen consumption in proportion to the decrease of left ventricular systolic stress-time integral. The inotropic effect of UDCG-115 does not result in significant extra energy expenditure of the myocardium. UDCG-115, therefore, may be a promising alternative to catecholamines or current phosphodiesterase inhibitors in the therapy of acute myocardial failure, in particular if the energy supply of the heart is critical. Long-term studies should be performed to evaluate whether the beneficial hemodynamic effect of UDCG-115 is sustained during long-term administration and whether UDCG-115 is advantageous compared to other inotropic agents or vasodilators. It should be a challenge to pharmacologists to develop new inotropic agents which exclusively influence the calcium affinity of the contractile proteins.

References

1. Berger C, Meyer W, Scholz H, Starbatty (1985) Effects of the benzimidazole derivatives pimobendan and 2-(4-Hydroxy-phenyl)-5-(5-methyl-3-oxo-4,5-dihydro-2H-6-pyridazinyl) benzimidazole-HCL on phosphodiesterase activity and force of contraction in guinea pig hearts. Arzneim-Forsch/Drug Res 35(II):1668–1673
2. Colucci WS, Wright RF, Braunwald E (1986) new positive inotropic agents in the treatment of congestive heart failure. Mechanisms of action and recent developments. N Engl J Med 341:290–296
3. Dunckker DJ, Hartog JM, Levinsky L, Verdouw PD (1987) Systemic haemodynamic actions of pimobendan (UD-CG 115 BS) and its o-demethyl metabolite (UD-CG 212 CL) in the conscious pig. Br J Pharmacol 91:609–615
4. Fritsche R, Scheld HH, Mulch J, Trach V, Gerstenberg W, van Meel JAC, Hehrlein W (1986) Effects of pimobendan on calcium sensitivity of skinned fibers isolated from human papillary muscles. Br J Pharmacol 89:751
5. Hasenfuss G, Holubarsch Ch, Heiss HW, Meinertz Th, Bonzel T, Wais U, Lehmann M, Just H (1989) Myocardial energetics in patients with dilative cardiomyopathy. Influence of nitroprusside and enoximone. Circulation (in press)
6. Herzig JW, Feile K, Rüegg JC (1981) Activating effects of ARL-115 BS on the Ca^{++} sensitive force, stiffness and unloaded shortening velocity (Vmax) in isolated contractile structures from mammalian heart muscle. Arzneim-Forsch/Drug Res 31:188–191
7. Hoh JFY, Rossmanith GH, Kwan LJ, Hamilton AM (1988) Adrenaline increases the rate of cycling of cross bridges in rat cardiac muscle as measured by pseudo-random binary noise-modulated perturbation analysis. Circ Res 62:452–461
8. Honerjäger P, Heiss A, Schäfer-Korting M, Schonsteiner G, Reiter M (1984) UDCG-115 – a cardiotonic pyridazinone which elevates cyclic AMP and prolongs the action potential in guinea pig papillary muscles. Naunyn-Schmiedeberg's Arch Pharmacol 325:259–269
9. Jaquet K, Heilmeyer LMG (1987) Influence of association and of positive inotropic drugs on calcium binding to cardiac troponin C. Biochem Biophys Res Comm 145:1390–1396
10. Katz AM (1983) Cyclic adenosine monophosphate effects on the myocardium: a man who blows hot and cold with one breath. J Am Coll Cardiol 2:143–149
11. Katz AM (1986) Potential deleterious effects of inotropic agents in the therapy of chronic heart failure. Circulation 73 (suppl III):184–190
12. Kitada Y, Narimatsu A, Matsumura N, Endo M (1987) Increase in Ca^{++} sensitivity of the contractile system by MCI-154, a novel cardiotonic agent, in chemically skinned fibers from the guinea pig papillary muscles. J Pharmacol Exp Ther 243:633–638
13. Leyen von der H, Brunkhorst D, Meyer W, Scholz H (1987) Positive inotropic and chronotropic effects and phosphodiesterase III inhibition of new cardiotonic agents. Circulation 76 (suppl IV):155
14. Lues I, Siegel R, Harting J (1988) Effects of isomazole on the responsiveness to calcium of the contractile elements in skinned cardiac muscle fibers of various species. Europ J Pharmacol 146:145–153
15. Meel van JCA (1985) Cardiovascular affects of the positive inotropic agents pimobendan and sulmazole in vivo. Arzneim-Forsch/Drug Res 35:284–288
16. Meel van JCA, Gersternberg W, Boss H, Mrwa U (1986) Effects of some cardiotonics on calcium sensitivity of skinned myocardial fibres. Br J Pharmacol 87:102

17. Mirsky I (1979) Elastic properties of the myocardium: a quantitative approach with physiological and clinical applications. In: Berne RM (ed) Handbook of physiology. The cardiovascular system. Washington DC: American Physiological Society, pp 497–531
18. Mügge A (1985) Alpha-adrenozeptoren am Myokard: Vorkommen und funktionelle Bedeutung. Klin Wochenschr 63:1087–1097
19. Müller-Beckmann B, Freund P, Honerjäger P, Kling L, Rüegg JC (1988) In vitro investigations on a new positive inotropic and vasodilating agent (BM 14.478) that increases myocardial cyclic AMP content and myofibrillar calcium sensitivity. J Cardiovasc Pharmacol 11:8–16
20. Packer M, Medina N, Yushak M (1984) Hemodynamic and clinical limitations of long term inotropic therapy with amrinone in patients with severe chronic heart failure. Circulation 70:1038–1047
21. Rackley CE, Dodge HT, Coble YD, Hay RE (1964) A method for determining left ventricular mass in man. Circulation 29:666–671
22. Rau G (1968) Messung der Koronardurchblutung mit der Argon-Fremdgasmethode. Arch Kreisl-Forsch 58:322–398
23. Rüegg JC, Pfitzer G, Eubler D, Zeugner C (1984) Effect on contractility of skinned fibers from mammalian heart and smooth muscle by a new benzimidazole derivative, 4,5-dihydro-6-(2-(4-methoxyphenyl)-1H-benzimidazole-5-yl)-5-methyl-3(2H)-pyridazinone. Arzneim-Forsch/Drug Res 34:1736–1738
24. Rüegg JC (1986) Effects of new inotropic agents on Ca^{++} sensitivity of contractile proteins. Circulation 73(suppl III):78–84
25. Rüegg JC (1986) The vertebrate heart: modulation of calcium control. In: Rüegg JC (ed) Calcium in muscle activation. Springer, Berlin Heidelberg New York London Paris Tokyo, pp 165–200
26. Salzmann R, Bormann G, Herzig JW, Markstein R, Scholtysik G (1985) Pharmacological actions of APP 201-533, a novel cardiotonic agent. J Cardiovasc Pharmacol 7:588–596
27. Sandler H, Dodge HT (1968) The use of single plane angiocardiograms for the calculation of left ventricular volume in man. Am Heart J 75:325–334
28. Scholtysik G, Salzmann R, Berthold R, Herzig JW, Quast JW, Markstein R (1985) DPI 201.106, a novel cardiotonic agent. Combination of cAMP-independent positive inotropic, negative chronotropic, action potential prolonging and coronary dilatory properties. Naunyn-Schmiedeberg's Arch Pharmacol 329:316–321
29. Scholz H, Meyer W (1986) Phosphodiesterase-inhibiting properties of newer inotropic agents. Circulation 73(suppl III):99–108
30. Scholz H (1984) Inotropic drugs and their mechanisms of action. J Am Col Cardiol 4:389–397
31. Fujino K, Sperelakis N, Solaro RJ (1988) Sensitization of dog and guinea pig heart myofilaments to Ca^{2+} activation and the inotropic effect of pimobendan: comparison with milrinone. Circ Res 63:911–922
32. Wallenstein S, Zucker CL, Fleiss JL (1980) Some statistical methods useful in circulation research. Circ Res 47:1–9

Author's address:

Gerd Hasenfuss, M.D., Medizinische Universitätsklinik, Innere Medizin III, Kardiologie, Universität Freiburg, Hugstetter Strasse 55, D-7800 Freiburg, FRG

IV. Energetic aspects in the intact heart

Myocardial energetics: experimental and clinical studies to address its determinants and aerobic limit

K. T. Weber, J. S. Janicki, P. Sundram

Cardiovascular Institute, Michael Reese, Hospital,
University of Chicago Pritzker School of Medicine, Chicago, Illinois, USA

Summary

The major determinants of myocardial oxygen consumption ($M\dot{V}O_2$) were examined in the isolated, servo-regulated, canine heart in which coronary perfusion pressure, heart rate, ventricular volume and pressure could be individually monitored and controlled. We found that the integral of systolic wall force and the time derivative of systolic force development were major determinants of $M\dot{V}O_2$. Net $M\dot{V}O_2$, seen in response to increments in contractility (dobutamine) and heart rate, were the result of the relative increments in each of these determinants. Additional studies were performed to assess the heart's metabolic reserve and aerobic limit (i.e., before the onset of lactate production). We found that with increments in left ventricular work, mediated by increments in filling volume, heart rate, and contractility (dobutamine), myocardial lactate production could be induced, but was dependent on the level of coronary perfusion pressure. When the aerobic limit of the myocardium was exceeded, its performance declined and pulsus alternans appeared.

In patients with cardiomegaly and advanced heart failure given the phosphodiesterase inhibitors enoximone and piroximone we did not observe a rise in $M\dot{V}O_2$ or the appearance of lactate production in the majority of patients. When patients with documented idiopathic (dilated) cardiomyopathy and marked heart failure received hemodynamically significant doses of dobutamine alone or in combination with amrinone, there again was no evidence of lactate production or a rise in $M\dot{V}O_2$, while a marked improvement in ventricular function was noted. Thus, these agents served to improve the efficiency of the dilated failing heart. Hence we would conclude that in most cases, the dilated failing heart has an adequate metabolic reserve. Its performance, and indeed its efficiency, can be improved with pharmacologic agents having positive inotropic properties without adversely altering myocardial energetics.

Introduction

Myocardial energetics refers to the energy utilized by the myocardium in sustaining cell function and viability, the contraction and relaxation of the myocardium, as well as the heat it produces. In the past, the methodologies available to monitor energy utilization were confined solely to the measurement of myocardial oxygen consumption ($M\dot{V}O_2$), derived from the product of coronary blood flow and the arteriovenous oxygen difference across the heart. Today's technological advances permit the monitoring of high energy phosphate consumption [1] and heat production [2]. In this report studies will be reviewed that we conducted over the pase decade, both in the experimental laboratory and at the bedside, to assess $M\dot{V}O_2$ [3–8]. We will specifically consider our studies in which we wished to a) identify the components of myocardial contraction that represent the major determinants of $M\dot{V}O_2$, b) determine if the heart has an aerobic limit beyond which additional increments in myocardial work would lead to myocardial lactate production, and

c) assess the response of the dilated failing myocardium to pharmacologic agents having positive inotropic and vasodilator properties.

Experimental studies: determinants of myocardial oxygen consumption

Studies of $M\dot{V}O_2$ in the isolated heart

The heart's consumption of oxygen serves to reflect its energy requirements. Calculated from the product of steady state coronary blood flow and arteriovenous oxygen difference across the myocardium, oxygen consumption ($M\dot{V}O_2$) was measured in our isolated heart preparation. We wished to determine which components of ventricular contraction represented the major determinants of $M\dot{V}O_2$. The isolated canine heart, in which a compliant balloon had been placed in the left ventricle and attached to a servo control system, permitted us to precisely monitor and control ventricular volume and pressure in both isovolumetric and ejecting contractions. As shown in Fig. 1, atrial pacing and a coronary perfusion reservoir provided control of heart rate and coronary perfusion pressure, respectively. Contractile state could be varied by the infusion of the synthetic catecholamine dobutamine or the beta adrenergic receptor antagonist propranolol into the perfusion conduit.

The integral of systolic wall force as a determinant of $M\dot{V}O_2$

To address the components of contraction that were most responsible for $M\dot{V}O_2$, we examined a series of isovolumetric and ejecting contractions (Fig. 2). Our intent was to first assess the relationship between generated force and $M\dot{V}O_2$. We considered the left ventricle to be a thick-walled sphere with internal radius (R). The volume of the sphere equals

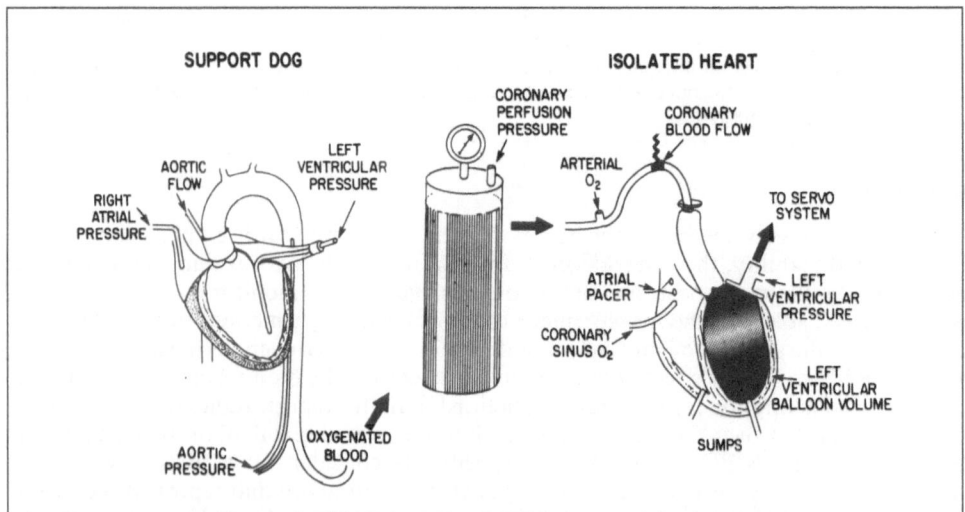

Fig. 1. Isolated canine heart preparation maintained in cross circulation with another dog

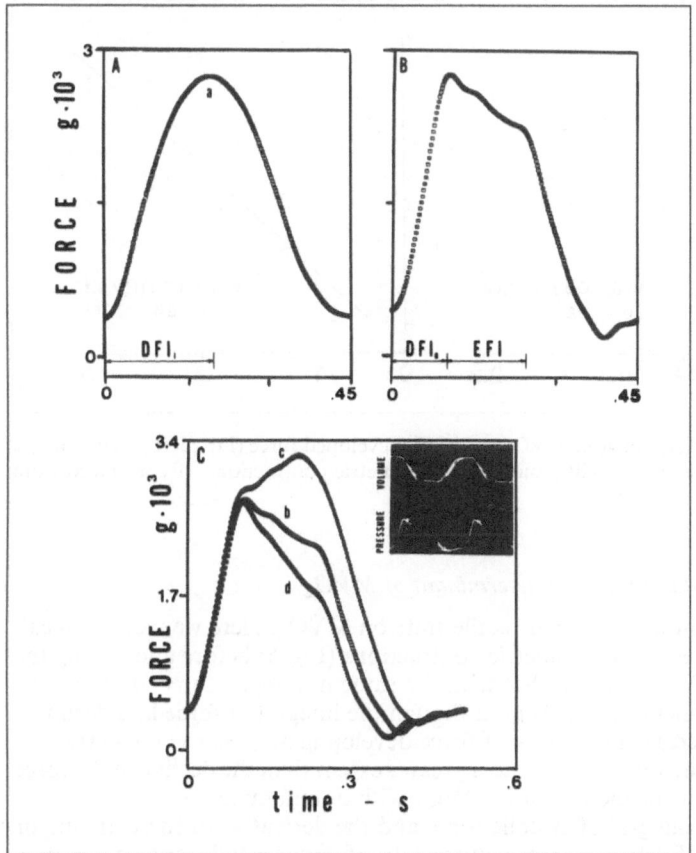

Fig. 2. A series of isovolumetric and ejecting contractions that were studied. Note that onset ejection force for the ejecting beat was similar to peak force in the isovolumetric beat. (Reproduced from [3] with permission.) DFI_i, DFI_e, and EFI = developed force integral in isovolumetric (i) and ejecting (e) beats and the ejection force integral

$4/3$ πR^3. Wall force was calculated from the product of chamber pressure and cross-sectional area (πR^2). We found that in the isovolumetrically beating ventricle there was a direct linear relationship between the peak force that was developed and $M\dot{V}O_2$ (Fig. 3). This was also true for the time integral of developed force.

Ejecting beats having the same developed force (Fig. 2) were than compared to the isovolumetric contractions. Here, however, we varied the integral of force during the ejection period to create different integrals of systolic force. We found that the ejecting beats consumed more oxygen than the isovolumetric contractions (Fig. 4). A linear correlation existed between the integral of systolic force (from onset contraction to end ejection) and $M\dot{V}O_2$ (Fig. 4). In contractions where force continued to rise during ejection there was a greater increment in $M\dot{V}O_2$ than that seen where ejection force declined. Thus the force that has to be developed and sustained during a contraction is directly related to the amount of energy the myocardium will consume.

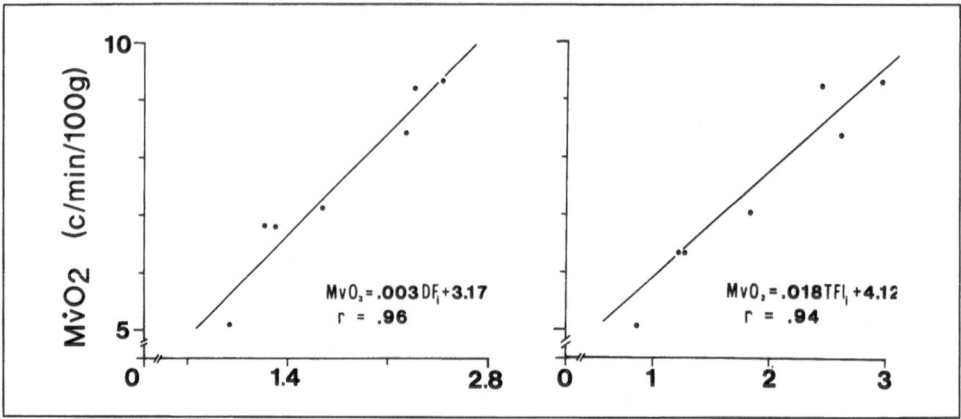

Fig. 3. A linear correlation exists between $M\dot{V}O_2$ and peak developed force (DF_i; left panel) and the time integral of systolic force (TFI_i; right panel) in isovolumetric contractions. (Reproduced from [3] with permission)

The derivative of systolic wall force as a determinant of $M\dot{V}O_2$

We next determined the influence of contractile state on $M\dot{V}O_2$. Here we matched peak developed force for a series of isovolumetric contractions (Fig. 5) before and during the infusion of dobutamine. We observed that with the same developed force, dobutamine led to an additional increment in $M\dot{V}O_2$ even though the integral of force had declined. This was related to an increase in the rate of force development. Because the derivative of force with respect to time was increased to a greater extent than the decline in the force integral, this accounted for the increment in $M\dot{V}O_2$ with dobutamine.

Thus, both the timed integral of systolic force and the derivative of force are major determinants of $M\dot{V}O_2$. Each of these components of myocardial contraction may change in opposite directions. The dominant effect will determine $M\dot{V}O_2$. One could therefore envisage that if the integral of force is reduced to a greater extent than the increment in the derivative of force is increased, $M\dot{V}O_2$ would decline. Alternatively, and as discussed above and shown in Fig. 5, when the increment in the derivative of force was greater than the reduction in the force integral, the net effect would be an increment in $M\dot{V}O_2$.

Heart rate as a determinant of $M\dot{V}O_2$

Variations in atrial pacing rate were invoked to assess the role of heart rate on steady state $M\dot{V}O_2$ in the isolated canine heart. Here we again examined isovolumetric contractions having the same peak developed force. This required us to reduce left ventricular end diastolic volume to offset the force-treppe phenomenon. As shown in Fig. 6, $M\dot{V}O_2$ expressed in ml/min/100 g, was increased as atrial pacing rate was increased from 120 to 150 and then 180 bpm. In terms of the energy consumed per beat, however, there was a decline in $M\dot{V}O_2$ with these increments in heart rate. This could be explained by the fall in the integral systolic force that was greater than the increment in the derivative of force. Thus, once again, the trade-off between the force integral and the derivative of force will determine net $M\dot{V}O_2$.

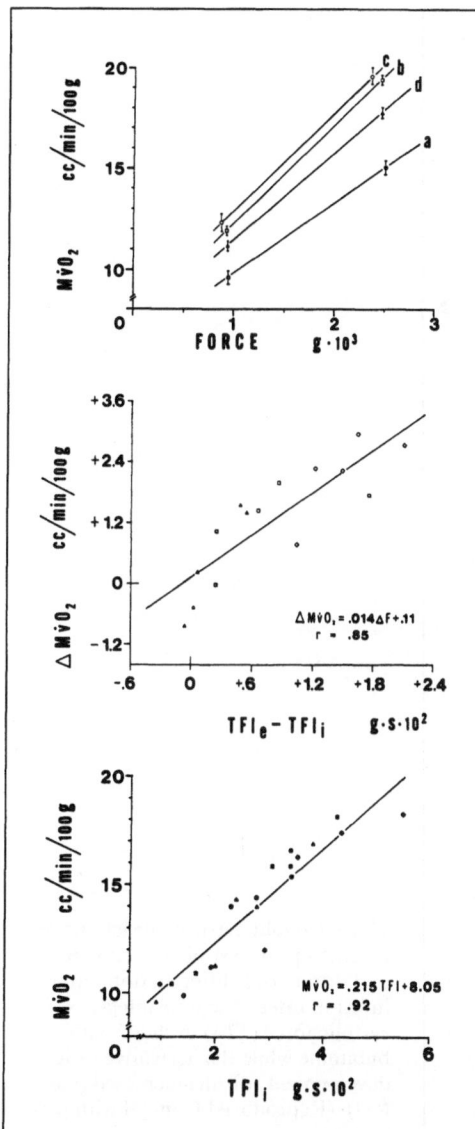

Fig. 4. Upper panel: ejection beats (see Fig. 2) consume a greater quantity of oxygen than isovolumetric contractions, in which onset ejection and peak isovolumetric force were matched. Middle panel: the difference in $M\dot{V}O_2$ ($\varDelta\, M\dot{V}O_2$) between ejecting and isovolumetric contractions was related to the greater integral of systolic force in the ejecting beats. Lower panel: $M\dot{V}O_2$ is most closely related to the integral of systolic force (TFI_i) in either ejecting or isovolumetric contractions. (Reproduced from [3] with permission)

Experimental studies: the aerobic limit of the myocardium

Using the isolated canine heart preparation we undertook a series of studies to determine if increments in the major determinants of $M\dot{V}O_2$ could lead to an imbalance in oxygen demand and supply with the heart becoming anaerobic with lactate production. For this purpose we utilized increments in left ventricular filling volume, graded infusion of dobutamine, and atrial pacing to raise $M\dot{V}O_2$ while monitoring transmyocardial lactate extraction. We further examined this aerobic limit of the myocardium when coronary per-

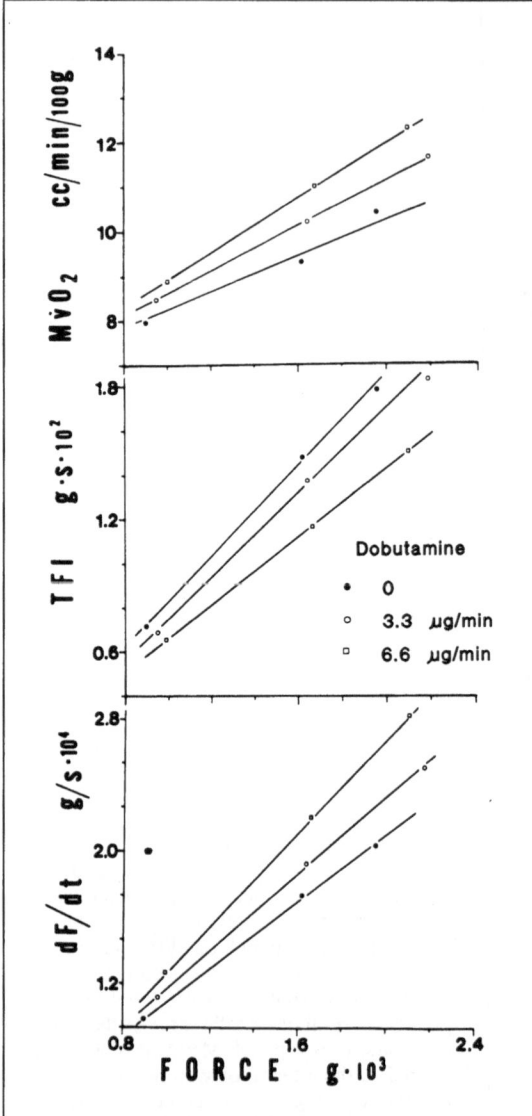

Fig. 5. Isovolumetric contractions of matched peak systolic force, before and during two different dobutamine infusion rates. The time integral of systolic force (TFI) declined with dobutamine while the derivative of force development (dF/dt) increased (see text). (Reproduced from [4] with permission)

fusion pressure was held constant at either 80 or 40 mm Hg to represent normal and ischemic conditons.

We found that normal lactate extraction averaged 15% or more for normal levels of coronary perfusion. When filing pressure was raised to 18 or more mm Hg, heart rate was increased to 180 or more bpm, and the infusion rate of dobutamine was 14 mcg/min, myocardial oxygen extraction exceeded 80% and negative lactate extraction, compatible with lactate production, was observed. With the appearance of anaerobic metabolism developed force declined and pulses alternans appeared. For ischemic conditions anaerobic

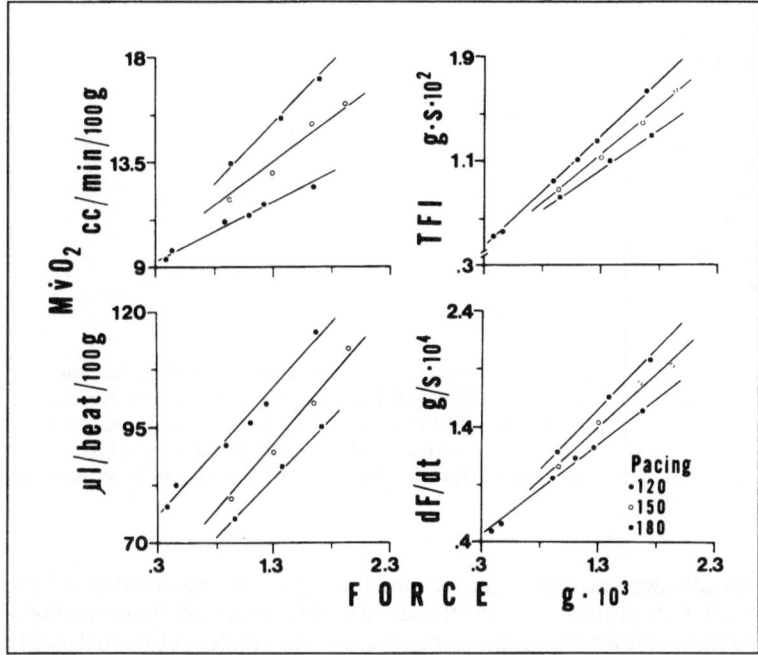

Fig. 6. Isovolumetric contractions of matched peak systolic force for heart rates of 120, 150, and 180 bpm. Note that MV̇O$_2$ per minute rises with heart rate while MV̇O$_2$ per beat declines (see text). (Reproduced from [4] with permission)

metabolism was again induced. Here, however, smaller increments were required. Filling pressure averaged 19 mm Hg, heart rate 150 bpm, and dobutamine 5 mcg/min for the onset of lactate production. At his point, oxygen extraction exceeded 80%, peak developed isovolumetric force declined, and pulsus alternans appeared.

These studies indicated that the heart normally has a remarkable metabolic reserve. It is only when marked increments in myocardial work are invoked that lactate production and active anaerobic metabolism occur. It should also be noted that when the heart's aerobic limit is exceeded, there is a decline in performance (developed force) and the appearance of pulsus alternans. Hence we would infer that in man with a dilated failing heart, any intervention that raises myocardial oxygen demand beyond oxygen availability would be accompanied by myocardial anaerobiosis, worsening heart failure, and pulsus alternans.

Clinical studies: response in MV̇O$_2$ of failing heart to inotropic agents

Bing et al. [9] were among the first to derive MV̇O$_2$ in man. For this purpose an inert gas was used to measure coronary blood flow and the arteriovenous difference of this gas across the heart was determined from coronary sinus catheterization. Today, various inert gases continue to be used for clinical determinations of coronary flow [10]. A thermodilution technique is also available [11]. We used the thermodilution principle to assess

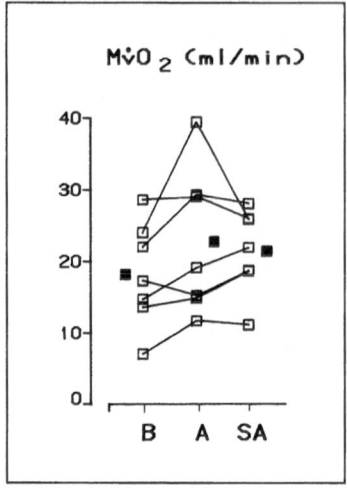

Fig. 7. The response in $M\dot{V}O_2$ to piroximone in patients with heart failure of diverse etiology. Shown are baseline $M\dot{V}O_2$ (B) and the responses to the initial (A) and fourth (SA) dose of piroximone (see text). Mean values are given as the solid squares. (Reproduced from [7] with permission)

$M\dot{V}O_2$ in patients with advanced cardiac failure secondary to previous myocardial infarction(s) or idiopathic (dilated) primary myocardial disease, who received pharmacologic agents having positive inotropic and vasodilator properties. The results of two such studies have been previously published [6, 7].

In the first study [6], we examined arterial-coronary sinus difference in oxygen and lactate, thermodilution coronary sinus blood flow, and calculated $M\dot{V}O_2$, and the extraction (arteriovenous difference/arterial concentration) of oxygen and lactate in response to graded doses of the type-III phosphodiesterase inhibitor enoximone. We have previously shown that this agent has both inotropic and vasodilator properties in our isolated heart preparation [12]. We found that with but two exceptions, $M\dot{V}O_2$ was not altered by enoximone in our study population. In two patients there was a rise in $M\dot{V}O_2$ that proved to be statistically significant, but which was modest and associated with a rise in left ventricular function. Myocardial lactate production was not observed. In the second study [7] involving similar patients, we administered another phosphodiesterase inhibitor piroximone. The response in $M\dot{V}O_2$ (Fig. 7) to piroximone was insignificant and there was no evidence of myocardial lactate production.

A similar experience was reported by Amin et al. [13] and Viquerat et al. [14], who examined the response in $M\dot{V}O_2$ to intravenous enoximone. Bennotti [15] and Monrad [16] and their co-workers found that other phosphodiesterase inhibitors, amrinone and milrinone, did not raise $M\dot{V}O_2$ or cause myocardial lactate production in their patients with advanced heart failure. Collectively, these findings underscore the fact that in the dilated, failing heart an inotropic agent that increases the function and emptying of the enlarged ventricle will reduce systolic wall stress and thereby offset any increment in $M\dot{V}O_2$ that might be mediated by an elevation in contractility. Moreover, and in keeping with our studies of the heart's aerobic limit, we did not find a decline in cardiac performance or the appearance of pulsus alternans which might be expected if the metabolic reserve of the myocardium had been exhausted. In fact, we noted [17] that pulsus alternans disappears with these agents. Thus, we would have to conclude that the available clinical studies in which $M\dot{V}O_2$ was measured directly does not support the claim that positive inotropic agents will create a detrimental imbalance in oxygen supply and demand. The physiologic basis for these findings must relate to a decline in the integral of systolic force as ventricular chamber size is reduced. This is in keeping with our use of digitalis and diuretics in patients with ischemic heart disease and cardiomegaly, who develop angina. Covell et al. [18] were able to demonstrate that with an enlarged left ventricle, an inotropic agent, like digitalis, could be given to improve pump function, reduce ventricular size, and thereby lower $M\dot{V}O_2$.

Clinical studies: does the failing myopathic heart have an aerobic limit?

In order to more critically evaluate the possibility that the failing heart is closer to its aerobic limit or that inotropic agents might adversely raise metabolic demand to shorten the threshold for the appearance of myocardial anaerobiosis, the following study was undertaken by one of us (P.S.) in our Cardiac Care Unit [8]. Patients with documented idiopathic (dilated) cardiopathy and no evidence of coronary or valvular heart disease and who were consecutively admitted to our unit with advanced heart failure were enrolled into the study after written consent was obtained. The coronary sinus was cannulated with a thermodilution catheter; and a flotation catheter was advanced into the pulmonary artery. An infusion of intravenous dobutamine was begun at 5 mcg/kg/min and raised by 5 mcg/min/kg increments to a maximum dose of 15 mcg/kg/min or until a plateau in the cardiac output response was observed. Thereafter, and with the maximal infusion rate of dobutamine continued (15 mcg/kg/min), we began intravenous amrinone. Here a 1–1.5 mg/kg loading dose was given over 5 min followed by a sustaining infusion of 15 mcg/kg/min. Hemodynamics, $M\dot{V}O_2$ and myocardial lactate extraction were measured at each dobutamine dose and for the combination of dobutamine with amrinone. We observed a dose-dependent improvement in ventricular function with dobutamine. Cardiac index rose from a baseline of 1.47 ± 0.44 l/min/m^2 to 2.89 ± 1.10 l/min/m^2 with the 15 mcg/kg/min dose of dobutamine. Baseline wedge pressure fell from 28 ± 7 mm Hg to 26 ± 8 mm Hg with the peak dose of dobutamine, but this was not statistically significant. Heart rate and mean arterial pressure were not significantly altered with dobutamine. With the addition of amrinone, a significant rise in cardiac index was seen to 3.64 ± 1.05 l/min/m^2 while wedge pressure fell significantly to 20 ± 6 mm Hg. Heart rate rose significantly with the combination of dobutamine and amrinone to 116 ± 18 bpm (from 102 ± 22 bpm baseline). Baseline $M\dot{V}O_2$ (18 ml/min) was invariant after dobutamine and the combination of inotropic agents. Myocardial lactate extraction rose from $21 \pm 10\%$ before drug intervention to $35 \pm 10\%$ with the combined drugs. No patient had evidence of lactate production. Thus, these findings suggest that in dilated cardiomyopathy, inotropic agents acting by different mechanisms may have additive, beneficial effects. There appears to be an adequate metabolic reserve in these severely failing hearts and that myocardial efficiency (work/$M\dot{V}O_2$) can be raised substantially without creating an imbalance in myocardial oxygen supply.

References

1. Bittl JA, Balschi JA, Ingwall JS (1987) Contractile failure and high-energy phosphate turnover during hypoxia: ^{31}P-NMR surface coil studies in living rat. Circ Res 60:871–878
2. Alpert NR, Mulieri LA (1982) Increased myothermal economy od isometric force generation in compensated cardiac hypertrophy induced by pulmonary artery constriction in rabbit: A characterization of heat liberation in normal and hypertrophied right ventricular papillary muscle. Circ Res 509:519–500

3. Weber KT, Janicki JS, Reeves RC, Hefner LL (1977) Myocardial oxygen consumption: the role of wall force and shortening. Am J Physiol 233:H421–H430
4. Weber KT, Janicki JS (1978) Interdependence of cardiac function, coronary flow and oxygen extraction. Am J Physiol 236:H784–H793
5. Weber KT, Janicki JS, Fishman AP (1980) Aerobic limit of the heart perfused at constant pressure. Am J Physiol 238:H118–H125
6. Martin JL, Likoff MJ, Janicki JS, Laskey WK, Hirshfield JW Jr, Weber KT (1984) Myocardial energetics and clinical response to the cardiotonic agent MDL 17,043 in advanced heart failure. J Am Coll Cardiol 4:875–883
7. Weber KT, Janicki JS, Jain MC (1987) Piroximone (MDL 19,205) in the treatment of unstable and stable chronic cardiac failure. Am Heart J 114:805–813
8. Sundram P, Reddy K, McElroy PA, Janicki JS, Weber KT (1988) Myocardial energetics and efficiency in patients with idiopathic cardiomyopathy response to dobutamine and amrinone (submitted)
9. Bing RJ, Hammond MM, Handelsman JC, Powers SR, Spencer FC, Eckenhoff JE, Goodale WT, Hafkenschiel JH, Kety SS (1949) The measurement of coronary blood flow, oxygen consumption and efficiency of the left ventricle in man. Am Heart J 38:1–24
10. Strauer BE, Beer K, Heitlinger K, Hofling B (1977) Left ventricular wall stress as a primary determinant of myocardial oxygen consumption: comparative studies in patients with normal left ventricular function with pressure and volume overload and with coronary heart disease. Basic Res Cardiol 306–313
11. Baim DS, Rothman MT, Harrison DC (1982) Simultaneous measurement of coronary venous blood flow and oxygen saturation during transient alterations in myocardial oxygen supply and demand. Am J Cardiol 49:743–752
12. Janicki JS, Shroff SG, Weber KT (1987) Physiologic response to the inotropic and vasodilator properties of enoximone. Am J Cardiol 60:15C–20C
13. Amin DK, Shah PK, Hulse S, Shellock FG, Swan HJC (1984) Myocardial metabolic and hemodynamic effects of intravenous MDL 17,043, a new cardiotonic drug, in patients with chronic severe heart failure. Am Heart J 108:1285–1292
14. Viquerat CE, Kereiakes D, Morris L, Daly PA, Wexman M, Frank P, Parmley WW, Chatterjee K (1985) Alterations in left ventricular function, coronary hemodynamics and myocardial catecholamine balance with MDL 17,043, a new inotropic vasodilator agent, in patients with severe heart failure. JACC 5:326–332
15. Bennotti J, Grossman W, Braunwald E, Carabello BA (1980) Effects of amrinone on hemodynamics and myocardial metabolism in patients with congestive heart failure from ischemic heart disease. Circulation 62:28–34
16. Monrad ES, Baim DS, Smith HS, Lanone AS, Braunwald E et al. (1985) Effects of milrinone on coronary hemodynamics and myocardial energetics in patients with congestive heart failure. Circulation 71:972–979
17. Weber KT, Janicki JS (1987) Chronic cardiac failure. In: Cardiopulmonary exercise testing; physiologic principles and clinical applications. Saunders, Philadelphia, pp 168–196
18. Covell JW, Braunwald E, Ross John Jr, Sonnenblick EH (1966) Studies on digitalis. XVI. Effects on myocardial oxygen consumption. J Clin Invest 10:1535–1542

Author's address:

Karl T. Weber, M.D., Michael Reese Hospital and Medical Center, Lake Shore Drive at 31st Street, Chicago, IL 60616, USA

Use of a conductance catheter to detect increased left ventricular inotropic state by end-systolic pressure-volume analysis

G. F. Leatherman, T. L. Shook, S. M. Leatherman, W. S. Colucci

Cardiovascular Division, Departments of Medicine, Brigham and Women's Hospital and Harvard Medical School, Boston, Massachusetts, USA

Summary

The slope of the left ventricular end-systolic pressure-volume relationship is thought to be a load-independent index of contractile state. However, clinical application requires a practical technique to simultaneously measure pressure and volume in man. We used a left ventricular conductance catheter to derive the left ventricular end-systolic pressure-volume relationship over a range of arterial pressures during nitroprusside infusion and washout in 14 patients, and to characterize the effect of dobutamine, a positive inotropic drug. End-systole was defined as the maximum pressure-to-volume ratio. Dobutamine (5 µg/kg/min) increased the slope of the end-systolic pressure-volume relationship in 11 of 14 patients, the mean slope increasing by 43% from 1.4 ± 0.1 to 2.1 ± 0.2 mm Hg/% end-diastolic volume ($p < 0.01$). In 12 of 14 patients there was a leftward and upward shift of the end-systolic pressure-volume relationship with dobutamine. To quantitate this shift, we derived left ventricular pressure at control end-systolic volume before and after dobutamine. Dobutamine increased the mean end-systolic pressure at control end-systolic volume in 12 of 14 patients, the average increased by 37% from 134 ± 10 to 189 ± 23 mm Hg ($p < 0.01$). We conclude that the conductance catheter can be used in man to detect a drug-induced change in left ventricular contractile state. This technique may be useful in the evaluation of drugs with positive inotropic actions, and in assessing the response of individual patients to positive inotropic agents.

Introduction

Although there is substantial evidence that the end-systolic pressure-volume relationship may provide an index of left ventricular contractile state that is relatively unaffected by preload and afterload changes over the physiologic range [22, 23, 25,8, 17, 13, 2, 20], the clinical application of this measure in man has been hampered by lack of a practical technique for the serial measurement of left ventricular volume. Contrast angiographic techniques were first used to demonstrate the potential clinical use of the end-systolic pressure-volume relationship [8, 17], but are limited to two or three assessments of only a few cardiac cycles each; they may be inaccurate when regional wall motion abnormalities are present, and the contrast agent itself may affect contractile state. Non-invasive techniques have also been used to estimate end-systolic volume [13, 2]. Despite relative ease of acquisition, echocardiographic assessment of left ventricular volume requires significant as-

Supported in part by NIH Training Grant HL07049. W.S.C. is an Established Investigator of the American Heart Association.

sumptions about ventricular geometry. Radionuclide ventriculography has been applied recently to end-systolic pressure-volume analysis in man [14, 9]. However, this method provides a time-averaged measure of ventricular volume, and therefore precludes beat-by-beat analysis.

The left ventricular conductance catheter [15, 1, 3, 10, 16] provides serial assessment of simultaneous pressure and volume for individual cardiac cycles, and therefore allows the continuous monitoring of the end-systolic pressure-volume relationship during interventions. The feasibility and validity of using the conductance catheter to determine the end-systolic pressure-volume relationship has been demonstrated [15, 1, 3, 10, 16]. However, the ability of this approach to detect a clinically relevant change in inotropic state in patients has not been evaluated. Therefore, the purpose of this study was to determine whether the left ventricular conductance catheter can be used to assess the end-systolic pressure-volume relationship in patients, and if so, whether this approach can be used to detect the positive inotropic effect of dobutamine [24].

Methods

Patient characteristics

The study population consisted of 14 patients (12 male, two female; mean age, 57 ± 3 years), all of whom had significant coronary artery disease by angiography. By left ventriculography and/or two-dimensional (2-D) echocardiography, segmental wall motion abnormalities (hypokinesia, akinesia) were present in 10 patients, and left ventricular enlargement was present in five patients. The mean left ventricular ejection fraction was 0.58 ± 0.03. Baseline hemodynamics are shown in Table 1. In addition, mean systemic arterial pressure was $133/71 \pm 9/3$ mm Hg, and mean pulmonary capillary wedge pressure was 14 ± 2 mm Hg. No patient had clinical, echocardiographic and/or angiographic evidence of significant mitral regurgitation. All studies were performed immediately following diagnostic catheterization. Patients were in the post-absorbtive state and were premedicated orally with diazepam (10 mg) and diphenhydramine (50 mg) 2 h prior to catheterization. Ten of the patients received atropine sulfate (1 mg, intravenously) during diagnostic angiography. The study protocol was approved by the Committee for the Protection of Human Subjects of the Brigham and Women's Hospital (October 10, 1984), and informed written consent was obtained.

Table 1. Hemodynamic effects of nitroprusside and dobutamine

	CON-1	NTP	CON-2	DOB	NTP & DOB
HR (bpm)	66 ± 3	$73 \pm 3^*$	64 ± 3	68 ± 4	$76 \pm 4^*$
SAP (mmHg)	130 ± 7	$101 \pm 6^*$	132 ± 6	$150 \pm 9^*$	$109 \pm 7^*$
LVEDP (mmHg)	16 ± 1	11 ± 2	15 ± 1	17 ± 2	9 ± 1

CON-1 and CON-2, control-1 and control-2; DOB, dobutamine; HR, heart rate; LVEDP, left ventricular end-diastolic pressure; NTP, nitroprusside; SAP, systolic arterial pressure.
$* = p < 0.01$ vs. CON-1. All values are the mean $\pm$ SEM in 14 subjects.

Conductance catheter

A 9F conductance catheter (Cardiac Pacemakers Inc., St. Paul, Minnesota, USA) was advanced to the left ventricle through a sidearm femoral artery sheath, and a stable position (assessed by frequent fluoroscopy) which did not stimulate ventricular ectopic activity was obtained with the catheter tip in the left ventricular apex. The conductance catheter used in this study is similar to that described by McKay et al. [15] except that there is a pigtail configuration at the catheter tip. Left ventricular pressure was measured through the central lumen using a low-volume pressure transducer (p50 Micron, Statham Instruments) mounted directly on the catheter manifold.

Study protocol

Nitroprusside (25–100 µg/min, intravenously) was infused to lower systolic arterial pressure by 20–30 mm Hg. When the systolic arterial pressure had decreased by 20–30 mm Hg, the nitroprusside infusion was discontinued. Systolic arterial pressure returned to baseline over the ensuing 3–8 min. During this washout period, a continuous strip chart recording (Honeywell Electronics for Medicine Physiologic Recorder) was made of the electrocardiogram, left ventricular volume signal and left ventricular pressure, with intermittant sampling every 15–30 s at a paper speed of 100 cm/s. After return of systolic arterial pressure to baseline, dobutamine (5 µg/kg/min, intravenously) was administered for 5 min. Nitroprusside was again infused at the same rate as previously; and after discontinuation of the nitroprusside infusion, the electrocardiogram, ventricular volume signal, and left ventricular pressure were recorded during nitroprusside washout as described above, without interruption of the dobutamine infusion.

Data analysis

Analog data recordings of left ventricular pressure and volume were digitized on a Summa Graphics MR1812 digitizing tablet, and the data were compiled on an IBM-XT computer. Simultaneous pressure and volume values were obtained for individual beats by gating from the electrocardiographic tracing. Relative ventricular volume was expressed as a percentage of the conductance catheter signal at end-diastole before drug infusions. Pressure-volume plots were constructed from 80–100 pressure and volume coordinates for each cardiac cycle (Fig. 1), and end-systole was defined as the time at which the ratio of pressure-to-volume was maximum [20]. During the nitroprusside washouts, the end-systolic pressure-volume ratio was assessed during approximately 15 beats under each condition (control, 15 ± 1 beats; dobutamine, 16 ± 1 beats). Least squares linear regression of left ventricular pressure and volume was performed to derive an equation describing the end-systolic pressure-volume relationship. From this equation we obtained, a) the slope of the end-systolic pressure-volume relationship (expressed as mm Hg/% of end-diastolic volume), and b) the pressure at a given volume.

Statistical analysis

The fit of the end-systolic pressure-volume coordinates to a linear equation was evaluated by calculating correlation coefficients and significance values. The slopes of the end-sys-

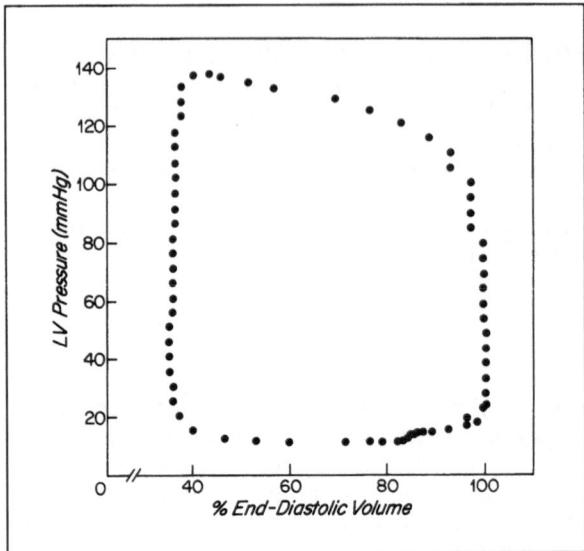

Fig. 1. A representative left ventricular pressure-volume loop (volume is expressed as percent of end-diastolic volume) generated from digitization of left ventricular pressure and volume signals

tolic pressure-volume relationship, and left ventricular pressure at control end-systolic volume before and during dobutamine were compared using a two-tailed paired *t*-test, and differences were considered significant at the $p < 0.05$ level. Linear regression and statistical analysis were performed on a VAX 8200 computer using the RS/1 and SAS statistical programs. All data are expressed as mean ± SEM.

Results

Baseline end-systolic pressure-volume relationship

Nitroprusside infusion (mean infusion rate, 51 ± 5 µg/min) decreased systolic arterial pressure by 29 ± 3 mm Hg and left ventricular end-diastolic pressure by 5 ± 1 mm Hg. These effects were associated with an $12 \pm 2\%$ increase in heart rate from 66 ± 3 to 73 ± 3 beats per min ($p < 0.01$) (Table 1). During nitroprusside washout, the end-systolic pressure-volume relationship was fit by a linear model in all cases, the correlation coefficients for the regression lines ranging from 0.80 to 0.99 (mean, 0.92 ± 0.01). Following nitroprusside washout, systolic arterial pressure, left ventricular end-diastolic pressure and heart rate returned to baseline values (Table 1).

Effect of dobutamine infusion on end-systolic pressure-volume slope

Infusion of dobutamine (5 µg/kg/min) caused no significant changes in heart rate or left ventricular end-diastolic pressure. Systolic arterial pressure was increased by 15% ($p < 0.01$). Nitroprusside infusion during continued dobutamine infusion caused changes

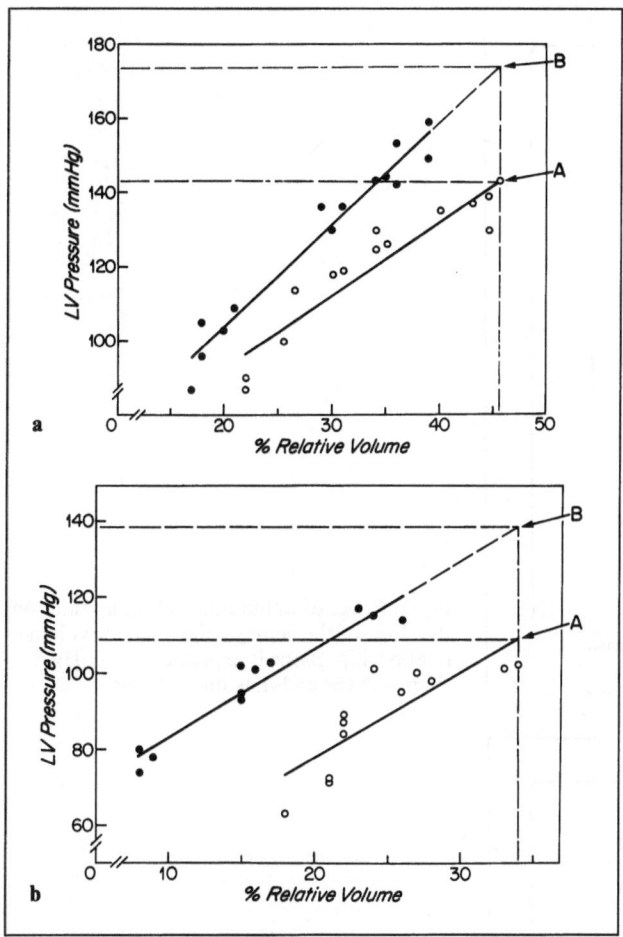

Fig. 2. Effect of dobutamine on the end-systolic pressure-volume relationship during nitroprusside washout in two representative patients (o, control; ●, dobutamine). Point A is the left ventricular pressure and relative volume at control end-systolic volume. Point B is left ventricular pressure and relative volume at control end-systolic volume during dobutamine infusion, extrapolated from linear regression of the end-systolic pressure-volume relationship. Panel A: Patient in whom there was an increase in slope. Panel B: Patient in whom there was an upward and leftward shift without an apparent change in slope

in heart rate, left ventricular end-diastolic pressure and systolic arterial pressure that were comparable to those caused by the first nitroprusside infusion (Table 1). The end-systolic pressure-volume relationships during dobutamine infusion and nitroprusside washout were fit by a linear model in all cases, with correlation coefficients ranging from 0.77 to 0.97 (mean, 0.92 ± 0.02). In 11 of 14 subjects the slope of the end-systolic pressure-volume regression line was increased (Fig. 2a). Dobutamine increased the slope of the end-systolic pressure-volume regression lines by an average of 43% from 1.4 ± 0.1 to 2.0 ± 0.2 mm Hg/% end-diastolic volume ($p < 0.01$) (Fig. 3).

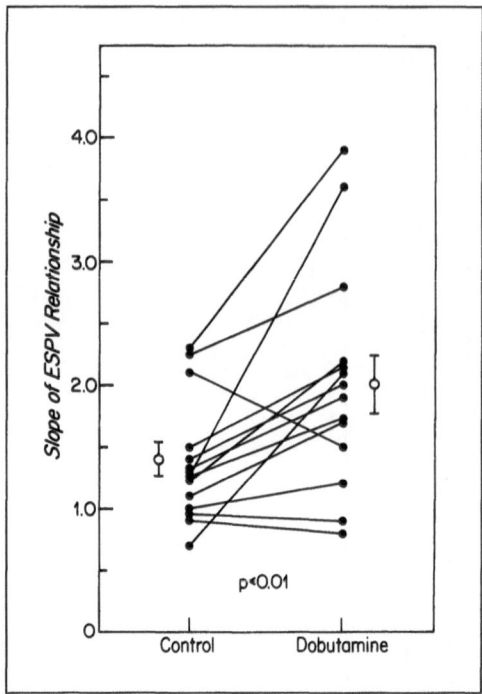

Fig. 3. Effect of dobutamine (5 µg/kg/min) on the slope of the end-systolic pressure-volume relationship. Slope is expressed as mmHg/% change in the end-diastolic volume.

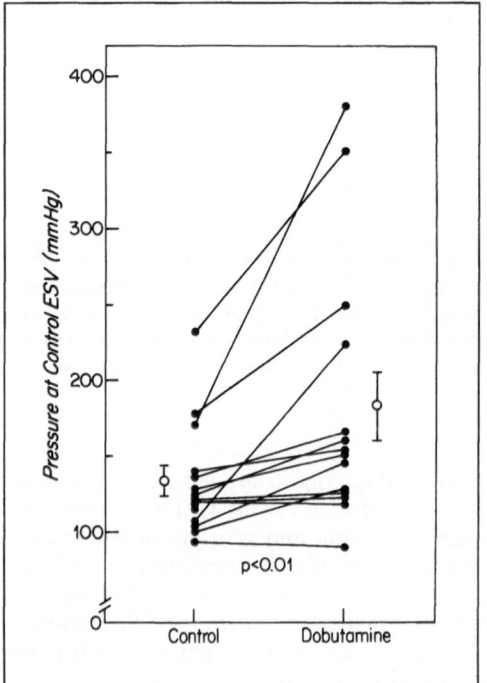

Fig. 4. The effect of dobutamine infusion on left ventricular pressure at control end-systolic volume, calculated as shown in Fig. 2

Effect of dobutamine on left ventricular pressure

In 12 of 14 subjects an upward and leftward displacement of the end-systolic pressure-volume relationship was also apparent during dobutamine infusion. Figure 2b demonstrates this shift of the regression line in a patient in whom no change in slope was detected. To quantify this shift of the end-systolic pressure-volume relationship, the equation of the pre-dobutamine regression line was used to calculate left ventricular pressure at end-systolic volume before nitroprusside infusion (Point A, Fig. 2a and b). The regression line during dobutamine infusion was then extrapolated to the same end-systolic volume to calculate the predicted left ventricular pressure at that volume (Point B, Fig. 2a and b). This predicted left ventricular pressure at control end-systolic volume was increased in 12 of 14 subjects during dobutamine infusion (Fig. 4). The mean left ventricular pressure at control end-systolic volume for the group increased by 37% from 134 ± 10 mm Hg at control to 183 ± 23 mm Hg with dobutamine ($p < 0.01$).

Discussion

This study demonstrates that the left ventricular conductance catheter, as recently described in detail by McKay et al. [14–16], Baan et al. [1, 3], and Kass et al. [10], can be used to derive the left ventricular end-systolic pressure-volume relationship in patients, and that this technique can be used to detect a drug-induced positive inotropic effect. A left ventricular conductance catheter similar to the one used in this study has been shown to measure accurately left ventricular volume in isolated hearts, animals and man [15, 1, 3], and McKay et al. [15] used the catheter in man to delineate left ventricular pressure-volume loops [15]. In animals, it has been shown that positive inotropic agents cause an increase in the slope of the end-systolic pressure-volume relationship as assessed with the conductance catheter [10, 16]. However, it is not known whether this approach can be used to detect the effect of a positive inotropic agent in man.

These data show that the conductance catheter can be used to derive the end-systolic pressure-volume relationship in patients, and that this technique is adequately sensitive to detect the positive inotropic effects of a modest dose of dobutamine (5 μg/kg/min). In 12 of 14 patients there was an upward/leftward shift in the end-systolic pressure-volume relationship, and in 11 of 14 patients there was an increase in the slope of this relationship. From these data we cannot draw conclusions regarding the sensitivity or specificity of this technique as compared to other measures of contractility such as left ventricular $+dP/dt$. However, the average magnitude of dobutamine-stimulated change in both of the end-systolic indices in this study (37%–43%) is larger than the magnitude of increase in left ventricular $+dP/dt$ (18%) caused by the same dose of dobutamine in patients with congestive heart failure [4], and is similar in magnitude to the increase (approximately 40%) observed in normal subjects in response to a comparable dose of dobutamine administered by the intracoronary route [5].

In two patients there was no apparent change in either end-systolic pressure-volume slope or pressure at control end-systolic volume with dobutamine. Since we had not independent measure of inotropic effect and did not attempt to use higher doses of dobutamine, we do not know whether this reflects lack of an inotropic effect in these patients as opposed to a technical limitation of the method. It has shown that the response to dobutamine can vary greatly, such that as many as four of 15 patients with congestive heart failure may fail to increase $+dP/dt$ in response to a dose of 5 μg/kg/min [4]. Future modification of the conductance catheter to allow simultaneous micromanometer measurement of pressure may allow a comparison to $+dP/dt$ be made over a range of inotropic drug doses, and under a variety of conditions.

An increase in left ventricular pressure at control end-systolic volume provided a second convenient means of quantifying the effect of dobutamine. It has been observed previously that an increase in contractile state can cause a parallel upward shift in the end-systolic relationship without an increase in slope. Noble postulated that increased pressure at a given volume would indicate increased contractility even if there were no change in the slope of the end-systolic pressure-volume relationship [18]. In a preliminary examination of end-systolic pressure-volume analysis in intact dogs, Sugawa et al. demonstrated that isoproterenol caused a parallel shift of the end-systolic pressure-volume relationship without an increase in slope in two of four animals [20]. Likewise, Mahler et al. found that digoxin or isoproterenol caused a parallel upward shift of the end-systolic pressure-dimension relationship in conscious dogs [12]; and Sasayama and Kotoura found that isoproterenol caused an upward parallel shift in the end-systolic pressure-diameter relationship in man [21]. A potential limitation of the use of left ventricular pressure at control end-systolic volume is that this value is based on extrapolation of the measured data. However, the extrapolated volume falls within the range measured in each patient, and it has been shown that the pressure-volume relationship is linear when pressure is increased over this range by pressor agent such as methoxamine or angiotensin-II. From these limited data, we hesitate to draw mechanistic conclusions regarding the relative effects of a positive inotropic intervention on end-systolic pressure-volume slope vs pressure at a given volume. Nevertheless, it appears that under the conditions of our study, an increase in pressure at a given volume is at least as sensitive as an increase in slope for detecting the positive inotropic effect of dobutamine.

In this study we have taken a practical, clinically feasible approach to the assessment of a positive inotropic drug effect. A number of potential theoretical limitations must be recognized. First, we did not utilize autonomic blockade or cardiac pacing, methods that might minimize the possible effects of reflex activation by nitroprusside. The use of sympathetic blockade was precluded since we wished to assess the effects of a beta-adrenergic agonist. Although the small (12%) increase in heart rate at the peak of nitroprusside effect indicates that there was a small degree of sympathetic activation which potentially could have altered the contractile state, it is unlikely that this effect significantly affected the interpretation of our data. First, heart rate increased to a similar degree during both nitroprusside infusions (i.e., control and dobutamine). Second, although large (i.e., 100%) increases in heart rate per se affect the end-systolic pressure-volume relationship, it is unlikely that such small changes exert a significant effect [22].

A second potential limitation of this method is the possibility that nitroprusside per se may cause a parallel rightshift in the end-systolic pressure-volume relationship [7]. However, our data are based on relative changes in volume, and nitroprusside was infused under *both* control and dobutamine conditions. Thus, it is unlikely that an effect of nitroprusside significantly confounded detection of an inotropic response to dobutamine. Nevertheless, the use of alternative methods to vary volume, such as inferior vena cava balloon occlusion, may be of value in future studies [16, 6, 11].

The present study expressed left ventricular volume as a relative, rather than absolute, change in volume [16, 6]. This simplification avoids technical rigors involved in the determination of absolute left ventricular volume. Changes in the conductance catheter signal are linearly related to changes in stroke volume as measured by thermodilution and agree with time-activity curves of the left ventricle by radionuclide ventriculography [15]. Although methods are available to calibrate the conductance catheter volume signal [1, 11], the purpose of this study was to detect *changes* in contractile state, rather than to assess the absolute level of contractility, and therefore we opted for the simplification afforded by the use of relative volume changes [16, 6].

These data show the potential utility of the conductance catheter in man as a method for detecting changes in the left ventricular end-systolic pressure-volume relationship indicative of a positive inotropic drug effect. This method was sensitive enough to detect the effect of a modest dose of dobutamine, thereby indicating that inotropic effects well within the clinically relevant range can be monitored. This approach may be of value in evaluating the direct myocardial actions of new positive or negative inotropic drugs in man.

Acknowledgements. We thank Diane F. Gauthier R. N. for technical assistance, Elizabeth Hawk and Michelle Somers for manuscript preparation, and Donald Kraiser Ph. D. for statistical consultation.

References

1. Baan J, van der Velde ET, de Bruin HG, Smeenk GJ, Koops J, van Dijk AD, Temmerman D, Senden J, Buis B (1984) Continuous measurement of left ventricular volume in animals and humans by conductance catheter. Circulation 870:812–823
2. Borow KM, Neumann A, Wynne J (1982) Sensitivity of end-systolic pressure-dimension and pressure-volume relations to the inotropic state in humans. Circulation 65:988–997
3. Burkhoff D, van der Velde E, Kass D, Baan J, Maughan WL, Sagawa K (1985) Accuracy of volume measurement by conductance catheter in isolated, ejecting canine hearts. Circulation 72:440–447
4. Colucci WS, Wright RF, Jaski BE, Fifer MA, Braunwald E (1986) Milrinone and dobutamine in severe heart failure: differing hemodynamic effects and individual patient responsiveness. Circulation 73(Suppl III):III-175–III-183
5. Colucci WS, Deniss AR, Leatherman GF, Quigg RJ, Ludmer PL, Gauthier D (1988) Intracoronary infusion of dobutamine to patients with and without severe congestive heart failure: dose-response relationships, correlation with circulating catecholamines, and effect of phosphodiesterase inhibition. J Clin Invest (in press)
6. Diver DJ, Safien RD, Leeman DE, Smith HS, Morgan JP, Aroesty JM, Lorell BH, McKay RG (1986) Assessment of ventricular mechanics with inferior vena cava occlusion. Circ 74(Suppl II):II-167 (abstr)
7. Freeman GL, Little WC, O'Rourke RA (1986) The effect of vasoactive agents on the left ventricular end-systolic pressure-volume relation in closed-chest dogs. Circ 74:1107–1113
8. Grossman W, Braunwald E, Mann T, McLaurin LP, Green LH (1977) Contractile state of the left ventricle in man as evaluated from end-systolic pressure-volume relations. Circulation 56:845–852
9. Herrmann HC, Ruddy TD, Dex GW, Strauss HW, Boucher CA, Fifer MA (1987) Inotropic effect of enoximone in patients with severe heart failure: Demonstration by left ventricular end-systolic pressure-volume analysis. J Am Coll Cardiol 9:1117–1123
10. Kass DA, Yamazaki T, Burkhoff D, Maughan WL, Sagawa K (1986) Determination of left ventricular end-systolic pressure-volume relationships by the conductance (volume) catheter technique. Circulation 73:586–595
11. Kass D, Midei M, Graves W, Maughan WL (1987) Measurement of the LV end-systolic pressure-volume relationship in patients by conductance catheter and IVC balloon occlusion. J Am Coll Cardiol 9:201A (abstr)
12. Mahler F, Covell JW, Ross J (1975) Systolic pressure-diameter relations in the normal conscious dog. Cardiovasc Res 9:447–455
13. Marsh JD, Green LH, Wynne J, Cohn PF, Grossman W (1979) Left ventricular pressure-dimension and stress-length relations in normal human subjects. Am J Cardiol 44:1311–1317

14. McKay RG, Aroesty JM, Heller GV, Royal H, Parker A, Silverman KJ, Kolodny GM, Gross-
    man W (1984) Left ventricular pressure-volume diagrams and end-systolic pressure-volume re-
    lations in human beings. J Am Coll Cardiol 3:310–312
15. McKay RG, Spears R, Aroesty JM, Baim DS, Royal HD, Heller GV, Lincoln W, Salo RW,
    Braunwald E, Grossman W (1984) Instantaneous measurement of left and right ventricular
    stroke volume and pressure-volume relationships with an impedance catheter. Circulation
    69:703–710
16. McKay RG, Miller MJ, Ferguson JJ, Momura S-I, Sahagian P, Grossman W, Pasternak RC
    (1986) Assessment of left ventricular end-systolic pressure-volume relations with an impedance
    catheter and transient inferior vena cava occlusion: use of this system in the evaluation of the
    cardiotonic effects of dobutamine, milrinone, posicor and epinephrine. J Am Coll Cardiol
    8:1152–1160
17. Mehmel HC, Stockins B, Ruffman K, Olshausen KV, Schuler G, Kubler W (1981) The linearity
    of the end-systolic pressure-volume relationship in man and its sensitivity for assessment of left
    ventricular function. Circulation 63:1216–1222
18. Noble MIM (1979) Left ventricular load, arterial impedance and their interrelationship. Car-
    diovasc Res 13:183–198
19. Sagawa K, Suga H, Shoukas AA, Bakalar KM (1977) End-systolic pressure-volume ratio: a new
    index of ventricular contractility. Am J Cardiol 40:748–753
20. Sagawa K (1981) The end-systolic pressure-volume relation of the ventricle: definition, modifi-
    cations and clinical use. Circulation 63:1223–1227
21. Sasayama S, Kotoura H (1979) Echocardiographic appraoch for the clinical assessment of left
    ventricular function: the analysis of end-systolic pressure (wall stress)-diameter relation and
    force-velocity relation of ejecting ventricle. Jpn Circ J 43:357–366
22. Suga H, Sagawa K, Shoukas A (1973) Load independence of the instantaneous pressure-volume
    ratio of the canine left ventricle and effects of epinephrine and heart rate on the ratio. Circ Res
    32:314–322
23. Suga H, Sagawa K (1974) Instantaneous pressure-volume relationships and their ratio in the ex-
    cised, supported canine left ventricle. Circ Res 35:117–125
24. Tuttle RR, Mills J (1975) Dobutamine-development of a new catecholamine to selectively in-
    crease cardiac contractility. Circ Res 36:185–196
25. Weber KT, Janicki JS, Hefner LL (1976) Left ventricular force-length relations of isovolumic
    and ejecting contractions. Am J Physiol 231:337–343

Author's address:

Wilson S. Colucci, M.D., Cardiovascular Division, Brigham and Women's Hospital, 75 Francis
Street, Boston, MA 02115, USA

Separation between vasodilation and positive inotropism by assessment of myocardial energetics in patients with dilated cardiomyopathy

Ch. Holubarsch, G. Hasenfuss, M. Allgeier, H. W. Heiss, Hj. Just

Medizinische Universitätsklinik, Innere Medizin III, Kardiologie,
Universität Freiburg, FRG

Summary

Phosphodiesterase inhibitors have vasodilating and positive inotropic properties, and these compounds may have energy saving effects due to vasodilation and energy consuming effects due to inotropism. In order to differentiate between the effects, it is necessary to relate myocardial oxygen consumption to its hemodynamic determinants. Myocardial oxygen consumption per beat was related to the following parameters: dp/dt_{max}, mean velocity of fiber shortening, pressure-volume work, peak developed wall stress, and stress-time integral. The best linear relationship was found between myocardial oxygen consumption per beat and the corresponding stress-time integral ($r = 0.71$; $p < 0.001$) in patients with idiopathic dilative cardiomyopathy. Using i.v. nitroprusside as a pure vasodilator, myocardial oxygen consumption per beat and stress-time integral decreased along this established relationship. In contrast, the phosphodiesterase inhibitor enoximone given intravenously decreased the stress-time integral significantly more than the myocardial oxygen consumption per beat.

We conclude from these data that phosphodiesterase inhibitors possess vasodilating properties which reduce the myocardial oxygen demand. In addition, they do have positive inotropic effects which increase the myocardial oxygen demand. Myocardial oxygen consumption always reflects the sum of both effects. The balance between the energy saving and the energy consuming effects may determine the efficacy of phosphodiesterase inhibitors, especially in the long-term treatment of chronic heart failure.

Introduction

The two prinicples of pharmacological therapy in chronic heart failure are vasodilation and positive inotropism. Whereas vasodilating substances decrease myocardial afterload and thereby myocardial energy consumption, stimulation of the myocardium by positive inotropic compounds may increase myocardial energy demand in an unknown manner.

In recent years, phosphodiesterase inhibitors have been developed which combine vasodilating and positive inotropic properties and are therefore thought to be useful for short- and long-term treatment of chronic heart failure [1–4].

However, some authors have raised doubts about the advantage of myocardial stimulation in patients with chronic heart failure [5, 6], because this may lead to an excess energy demand and thereby deterioration of myocardial cells. It is therefore essential to measure the oxygen consumption of the myocardium after application of these new drugs.

Three different scientific groups measured myocardial oxygen consumption in patients with severe heart failure (NYHA III and NYHA IV) before and after intravenous application of enoximone and found quite different results: Myocardial oxygen consumption increased significantly by 19% or 7% [7, 8] or decreased significantly by 18% [9]. This discrepancy is quite surprising, because the studies were performed in relatively homogenous groups of patients. However, it may not be sufficient to simply measure myocardial oxygen consumption. Phosphodiesterase inhibitors may exert vasodilating effects which lead to a reduction of the left ventricular load and thus myocardial oxygen consumption. In addition, these substances have positive inotropic effects which may alter the relation between myocardial contractile force and energy demand resulting in excess myocardial oxygen consumption.

Therefore, myocardial energy consumption must be analyzed in relation to its hemodynamic determinants in order to characterize the metabolic effects of pharmacological therapy on the myocardium in vivo. In the present study, we
1) established the relationship between myocardial oxygen consumption per beat and the most imported hemodynamic determinants in patients with idiopathic dilated cardiomyopathy. The best linear correlation was found between MVO_2 and the stress-time integral [10];
2) This relation was then tested under conditions of pure vasodilator nitroprusside;
3) Using this concept, the effect of the phosphodiesterase inhibitor enoximone on the relation between myocardial oxygen consumption and the stress-time integral was studied in patients with idiopathic dilated cardiomyopathy.

Methods

Patients

The data presented in this study was obtained from 29 patients suffering from idiopathic dilated cardiomyopathy NYHA II–III. The diagnosis of idiopathic dilated cardiomyopathy was accepted if left ventricular volume was increased (>220 ml) and left ventricular ejection fraction was reduced ($<55\%$); coronary heart disease, valvular heart disease, as well as arterial hypertension were excluded.

Study protocol

All patients were in sinus rhythm. Previous medications were withheld at least 48 h before catheterization. After coronary angiography, left ventriculography with simultaneous pressure measurement was performed using a Millar microtip catheter pressure transducer (PC-485 A). Care was taken to obtain a series of ventricular contractions without ventricular premature beats. The film speed was 50 frames per s. Myocardial blood flow was measured using the argon method [11]. Blood samples were taken from the aorta and the sinus coronarius for oxygen saturation measurement before and after measurement of myocardial blood flow. After control measurements were performed (basal state), in 10 patients nitroprusside was applied intravenously at an average dose of 71 ± 58 µg/min, in order to analyze the effect of a pure vasodilator on the relation between myocardial oxygen consumption and the mechanical parameters.

In 12 other patients with idiopathic dilated cardiomyopathy we gave enoximone intravenously at an average dose of 1.42 ± 0.52 mg/kg at 12.5 mg/min.

Left ventriculography with simultaneous pressure measurements was performed again and myocardial blood flow and oxygen consumption measurements were repeated.

Determination of pressure-volume work and stress-time integral

Left ventricular angiography was performed by power injection of 40 ml nonionic contrast solution. The projection was a 10° caudally angulated 45° right anterior oblique view. Heart rate was taken from the continuously recorded electrocardiogram. During one complete cardiac cycle, left ventricular volumes were calculated from each frame at intervals of 20 ms according to [12]. Left ventricular muscle mass was calculated according to a modification of the method of [13]. Instantaneous wall thickness was determined at intervals of 20 ms using our own computer program [14] which calculates instantaneous wall thickness from wall mass, instantaneous volume, and major and minor hemiaxes. Instantaneous circumferential wall stress was calculated according to [15] using an ellipsoidal model. By integrating the stress-time curve from enddiastole to endsystole (see Fig. 1), the stress-time integral was obtained. Peak systolic stress could easily by chosen

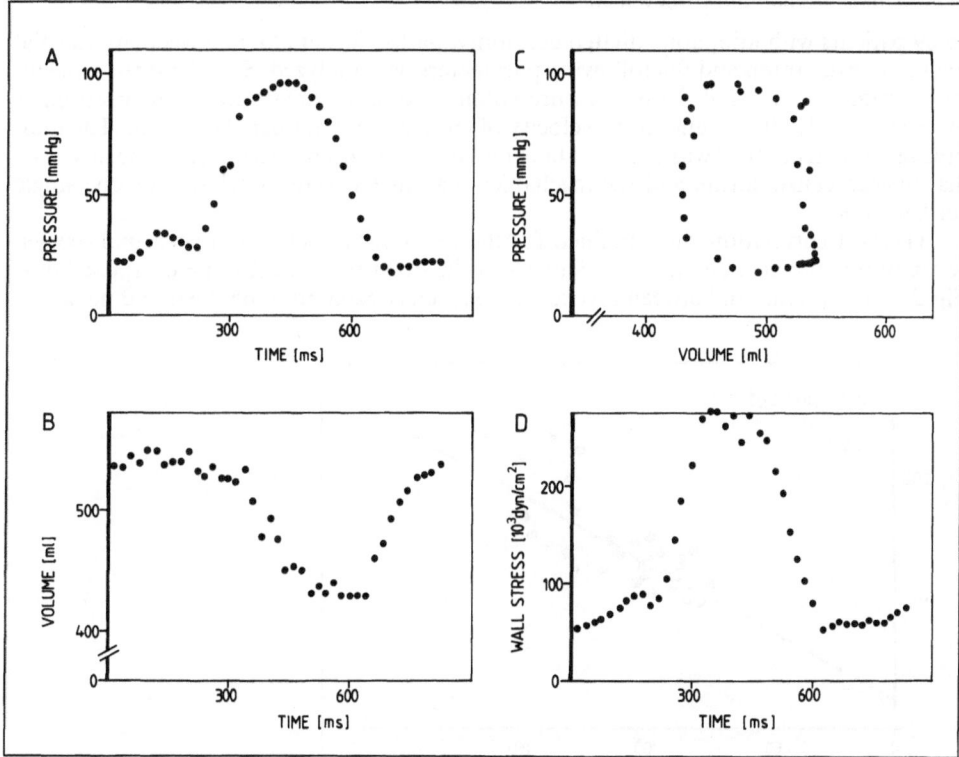

Fig. 1 A–D. Analysis of the mechanical parameters for one representative patient with dilated cardiomyopathy; **A** Pressure vs time plot; **B** left ventricular volume vs time plot; **C** pressure-volume plot. The pressure-volume loop represents left ventricular work; and **D** the left ventricular circumferential wall stress as a function of time, (for calculation see text)

from the stress-time curve (see Fig. 1). Pressure-volume work was calculated as the area of the pressure-volume loop obtained from corresponding pressure-volume values which were normalized for 100 g wall mass (Fig. 1). dp/dt_{max} (maximal rate of left ventricular pressure rise) was determined from continuous differentiation of left ventricular pressure tracings obtained by means of a Millar microtip catheter pressure transducer (PC-485 A).

Statistics

The relationship between myocardial oxygen consumption and mechanical parameters was investigated using multiple regression analysis (SPSS: PC + computer programme). The paired *t*-test was used to compare the values obtained before and after drug application. P values of less than 0.05 were accepted as significant.

Results

Relationship between myocardial oxygen consumption and mechanical parameters in patients with idiopathic dilated cardiomyopathy

In 29 patients with idiopathic dilated cardiomyopathy the coupling between myocardial oxygen consumption and the following parameters was analyzed: Systolic stress-time integral, peak systolic wall stress, pressure-volume work, maximal rate of left ventricular pressure rise (dP/dt_{max}), and mean velocity of circumferential fiber shortening. The multiple regression analysis was used. To take into account differences in heart rate, myocardial oxygen consumption and the mechanical parameter were calculated for one single cardiac cycle.

The best correlation was obtained for the relationship between myocardial oxygen consumption per beat and left ventricular systolic stress-time integral ($r = 0.71$; $p < 0.001$; Fig. 2). The correlation between myocardial oxygen consumption per beat and peak sys-

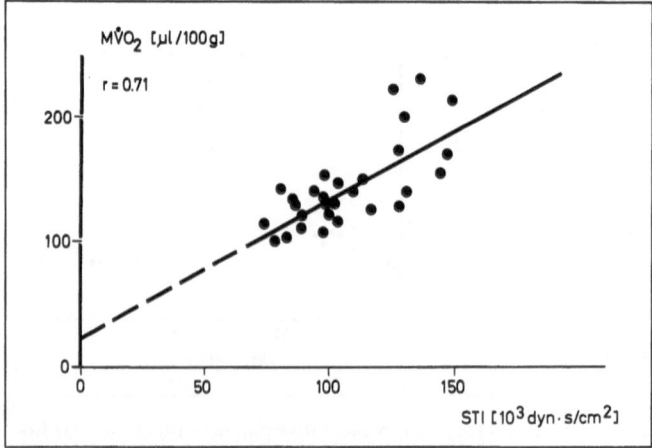

Fig. 2. Myocardial oxygen consumption as a function of the stress-time integral obtained from 29 patients with idiopathic dilated cardiomyopathy ($MVO_2 = 1.11 \int \sigma \cdot t + 23.6$)

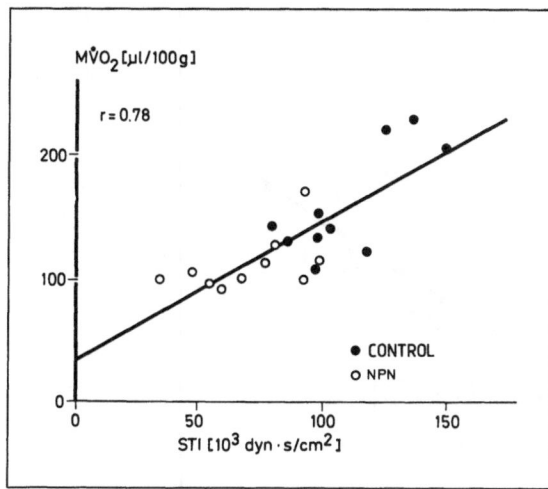

Fig. 3. Myocardial oxygen consumption as a function of the stress-time integral. In 10 patients with idiopathic dilated cardiomyopathy, the pure vasodilator nitroprusside was applied intravenously. The linear regression was obtained from n = 20 data points ($MVO_2 = 1.02 \int \sigma \cdot t + 37.9$)

tolic wall stress was less relevant although significant (r = 0.55; p < 0.005). No significant correlation was found for the relationship between myocardial oxygen consumption per beat and pressure-volume work, left ventricular pressure rise (dP/dt_{max}) and mean velocity of circumferential fiber shortening. Consideration of pressure-volume work, maximal rate of left ventricular pressure rise, or mean velocity of circumferential fiber shortening did not significantly improve the relation between myocardial oxygen consumption per beat and systolic stress-time integral.

Validation of the relationship between myocardial oxygen consumption and stress-time integral using nitroprusside

In order to test the relationship between myocardial oxygen consumption and the stress-time integral which was obtained from an interindividual analysis in 29 patients with idiopathic dilative cardiomyopathy, we applied nitroprusside intravenously in 10 patients with enlarged ventricles. The oxygen consumption and the stress-time integral before and after application was analyzed. This relationship is shown in Fig. 3. The linear regression analysis yielded a function which was not significantly different from that in Fig. 2. This analysis shows that the relation between myocardial oxygen consumption and the stress-time integral is also valid for intraindividual comparisons.

Influence of the phosphodiesterase inhibitor enoximone on the relationship between myocardial oxygen consumption and stress-time integral

With enoximone, systolic stress-time integral decreased by 60%, whereas myocardial oxygen consumption decreased by only 19%. As shown in Fig. 4, myocardial oxygen consumption should have decreased along the linear relationship, if enoximone would have only vasodilating effects. The deviation of the enoximone data points from the established myocardial oxygen consumption vs stress-time integral plot indicates that enoxi-

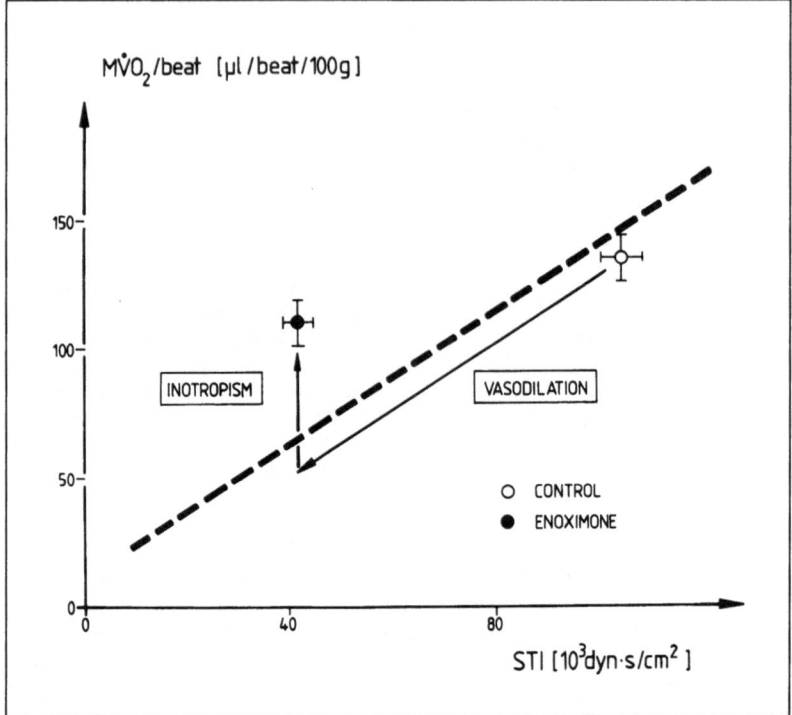

Fig. 4. Myocardial oxygen consumption as a function of the stress-time integral. The linear relationship indicates the function obtained from Fig. 2 (broken line: 29 patients with idiopathic dilated cardiomyopathy). If enoximone would have only vasodilating effects, a decrease of the MVO_2 would have been expected as indicated by the arrow. The deviation of the enoximone data point from the linear relationship indicates the additional energy demand due to direct myocardial stimulation

mone increases the myocardial oxygen consumption in addition to the expected decrease of myocardial oxygen consumption.

Discussion

Positive inotropic drugs exert inotropic and vasodilating effects. The vasodilating effects may be directly mediated by the substance itself or be a result of neurohumoral reflex mechanisms. Vasodilation reduces pre- and afterload, and therefore, is supposed to decrease myocardial oxygen consumption due to altered myocardial working conditions. The inotropic effect may increase myocardial oxygen by altered activation, relaxation and contraction processes. Measured myocardial oxygen consumption is the sum of both effects. If the vasodilating effect dominates, myocardial oxygen consumption will decrease [9]. If the positive inotropic effect is more pronounced, myocardial oxygen consumption will increase [7, 8]. In the present study the energetic consequences of vasodilation and positive inotropism are separated by relating myocardial oxygen consumption to its hemodynamic determinants.

What are the hemodynamic parameters of myocardial oxygen consumption? It was shown that heart rate, wall stress, tension-time index, and velocity of myocardial contraction [16–18] are closely related to myocardial oxygen consumption, whereas the influence of myocardial work on energy consumption is discussed controversely [19, 20].

In the present paper, heart rate was taken into account by calculating myocardial oxygen consumption and the hemodynamic parameters for one cardiac cycle. In 29 patients with idiopathic dilated cardiomyopathy, the best relationship was found between myocardial oxygen consumption and the systolic stress-time integral ($r = 0.71$), whereas the relationship between myocardial oxygen consumption and systolic peak stress was less significant ($r = 0.55$). No significant relation was found between myocardial oxygen consumption and pressure-volume work, maximal rate of pressure rise or mean velocity of circumferential fiber shortening. Because the highly significant correlation between myocardial oxygen consumption and the systolic stress-time integral could not be improved by additional consideration of velocity parameters or pressure-volume work using the multiple regression analyses, we consider the stress-time integral as the most powerful parameter for predicting myocardial oxygen consumption in patients with idiopathic dilated cardiomyopathy.

In order to test this relationship between myocardial oxygen consumption and stress-time integral by means of intraindividual observations, the pure vasodilator nitroprusside was applied in 10 patients with idiopathic dilated cardiomyopathy. A proportional decrease of myocardial oxygen consumption and stress-time integral was observed, as would have been expected from the established relationship.

Compared to nitroprusside, enoximone induced a more pronounced decrease of stress-integral and a less pronounced decrease of myocardial oxygen consumption. In other words, the deviation of the enoximone data points from the established myocardial oxygen consumption vs stress-time integral plot represents the additional energy demand due to direct myocardial stimulation by enoximone.

Whereas the energy-saving effect of enoximone is clearly due to the vasodilating properties of the substance, the energy consuming effect of enoximone has to be discussed. Enoximone increases the myoplasmic cyclic adenosinemonophosphate concentration [21] which may have two effects: Firstly, the number of calcium ions delivered into the cytoplasma available for the contractile proteins is increased. Therefore, the energy needed for the calcium transport during one cardiac cycle is enhanced. Secondly, the cyclic adenosine monophosphate may decrease the energy transduction efficiency of the contractile proteins. Hoh et al. [22] recently demonstrated that catecholamines increases the cycling rate of cross-bridges. Because one molecule of adenosine triphosphate is hydrolyzed during each cross-bridge cycle, augmented energy consumption of the contractile proteins may result. This was clearly shown in recent studies using the myothermal method for papillary muscles of rats and guinea pigs [23–25].

Therefore, it is postulated that cyclic adenosine monophosphate increases the number of calcium ions cycling during one twitch and speeds up the cycling frequency of the cross-bridges. Both effects will increase the energy demands of the myocardium.

The presented data imply that phosphodiesterase inhibitors have vasodilating effects which decrease myocardial oxygen consumption. Secondly, they have positive inotropic effects which increase myocardial oxygen consumption by an increased number of cycling calcium ions and an increased rate of cycling actomyosin crossbridges. The balance between the energy-consuming and the energy-saving effects of the phosphodiesterase inhibitors may determine the efficacy of phosphodiesterase inhibitors in long-term treatment in chronic heart failure.

References

1. Alousi AA, Canter JM, Montenero MJ, Fort DJ, Ferrari RA (1983) Cardiotonic activity of milrinone a new and potent cardiac bipyridine on the normal and failing heart of experimental animals. J Cardiovasc Pharmacol 5:792–803
2. Baim DS, McDowell AV, Cherniles J, Monrad ES, Parker JA, Edelson J, Braunwald E, Grossman W (1983) Evaluation of a new bipyridine inotropic agent – milrinone in patients with severe congestive heart failure. N Engl J Med 309:748–756
3. Dage RC, Roebel LE, Hsieh CP, Weiner DL, Woodward JK (1982) Cardiovascular properties of a new cardiotonic agent: MDL 17,043 (1,3 dihydro-4-methyl-5(4-(methylthio)-benzoyl)2H-imidaxol-2-one). J Cardiovasc Pharmacol 4:500–508
4. Uretzky BF, Generalovich T, Verbalis JG, Valdes Am, Reddy PS (1985) MDL 17,043 therapy in severe congestive heart failure: characterization of the early and late hemodynamic, pharmacokinetic, hormonal and clinical response. J Am Coll Cardiol 5:1414–1421
5. Katz AM (1986) Potential deleterious effects of inotropic agents in the therapy of chronic heart failure. Circulation 73, Suppl II:184–190
6. Packer M, Medina N, Yushak M (1984) Hemodynamic and clinical limitations of long-term inotropic therapy with amrinone in patients with severe chronic heart failure. Circulation 70:1038–1047
7. Viquerat CE, Kereiakes D, Morris DL, Daly PA, Wexman M, Frank P, Parmley WW, Chatterjee K (1985) Alterations of left ventricular function, coronary hemodynamics and myocardial catecholamine balance with MDL 17043, a new inotropic vasodilator agent, in patients with severe heart failure. J Am Coll Cardiol 5:326–332
8. Martin JL, Likoff MJ, Janicki JS, Laskey WK, Hirshfeld JW, Weber KT (1984) Myocardial energetics and clinical response to the cardiotonic agent MDL 17043 in advanced heart failure. J Am Coll Cardiol 4:875–883
9. Amin DK, Shah PK, Hulse S, Shellock FG, Swan HJC (1984) Myocardial metabolic and hemodynamic effects of intravenous MDL-17,043, a new cardiotonic drug, in patients with chronic severe heart failure. Am Heart J 108:1285–1292
10. Hasenfuss G, Holubarsch Ch, Heiss WH, Meinertz Th, Bonzel T, Wais U, Lehmann M, Just H (1989) Myocardial energetics in patients with dilative cardiomyopathy. Influence of nitroprusside and enoximone. Circulation (in press)
11. Bretschneider HJ, Cott L, Hilgert G, Probst R, Rau G (1966) Gaschromatographische Trennung und Analyse von Argon als Basis einer neuen Fremdgasmethode zur Durchblutungsmessung von Organen. Verh Dtsch Ges Herzkreislaufforsch 32:267–273
12. Sandler H, Dodge HT (1968) The use of single plane angiocardiograms for the calculation of left ventricular volume in man. Am Heart J 75:325–334
13. Rackley CE, Dodge HT, Coble YD, Hay RE (1965) A method for determining left ventricular mass in man. Circulation 29:666–671
14. von Herrath M, Hasenfuss G, Holubarsch Ch, Hofmann Th, Heiss WH, Just H (1989) Continuous calculation of left ventricular wall thickness from mass and volume during one cardiac cycle for the determination of left ventricular wall stress. Clin Card (in press)
15. Mirsky I (1979) Elastic properties of the myocardium: a quantitative approach with physiological and clinical applications. In: Berne RM (ed) Handbook of physiology. The cardiovascular system, American Physiological Society, Washington DC, p 497
16. Braunwald E (1971) Control of myocardial oxygen consumption. Physiologic and clinical considerations. Am J Cardiol 27:416–432
17. Strauer BE (1979) Myocardial oxygen consumption in chronic heart disease: role of wall stress, hypertrophy and coronary reserve. Am J Cardiol 44:730–740
18. Sarnoff SJ, Braunwald E, Welch GH, Case RB, Stainsly WN, Marcuz R (1958) Hemodynamic determinants of oxygen consumption of the heart with special reference to the tension-time-index. Am J Physiol 192:148–156
19. Coleman HN, Sonnenblick EH, Braunwald E (1969) Myocardial oxygen consumption associated with external work: the Fenn effect. Am J Physiol 217:291–296
20. Weber KT, Janicki JS (1977) Myocardial oxygen consumption: the role of wall force and shortening. Am J Physiol 233:H421–H430

21. Roebel LE, Dage RC, Cheng HC, Woodward JK (1982) Characterization of the cardiovascular activities of a new cardiotonic agent, MDL 17,043. J Cardiovasc Pharmacol 4:721–729
22. Hoh JFY, Rossmanith GH, Kwan LJ, Hamilton AM (1988) Adrenaline increases the rate of cycling of cross-bridges in rat cardiac muscle as measured by pseudo-random binary noise-modulated perturbation analysis. Circ Res 62:452–461
23. Holubarsch Ch, Hasenfuss G, Blanchard E, Alpert NR, Just H (1986) Myothermal economy of rat myocardium, chronic adaptation versus acute inotropism. Bas Res Cardiol 81, Suppl I:95–102
24. Holubarsch Ch, Hasenfuss G, Heiss WH, Meinertz T, Just H (1987) Acute and chronic changes of myocardial energetics in the mammalian and human heart. Bas Res Cardiol Vol 82, Suppl II:377–388
25. Holubarsch Ch, Hasenfuss G, Just H, Blanchard E, Mulieri LA, Alpert NR (1989) Modulation of myothermal economy of isometric force generation by positive inotropic interventions in the guinea pig myocardium. Am J Cardiol (in press)

Author's address:

PD Dr. Chr. Holubarsch, Medizinische Universitätsklinik, Innere Medizin III, Kardiologie, Universität Freiburg, Hugstetter Strasse 55, D-7800 Freiburg, FRG

Subject index